DAILY DISCIPLES

BUILDING DIVINE

Devotional
TIME

Tonilee Adamson & Bobbye Brooks

Daily Disciples Publishing
Carlsbad, California 92013

Building Divine Devotional Time
By Tonilee Adamson and Bobbye Brooks

Copyright © 2010 by Tonilee Adamson and Bobbye Brooks

All rights reserved. Under International Copyright Law, no part of this publication may be reproduced, stored, transmitted by any means – electronic, mechanical, photographic (photocopy), recording, or otherwise – without written permission from the publisher.

Library of Congress Catalog Card Number: T.B.A.
International Standard Book Number 978-0-9840913-3-1

Printed in the United States of America.

Scripture quotations taken from the New King James Version. Copyright © 1979, 1980, 1982 by Thomas Nelson, Inc. Used by permission. All rights reserved.

Any emphases or parenthetical comments within Scripture are the authors' own.

From Our Hearts to Yours

We are pleased to present this cherished collection of Bible readings and daily devotions in a 365-day devotional book. This book *Building Divine Devotional Time* is the second in our "Building Divine Intimacy" series, designed to encourage and inspire you in God's hope and love for you personally. Romans 10:17 says, "Faith comes by hearing and hearing by the Word of God." We know that time spent with the Lord in His Word *every day* will make a huge difference in your life. It is our privilege to share these writings from our hearts with you. Every devotional was written by us personally, most of them reflecting our own intimate thoughts and experiences. We desire to be honest and open in our writings, with a hope that these words will minister to you and give you daily nuggets of God's wisdom.

Daily Disciples Ministries was founded on Bible studies and daily devotionals. The heart of this ministry is to evangelize the lost and awaken the saved to experience lives empowered by the Word of God through His Spirit. We teach and train those who desire to grow, to serve and to have all God has for them, encouraging everyone to be a daily disciple of Jesus Christ.

A daily reading plan in the Word of God is a key component of walking in faith and victory. For some, reading the Bible is difficult because they find it obscure and hard to understand. This devotional removes the difficulty by leading you to take steps, one day at a time, to gain insight into God's personal message for your life.

In this devotional book, each day begins with the daily reading plan. The reading plan is designed to help you read through the Bible in one year. We sincerely encourage you to take the time each day to read these Scriptures, so God, through His Word and Holy Spirit, can communicate with you, build your faith and grow your walk. Plan your time and pray for God's protection against any distractions so you can gain intimate time with the Lord.

Grab your Bible, pen and your *Building Divine Devotional Time* book and spend some life-enriching moments with the Lord. As an added building block to help you in your daily walk in the Word, the pages around the 30th of each month are blank. These days are your personal devotional days. Use the Scriptures from the daily reading plan at the top of the page and try writing your own devotional. This exercise will help you learn to apply the Bible to your daily life. What verse seems to jump off the page at you? What is weighing heavily on your heart? Try writing your thoughts down with the verses given and see how the Lord guides you with a message just for you personally. As you continue to be diligent in seeking the Lord, He will be faithful to reveal Himself to you. Remember that with meditation comes revelation.

May God bless you as you seek Him and grow intimately acquainted with Him, as He is with you.

January 1

Today's Reading: Genesis 1-3; Matthew 1
Today's Thoughts: Back to the Beginning

In the beginning God created the heavens and the earth. The earth was without form, and void; and darkness was on the face of the deep. And the Spirit of God was hovering over the face of the waters. Then God said, "Let there be light"; and there was light. And God saw the light, that it was good; and God divided the light from the darkness. **Genesis 1:1-4**

Sometimes I need to go back to the beginning of the Bible to get a new perspective on things. The book of Genesis is a wonderful book filled with all kinds of new beginnings. From the creation of the universe to the beginnings of civilization, I am always blessed when I start reading Genesis. Let's look at the first four verses to see how they can give us a fresh and new perspective for today.

I love the first four words: "In the beginning God...." That says it all. Sometimes I just need to be reminded that God has always existed. It is hard to comprehend the concept that before "the beginning," there was God, but it is a truth that we can embrace with love and respect. He is God and there is none other. "God," as used in this passage, is the Hebrew word "Elohim," which is plural (meaning more than one). Elohim refers to the Father, Son and Holy Spirit, existing since the beginning of creation. It brings great comfort in being reminded that God is all-knowing, all-powerful and ever-present, from beginning to end (Alpha and Omega), while taking care of us in the middle too.

The next amazing part of these verses describes how God separated the light from the darkness. The earth was dark and had no form or substance. But God is light and in Him, there is no darkness at all (1 John 1:5). God saw the darkness and knew that the whole earth needed His light. Thus, God brought forth His light for us. When we feel as though darkness surrounds us, we look upon Jesus and He is still bringing forth His light for us today. These verses remind us that God understands darkness and never intended for us to live in it. He created the light, separated the light from darkness, and said, "it was good."

When man sinned, darkness entered into our hearts and the light of God went out. God sent His Son, Jesus, to bring light into the world so that all who believe in Him might be saved from the eternal darkness. For those of us today who have accepted Jesus as our Savior, we are filled with His eternal presence through the indwelling of His Holy Spirit. Everywhere we go, we shine His light. Isn't it great to go back to the beginning?

January 2

Today's Reading: Genesis 4-6; Matthew 2
Today's Thoughts: Master Your Anger

So the Lord said to Cain, "Why are you angry? And why has your countenance fallen? If you do well, will you not be accepted? And if you do not do well, sin lies at the door. And its desire is for you, but you should rule over it." **Genesis 4:6-7**

The classic story of Cain and Abel unfolds in Genesis chapter four. Two brothers (from the real first family) each presented the Lord with their offerings. Cain offered the Lord fruit and Abel offered a first-born from his flock. The Lord accepted Abel's offering but rejected Cain's. Cain became angry and jealous of his brother's favor with the Lord. Today's verses give us insight as to how the Lord responded to Cain's anger. In essence, He warned Cain to be careful because sin was lying in wait for him. The Lord was giving Cain a choice to let it go, try again and be accepted, or to give in and allow sin to control him. Cain chose the latter option, which resulted in him killing his brother Abel and running from the Lord for the rest of his life. Sin did come to control him as the Lord warned!

Has anger ever led you to do things you would later regret? Has jealousy or envy stirred you to the point of sinning against someone close to you? These sins will eventually lead to destruction, unless we stop them from taking control of us. We must all take notice of the Lord's advice and beware of the sin that lies at our door. Sin leads to death if left to its own desires. If we allow sin to rule over us, then our actions will result in serious consequences. The Lord gives us this warning as a reminder that we "should rule over it." As Christians, we have the Holy Spirit who will make us aware of these dangers and will lead us away from dangerous temptations. He will help us to "rule over sin."

No one is immune to the emotions involving jealousy, envy or covetousness. But even if we are tempted with them, we do not have to be ruled or mastered by them. When we find ourselves in a situation that stirs these types of feelings or thoughts, we must start praying immediately. We must confess our thoughts to the Lord and be honest with Him, then ask Him to help us to let go of them. When we turn those things over to the Lord, He will give us the strength through His Spirit to turn the tables and rule over them. Left on its own, sin destroys. Do not give in to its desires. Listen to the Lord and let Him be your Master.

January 3

Today's Reading: Genesis 7-9; Matthew 3
Today's Thoughts: Press Into Him

So it was, as the multitude pressed about Him to hear the word of God, that He stood by the Lake of Gennesaret. **Luke 5:1**

Have you ever been in the presence of someone famous and had an opportunity to get close to them – close enough to touch them? Fans at concerts get as close as they can to the stage and others wait in lines for hours just to get a glimpse of someone they want to see in person. People in these situations will keep pressing in until they see or hear that special person they desperately want to meet. Jesus had become such a person of interest that many people came from miles around to see and hear Him speak. They "pressed about Him" to get as close as they could. Was it because He looked good or had a great voice? Was there something about Him physically that attracted people? This verse says that they came to "hear the word of God." When was the last time you pressed into church and sat as close as you could to the pastor just to hear the word of God?

When we fall in love with the word of God, we fall in love with the Lord. In that place of love, we cannot get close enough to Him. We find ourselves on our knees, on our faces and in our closets praying to see His face and hear His voice. We desire an intimacy that is so special and so personal that nothing else matters. James 4:8 says that if we draw near to God, then He will draw near to us. Press into Jesus today. Draw close to Him. Find a quiet place to be alone with Him and ask His Spirit to fill you with His love and presence. Open your Bible and start reading His Word. Soon you will understand why the multitude pressed about Him just to hear the word of God. Today, start your day off with a prayer to know Him more and do not stop pressing into Him.

January 4

Today's Reading: Genesis 10-12; Matthew 4
Today's Thoughts: It's Just That Simple

But as many as received Him, to them He gave the right to become children of God, to those who believe in His name. **John 1:12**

The "He" and the "Him" in these verses is Jesus Christ, the Son of God. He came to earth to save His people, but many of His own people rejected Him. For those who "received Him," Jesus gave them inheritance into the kingdom of God. All they had to do was "believe in His name." There was nothing else to do, not even their best religious works would be good enough to earn salvation. The same is true for us today as we have the Word of God to teach us about Jesus. The Gospel of John was written so that we would *believe* in Him, a simple message that requires simple faith. But is it really that simple?

The short answer is "yes." *It is just that simple.* All anyone has to do to be saved and inherit eternal life with God in heaven is to believe in His Son Jesus and the work He did on the cross. If it is that easy, then why doesn't everyone believe? Jesus would compare the faith required of us to the faith of a child. He desires for all of us to have childlike faith. Why? Because children just believe. They trust without proof of purchase. They do not carry baggage from years of pain, mistrust and betrayal. They do not know what it means to be skeptical, cynical or illogical. But we do. We get hardened by the world and its imprints on our lives. And, unfortunately, some have been hurt by Christians themselves. Even those of us who are Christians, who believe in Jesus, walk around without any power to change. Where is the victory that overcomes the world? In everyday life, it is not so easy to believe in a God we cannot see or hear or touch. When the world takes our time and attention, we have an even harder time believing beyond what we are dealing with at the moment.

If you feel overwhelmed with life, take a moment and set your eyes on Jesus. Ask Him to help you believe His promise to never leave you nor forsake you (Hebrews 13:5). Ask Him to help you believe that He has a future and a hope for you. (Jeremiah 29:11) Ask Jesus to fill you with His peace that He promised to leave with you (John 14:27). Stop trying to work it all out for yourself. And if you are having trouble believing, ask the Lord to help you even to believe (Mark 9:23-24). We spend so much time on other things; why not spend some of it with the Lord today?

January 5

Today's Reading: Genesis 13-15; Matthew 5:1-26
Today's Thoughts: Glorify Our Father

"Let your light shine before men in such a way that they may see your good works, and glorify your Father who is in heaven."
Matthew 5:16

I was meditating on this verse today and saw something different. Jesus has just finished teaching the beatitudes, "Blessed are the poor in spirit…. Blessed are those who mourn…. Blessed are the meek…." Then He summarizes this section by encouraging the believer to continue doing all those things that are making them poor in spirit, mournful, meek and persecuted because the prophets were treated the same way. This leads into being salt and light. Jesus is instructing the believer to remain salty and bright, to keep burning and stinging for Him. The transitions in the message flow beautifully and make perfect sense. However, when I would read Matthew 5:16, I wondered: "How would the 'men' glorify the Father in heaven as a result of seeing the believers' good works? If these same men are the ones causing the believer to be poor in spirit, meek and persecuted, how can they be hurting the believer while glorifying God?"

Today, I realized that the believer is the one who is instructed to glorify the Father. We, as believers, are to remain faithful to continue on regardless of the reaction of the people. We are to remain diligent, to stay sharp and be alive in living for Jesus. We don't hide who we are or apologize for what we believe. We stand strong and continue. Not only will men see our good works which will bring glory to our heavenly Father, but also we glory in our heavenly Father that we are able to endure in His good works.

There will be times that others will also glorify the Lord because of your good works. Jesus' ministry caused others to thank God. But we don't look at others for affirmation to know that the Lord is blessing… we look to God. We need to be faithful to things God has called us to do and leave the results with Him, as we glorify our Father who is in heaven.

January 6

Today's Reading: Genesis 16-17; Matthew 5:27-28
Today's Thoughts: Fall on your Face

When Abram was ninety-nine years old, the Lord appeared to Abram and said to him, "I am Almighty God; walk before Me and be blameless. And I will make My covenant between Me and you, and will multiply you exceedingly." Then Abram fell on his face, and God talked with him. Genesis 17:1-3

Notice what the last verse says: "Then Abram fell on his face." The Lord has appeared to Abram and has begun to speak to him. As soon as Abram hears the Lord, he falls on his face. It does not say that he knelt down or bowed to the ground. It says he "fell on his face." We are not told from the Scriptures how the Lord appeared to Abram or how He spoke to him, but we can assume that Abram recognized the appearance and voice of God Almighty. At that moment, Abram lost all composure. Then, God talked with him some more. As the chapter continues, God continues to speak to Abram. What an awesome experience to be in the presence of God!

How do you react to the presence of God? When was the last time you fell on your face before the Lord? Many Christians today have never heard the Lord speak to them and have no idea of what it means to be in God's presence. Just saying you heard the Lord speak at all can bring raised eyebrows and concerned looks from other believers. Does the Lord still speak to His people? Can we really be in His presence and know it is Him? The answers are yes and yes. Yes, we can fall on our faces in the presence of holy God and yes, He will talk to us.

We wrote and taught a study called "Practicing the Presence of God," and in the study we included two sections on how to Practice the Presence of Hearing God's Voice. The response from Christians was enlightening and encouraging. Those who had never experienced an intimacy with the Lord learned how to worship and pray. Those who were not sure how to know if they were hearing God's voice learned how to find confirmation in the Word as well as other ways that God confirms His message to us. Most of all, we learned how to fall on our face in worship, how to come to His throne in reverence, and how to know His presence.

Today, find time to fall on your face and worship the Lord. Ask Him to speak to you through His Word and to confirm His message to you through His Spirit. Your day will be blessed and nothing else will matter as much as it once did.

January 7

Today's Reading: Genesis 18-19; Matthew 6:1-18
Today's Thoughts: Our Daily Bread

Give us this day our daily bread. **Matthew 6:11**

Today's verse is taken from what is commonly called the Lord's Prayer. When Jesus' disciples asked Him how to pray, He gave them this prayer as a model. Matthew 6 and Luke 11 record this prayer in its entirety. Like many people, I learned the Lord's Prayer as a child and I have prayed it thousands of times. But how many of us really think about what we are praying when we recite this prayer? How often do I ask for my daily bread?

I guess a better question might be: what exactly is "our daily bread?" Certainly our daily bread consists of those things that we need to survive, to live on. We need food, water, clothing and shelter. We need financial resources to get the things we need. So, we must pray that the Lord will meet those needs and give us our daily provisions. There is another type of "daily bread" that applies here as well – our spiritual bread. We cannot live abundant spiritual lives without a daily portion of God's Word. Learning to feed on His Word every day will give us the spiritual nourishment we need to live a victorious Christian life.

Ask the Lord today to give you His bread of life: to meet your physical needs here on earth, and to also meet your spiritual needs as you walk with Him. When you allow the Word of God to penetrate your heart and mind daily, you will be transformed. Make time for reading your Bible today and let the Lord feed you with His amazing Word. Start the transformation today.

January 8

Today's Reading: Genesis 20-22; Matthew 6:19-34
Today's Thoughts: Seek God First

But seek first the kingdom of God and His righteousness, and all these things shall be added to you. **Matthew 6:33**

If you do not spend time with the Lord today, the devil will kill, steal and destroy any fruit of the Spirit. The most important thing that allows us to live a victorious Christian life seems to be the hardest thing to maintain on a consistent basis. We have to understand that the power of the Holy Spirit is released by abiding in the Word. If the devil can keep you from opening the Word, you will be ineffective and have little peace in knowing the will of God. The Lord wants to instruct you and keep you in the way you should go; but you must spend time with Him for Him to be able to lead you.

Repeatedly, the common excuse is that we are too busy. But by spending time with the Lord, He is then allowed in intervene in your life in such a way that gives you more time in your day. It is like tithing. You give and then you get. Time works the same way with the Lord.

No one was busier than Jesus. He had many things to do and many lessons to teach in a very short period of time. What did He do? Jesus made it a priority to spend time with the Father. He would rise early, before the dawn, just to pray. There was nothing that was more important than or as pressing as spending time with His Father. If Jesus needed to spend time alone with God, how much more do we need to spend time with Him? If the devil cannot make you bad, he will make you busy. Why? Because all those seemingly valuable things you are doing distracts you from the only thing that really matters for eternity. Even church-oriented events will not bear the fruit you are praying for without His counsel and His blessing.

Seek Him first – first thing in the morning, first before any plans are made and first in your thoughts and prayers. No other task can compare to finding God after seeking Him with your whole heart. Seek first His kingdom and His righteousness and all these things will be added unto you. (Matthew 6:33)

January 9

Today's Reading: Genesis 23-24; Matthew 7
Today's Thoughts: Adopted by a Great God!

For you have not received a spirit of slavery leading to fear again, but you have received a spirit of adoption as sons by which we cry out, "Abba! Father!" **Romans 8:15**

I drove carpool today. Boy, the things I learn while having a few more junior high kids in the car are so valuable to me. I am also thankful because the kids are still at a point of listening to instruction if I say it clearly, simply and in as few words as possible.

Two of the boys are struggling with the same issue under different circumstances. Both of them feel that they are being misrepresented by their teachers and both are very disheartened by the teachers' decision in their matters. They both could clearly state the problem, the areas that they were at fault and why they feel that a different decision should be made. However, they both dealt with fear in speaking to their teachers about it.

Many times, we find the same struggles in our prayer life with the Lord. We understand the issues at hand and even accept the blame and resulting consequence. But we don't know how to express how we feel about it to the Lord. As a result, we experience depression and frustration, waiting for time to play itself out.

God tells us that He is our Abba Father. He wants us to know that He is our Daddy, desiring to hear from us. He didn't create us to be robots, but His children who can reason with Him. He tells us to draw near to Him and He will draw near to us. Our opinions matter to the Lord regardless of the poor choices we may have made to reach our situation. The Lord loves us and if we include Him in our struggles and frustrations, He works with us to get through the difficulties, sometime miraculously. If we don't share with Him and draw near, we will miss out on His intervention and in the nearness of His presence to walk through every part of life with Him. Do not fear; you have been adopted by a great God!

January 10

Today's Reading: Genesis 25-26; Matthew 8:1-17
Today's Thoughts: Chronicle Your Blessings

Then Moses said, "This is the thing which the Lord has commanded: 'Fill an omer with it, to be kept for your generations, that they may see the bread with which I fed you in the wilderness, when I brought you out of the land of Egypt.' " **Exodus 16:32**

Do you ever forget things? Have you ever tried something out of the ordinary, maybe even silly, just to remember something of importance? Some people tie strings around their fingers, some write post-it notes and stick them in plain view, and others go so far as to take memory classes or use self-help tools for memory enhancement. Let's face it, we all deal with memory lapses at times, and most of us have forgotten at least one thing we wished we had remembered.

Think about how much information goes into our brains on a constant basis. We live in a world filled with constant stimuli with each piece fighting for a place of priority in our thoughts. Our days begin with so much to do that our organizers and calendars need more power. Sometimes I am amazed at just how soon I can forget something, regardless of the tactics, tools or tricks I might use to help me to not forget.

I am thankful that God understand, and He wants to help. The verses above demonstrate how important it was to the Lord that the children of Israel have a physical reminder of His faithfulness to them. The Lord instructed Moses on how to store the manna as a reminder of how He fed them in the wilderness for forty years. Sadly, as the Bible reveals their continuing story to us, these people would soon forget how good God was to them. But God did not want them to forget, and He does not want us to forget either. How do we remember the good things God does for us? There are so many awesome ways that the Lord demonstrates His love and faithfulness towards us such as: answered prayers, answers to prayers never prayed, surprise blessings, and all of those "must be a God-thing" events. How can we remember, especially when we find ourselves in a valley of darkness or a barren wilderness?

One of the best blessings in my life has been the habit of keeping a daily prayer journal. Writing out my prayers, my thoughts and feelings, and my personal evidence of God's hand in my life has strengthened my faith and trust in Him. I can look back over past journals and read about the amazing times He intervened in my circumstances. A prayer journal becomes our personal testimony of how God has worked in our lives. Let's commit ourselves to making this activity a habit and to keeping this daily journal as a testimony of how awesome our Lord and Savior is.

January 11

Today's Reading: Genesis 27-28; Matthew 8:18-34
Today's Thoughts: God's Promise

"Behold, I am with you and will keep you wherever you go, and will bring you back to this land; for I will not leave you until I have done what I have spoken to you." **Genesis 28:15**

Today's verse is a promise that God made to Jacob. As Jacob lay asleep, the Lord stood over him and reaffirmed the promise that He had given his father, Isaac, and his grandfather, Abraham. The Lord told Jacob of how his descendants would be as the dust of the earth, scattered in all directions and blessed to be in God's family. God kept His promise and the rest of the Old Testament tells the story of His people, a people known as the children of Israel. (Israel was the new name given to Jacob by God.)

The literal context of this verse is stated above but the spiritual context is applicable to us today. Did you know that the Lord has a promise for your life, even in this verse? How does something written so long ago become relevant to our lives now? Many people miss out on the Old Testament message because they see no relevance to our modern lives. But the Holy Spirit makes the message critically relevant to our lives right here and right now. The promise that God gave Jacob is also a promise that He gives us. Hebrews 13:5 says that the Lord will never leave us nor forsake us. For Jacob, God had given him a specific promise about the nation of Israel, a nation yet to be born at the time of the encounter. What promise has God given you, or do you know? For starters, He promises to never leave you and that He will be with you and keep you wherever you go. God promises to always be with you. What an awesome promise!

Sometimes we just need to open the Bible and start reading. Maybe someone reading this right now needs to hear that God promises to always be with them. The Holy Spirit takes the Word of God and speaks to our hearts personally, a message just for us. The Bible is filled with stories and literal accounts of historical events, but in every story and every account, there is a personal message just for you. Please do not miss out on the spiritual message. You will find what you need if you seek for it: a promise, a confirmation, an answer. You will love the Bible stories, but when they become real to you in your life, and apply to you in some way today, your life will change and you will have an insatiable appetite for God's Word.

January 15

Today's Reading: Genesis 36-38; Matthew 10:21-42
Today's Thoughts: The Power of Friendship

For if they fall, one will lift up his companion. But woe to him who is alone when he falls, for he has no one to help him up.
Ecclesiastes 4:9-10

It seems natural that with the New Year comes new goals and dreams. As I grow older however, I realize that my goals do not change much from year to year but my ability to achieve the goals change. My faith has not been the motivation to get me to change instead my faith provides the inspiration to accept the need for change. So, how can my goals to change be best accomplished? Over the years, I have come to discover that the peer pressure (or the friends) in my life are actually a very strong motivating factor to change. Peers are like deadlines dates. They both hold you accountable and have the ability to motivate you to complete necessary tasks. Think of reunions or house parties or even going out, we all make a special effort to try extra hard.

It is the same thing for me at work. It has not always been the inspiration of spiritual revelations that have kept me going. I really believe that it has been the physical tangible person, having a co-partner who has challenged me as well as encouraged me to continue. We could become resentful of the peer pressure in our lives or we can be thankful for the power of friendship to change our lives.

Gossip, negativity, criticism, lack of support and judgments can cause anyone to quit because change can be too hard. Personally, I do not blame anyone for quitting. The world is filled with people who lack support, finances, encouragement and personal abilities to break out of habitually bad patterns. But I would recommend that they find a friend or an accountability partner to talk things out and get perspective before making any big decisions. Sometimes, it is in the honest sharing of your heart that someone else can give you the wisdom to endure.

Just because we know we need to change does not mean that it will be easy. Just because others tell us that we need to change does not mean they can give you the tools to do it. Life can be confusing and tough. The power of friendship can cut through the random factors to keep you focused. You do not have to go at it alone; people have been pairing up for years. God is totally into fellowship, counsel, and unity and empowers us through friendship. Friendship is a powerful gift of life.

January 13

Today's Reading: Genesis 31-32; Matthew 9:18-38
Today's Thoughts: A Changed Heart

And Jacob saw the countenance of Laban, and indeed it was not favorable toward him as before. Then the LORD said to Jacob, "Return to the land of your fathers and to your family, and I will be with you." So Jacob sent and called Rachel and Leah to the field, to his flock, and said to them, "I see your father's countenance, that it is not favorable toward me as before; but the God of my father has been with me. And you know that with all my might I have served your father. Yet your father has deceived me and changed my wages ten times, but God did not allow him to hurt me. **Genesis 31:2-7**

God is God and He can change the heart of people for your advantage. Jacob knew that despite all his efforts and hard work, Laban's heart was changing towards him. Jacob described it as "not favorable toward him as before." This would seem like an injustice to Jacob since he had worked for Laban who had deceived him and changed his wages unfairly 10 times. And through it all, Laban was completely blessed because of Jacob. However, for some strange reason, Laban's heart changed towards Jacob and his work. With the unfair treatment and Laban's change of heart, Jacob wanted to move on. Why? Because God wanted Jacob to leave and when leaving, God did not want Jacob to ever look back and regret his decision. So God allowed Laban to test Jacob unfairly and to cause Jacob to desire to leave.

Has this ever happened to you? It happens in all kinds of circumstances where people form close ties that are mutually beneficial. A person's heart can change towards a friend when they fear they are losing something of value, but God can change a heart simply for His purposes, even when the innocent one is treated unfairly. It is hard being the one who is mistreated because the heart of a friend has changed against you. But God knows your heart as well as the heart of your former friend.

If you are finding yourself in this place, completely trust in the Lord. Even though you might not have ever left, He still may change circumstances and hearts to get you to change. Some of us pray for years to have our circumstances change. For others of us, we are so surprised when God intervenes and changes a heart. But if you really examine your prayers, the change should not catch you off guard. God is good like that. You can trust Him. He knows where you need to be and He will perform miracles to get you there.

January 14

Today's Reading: Genesis 33-35; Matthew 10:1-20
Today's Thoughts: Renew Your Mind

I beseech you therefore, brethren, by the mercies of God, that you present your bodies a living sacrifice, holy, acceptable to God, which is your reasonable service. And do not be conformed to this world, but be transformed by the renewing of your mind, that you may prove what is that good and acceptable and perfect will of God. **Romans 12:1-2**

Have you ever found yourself humming or singing a song that is in your mind but you are not sure where it came from? So often, I will walk out of a store with a familiar tune playing in my head. The same thing happens when we watch television. Images from movies and television shows repeatedly play over and over in our minds. The power of sound and visual imagery can dominate us and yet, in many ways, we are somewhat immune to their influence. The entire marketing industry knows that repetitive marketing slogans will lead us to desire and purchase them. Our minds are well-trained in the ways of the world.

As Christians, we need to be aware of the worldly influences that infiltrate our minds. Thoughts tend to build upon each other and that can lead us into ungodly habits and patterns. The devil will tell us that whatever is in our minds is just for us; no one will get hurt, and no one will ever know. But in time, out thoughts will begin to change our behaviors and ultimately, some of those hidden things in our minds will be exposed in our lives. We must guard against these influences from the start. We must be on the offensive at all times. The apostle Paul tells us that we are not to be conformed to this world but to be transformed by the renewing of our minds. Our minds must be renewed daily, moment by moment, and this training begins by keeping certain things out. Make a conscious choice to avoid watching certain programs or movies. Turn off the noise in the car and replace it with music of praise and worship. We need to put things in our minds that glorify our Lord, for He that is in us is greater than he that is in the world.

Is it a sacrifice? Yes. We must present our bodies and minds to be a sacrifice unto God. The best part is that our sacrifice is greatly rewarded in the blessings that come from having a sound mind filled with peace and joy. Ask the Lord to help you today, and pray for His power to overcome the enticing activities of the world. Retrain your mind and your life will change.

January 15

Today's Reading: Genesis 36-38; Matthew 10:21-42
Today's Thoughts: The Power of Friendship

For if they fall, one will lift up his companion. But woe to him who is alone when he falls, for he has no one to help him up.
Ecclesiastes 4:9-10

It seems natural that with the New Year comes new goals and dreams. As I grow older however, I realize that my goals do not change much from year to year but my ability to achieve the goals change. My faith has not been the motivation to get me to change instead my faith provides the inspiration to accept the need for change. So, how can my goals to change be best accomplished? Over the years, I have come to discover that the peer pressure (or the friends) in my life are actually a very strong motivating factor to change. Peers are like deadlines dates. They both hold you accountable and have the ability to motivate you to complete necessary tasks. Think of reunions or house parties or even going out, we all make a special effort to try extra hard.

It is the same thing for me at work. It has not always been the inspiration of spiritual revelations that have kept me going. I really believe that it has been the physical tangible person, having a co-partner who has challenged me as well as encouraged me to continue. We could become resentful of the peer pressure in our lives or we can be thankful for the power of friendship to change our lives.

Gossip, negativity, criticism, lack of support and judgments can cause anyone to quit because change can be too hard. Personally, I do not blame anyone for quitting. The world is filled with people who lack support, finances, encouragement and personal abilities to break out of habitually bad patterns. But I would recommend that they find a friend or an accountability partner to talk things out and get perspective before making any big decisions. Sometimes, it is in the honest sharing of your heart that someone else can give you the wisdom to endure.

Just because we know we need to change does not mean that it will be easy. Just because others tell us that we need to change does not mean they can give you the tools to do it. Life can be confusing and tough. The power of friendship can cut through the random factors to keep you focused. You do not have to go at it alone; people have been pairing up for years. God is totally into fellowship, counsel, and unity and empowers us through friendship. Friendship is a powerful gift of life.

January 16

Today's Reading: Genesis 39-40; Matthew 11
Today's Thoughts: John's Doubts

And when John had heard in prison about the works of Christ, he sent two of his disciples and said to Him, "Are You the Coming One, or do we look for another?" Matthew 11:2-3

John the Baptist was the forerunner of Jesus. Isaiah 40:3 and Malachi 3:1, he was prophesied as the messenger who would prepare the way for the Messiah. He was the one anointed in his mother's womb and filled with the Holy Spirit (Luke 1:41), and appointed by God to bear witness of Jesus, the Son of God. John the Baptist was the one who baptized Jesus and witnessed the Holy Spirit's descent upon Him like a dove (John 1:32). This same John the Baptist is now wondering if this Jesus is the real Messiah. Should he look for someone else? What happened to John? What went wrong with his faith?

The Messiah was expected to come to earth and to set up His kingdom. Jesus, therefore, was the long-awaited King of the Jews. Everyone who believed in Him fully expected the prophecies of His earthly reign to be fulfilled in their day. But Jesus would answer John's question showing a much different purpose for His coming than what John and the Jewish people expected. Jesus sent back the message that, "The blind see and the lame walk; the lepers are cleansed and the deaf hear; the dead are raised up and the poor have the gospel preached to them" (Matthew 11:5). Jesus did not come the first time to set up His kingdom; He humbled Himself and died on a cross for our sins. He came as the final sacrifice. Jesus came to serve, not to be served. So John's expectations were not met, thus, he began to question (or doubt) if Jesus was the true Messiah.

How often do we put Jesus in our box and look to Him to meet our expectations? What happens when Jesus does not meet us the way we think He should? It is in those moments that our faith is tested the most. As humans, we tend to have very short memories. We forget so quickly how real the Lord has been to us, only to experience those moments when we wonder if He was ever real at all. Pray that your faith is strengthened in times of testing. Pray that you do not forget all that Jesus has done in your life. Pray that you never doubt that He is the true Son of God.

January 17

Today's Reading: Genesis 41-42; Matthew 12:1-23
Today's Thoughts: Pray Without Ceasing

Pray without ceasing. **1 Thessalonians 5:17**

How do we do that? Well, I studied prayer for a while as if it were a science. I read about men who were known to be prayer warriors. But I noticed that by the end of their lives, many of them said that the one regret they had was that they didn't pray enough. Immediately I prayed, "Oh Lord, don't let that happen to me. Do not let anyone quote me saying that. Teach me to pray without ceasing. Teach me to tap into Your heart at the mall as much as when I am in trouble or in pain. Teach me that You are listening at a rock concert as well as at the beach. Teach me to include You in everything and take nothing for granted. And oh Lord, teach me to thank You for it all."

God has given us His Holy Spirit on earth and has a will for us that He wants fulfilled on earth as it is in heaven. So, what does it take to have God's heart so much in mine that He can fulfill His will on earth as it is in heaven with me? Verses like "Ask and it will be given," "Whatever you ask, it will be given," allowed me permission to ask for everything. I realized that He was with me wherever I went. He just wanted to be included. He heard my words but knew the desires of my heart. He might not answer my words but He would fulfill the desire. And as time passed, I realized that He placed a lot of those desires in my heart just so He could fulfill them. The key was learning to pinpoint my true motivation. God looks at my heart and He weighs the thoughts and motives of the heart. If I was asking out of selfishness, He didn't answer. If I told Him that I thought my motive was selfish, He answered. And He would answer those prayers just like I asked. So I got another piece in the puzzle. God rewards honesty.

I could not have learned these lessons without writing out my prayers. Prayer Journaling helps to clarify your thoughts, intents and motives. Writing allows you to focus and really understand who you are and what you want before the Lord. We have prayer journals that have helped us. And we have instructional CDs to teach you how to get started. Our heart's desire is that you may learn to pray God's heart and for you to see heaven open up on earth as He uses you, for we serve an Awesome and Faithful God.

January 18

Today's Reading: Genesis 43-45; Matthew 12:24-50
Today's Thoughts: What's in Your Heart?

So Judah and his brothers came to Joseph's house, and he was still there; and they fell before him on the ground. And Joseph said to them, "What deed is this you have done? Did you not know that such a man as I can certainly practice divination?" Then Judah said, "What shall we say to my lord? What shall we speak? Or how shall we clear ourselves? God has found out the iniquity of your servants; here we are, my lord's slaves, both we and he also with whom the cup was found." But he said, "Far be it from me that I should do so; the man in whose hand the cup was found, he shall be my slave. And as for you, go up in peace to your father."
Genesis 44:14-17

What kind of response do you think Joseph was looking for? He placed his cup into the pouch of his brother, Benjamin (the only brother that had the same mother). Did Joseph want to reveal his identity to Benjamin alone or was he testing the hearts of the brothers to see if they had changed after all of these years? Both Joseph and Benjamin were their father's favorite children because they were born by the mother Jacob loved most. Favoritism does bring about jealousy, resentment and bitterness but the brothers were still responsible for their own behavior before God. After all these years, Joseph probably wondered what his other brothers thought of their actions in selling him. Did they regret it? Would they do it again? Were they treating Benjamin the same way?

Now, that God had given Joseph the upper hand, Joseph patiently and wisely tested his brothers to discern what was in their hearts. Judah said, "God has found out the iniquity of your servants." In other words, he was acknowledging that the brothers deserve to be treated as slaves for the sin they had been hiding all these years in selling Joseph as a slave.

Because God exists and He is just, what goes around, comes around. We truly do reap what we sow and these boys knew it firsthand. God is willing to forgive and we find Joseph acting like the God he trusts in.

I pray that we all may have the heart of both types of men:
Joseph – being willing to forgive
The brothers – being willing to openly and honestly confess and repent for wrongdoing against God and others.

God tests our hearts because He wants us to know what is in them. He already knows and He is so gracious to forgive as soon as we are willing to confess.

January 19

Today's Reading: Genesis 46-48; Matthew 13:1-30
Today's Thoughts: God's Ways

Oh, that my ways were steadfast in obeying your decrees! Then I would not be put to shame when I consider all your commands.
Psalms 119:5-6

This psalmist dealt with the same issues of guilt that we have today. When we consider all that God requires, His standard is too high. How is one to follow His ways when God's ways are perfect and we are not? Like the psalmist, we cry out, "Oh, I wish my ways were consistent in following and obeying You. Then I would feel better about myself."

The problem with thinking this way is that God wants us to be thankful for His provision and not think of ourselves any higher than we ought. Jesus provides us with all that we need for life and godliness (2 Peter 1:3). It is because of Jesus that we can have a relationship with the Father. When our shortcomings turn our focus to ourselves, we have guilt and condemnation. When our eyes are focused on Jesus, despite our shortcomings, we are thankful to Him. God wants us thankful. He already knows our shortcomings and knows the extent of our ability to sin, but He saved us anyway. And He loves us anyway.

Lately, everywhere I turn in the Bible, I am seeing that God desires a sacrifice of praise and thanksgiving. He wants me to sing to Him in gratitude for all He has done and is doing. The word **sacrifice** reminds us that giving Him thanks and praise is not easy when we are feeling guilty, angry, frustrated, depressed or miserable. But by praising Him during those times, it releases our faith to look to Him instead of to ourselves.

January 20

Today's Reading: Genesis 49-50; Matthew 13:31-58
Today's Thoughts: God means for Good

When Joseph's brothers saw that their father was dead, they said, "Perhaps Joseph will hate us, and may actually repay us for all the evil which we did to him." So they sent messengers to Joseph, saying, "Before your father died he commanded, saying, 'Thus you shall say to Joseph: "I beg you, please forgive the trespass of your brothers and their sin; for they did evil to you." ' Now, please, forgive the trespass of the servants of the God of your father." And Joseph wept when they spoke to him. Then his brothers also went and fell down before his face, and they said, "Behold, we are your servants." Joseph said to them, "Do not be afraid, for am I in the place of God? But as for you, you meant evil against me; but God meant it for good, in order to bring it about as it is this day, to save many people alive. Now therefore, do not be afraid; I will provide for you and your little ones." And he comforted them and spoke kindly to them. Genesis 50:15-21

One of my favorite Bible stories is the story of Joseph. Joseph was one of twelve brothers, all born to Jacob (Israel). From these brothers came the nation of Israel, God's chosen people. God had a plan from the beginning to raise up a nation of His people, a people set apart to be holy and dedicated to Him. The fulfillment of God's plan of redemption, the Messiah, Jesus Christ, would come from this line of people. But early on, attempts would be made to thwart this plan. One such attempt came in the life of Joseph. Jealousy and selfishness led his brothers to sell him to a foreign people where Joseph would eventually end up as a prisoner in Egypt. But God's hand was upon Joseph and through the years of his captivity he found much favor with his superiors, being entrusted ultimately with governing the land of Egypt. But the day would come when the family would be reunited. How would Joseph receive his brothers who had done such an evil thing to him so many years earlier?

The brothers were afraid that Joseph would avenge their wrongdoing, but instead he forgave them. Verse 20 sums it up: what they meant for evil against Joseph, God meant for good so that many would be saved. God had a plan and that plan was carried out through Joseph. If Joseph had not been in Egypt, the famine that engulfed the land would have killed the entire family. Joseph was the one who would make sure there was enough grain to sustain the land. It was God's plan to spare this family. They were to become His chosen people. It was God's plan to use Joseph. It was God's plan to work for good the evil that Joseph had endured. For us today, God has a plan. Regardless of the evil that comes against us, God can work it together for good. Look to the Lord and ask Him to work all things together for good.

January 21

Today's Reading: Exodus 1-3; Matthew 14:1-21
Today's Thoughts: To Fear the Lord

But the midwives feared God, and did not do as the king of Egypt commanded them, but saved the male children alive... Therefore God dealt well with the midwives, and the people multiplied and grew very mighty. And so it was, because the midwives feared God that He provided households for them. Exodus 1:17, 20-21

We all know what it means to have fear. We struggle with the fear of failure, the fear of being alone, and the fear of rejection. These fears bring insecurity and slant our judgment. How can we differentiate between the fears of this life and the fear of the Lord? What is the difference? The term to "fear the Lord" means to "revere or to stand in awe of." We tend to fear the things that we see and touch, not a God who is invisible. So, we pay more respect to the things of this world than to the Lord who is in control of all these things in the first place. We have choices that we make every day. Are we trying to please people or are we trying to please the Lord?

From Exodus, we see how the midwives are excellent examples of making the right choice. They were given a direct order from their king, the ruler of all Egypt. These women should have had "the fear of the king" because on earth he was the position of authority. He could command whatever he wanted and had the power to kill those who were disobedient. However, these women chose to "fear the Lord" even though they knew they could be killed. They saw something in the Hebrew people that indicated the existence of a higher authority than Pharaoh. They realized that the Hebrew God is the one true God, in control of life and death. They chose to fear and honor the Lord, even if it meant their own death.

Sometimes we think that by trying to please people, we are pleasing the Lord. As Christians, are we not suppose to lay down our rights and become servants to all? This is true only when we first lay down our rights to God. We can only begin to understand this concept as we learn what it means to *fear the Lord*. He has given us His word to teach us. You may be put in a difficult situation in your home, in your work, or at school. Maybe your husband, or boss or principal has made a decision that goes against your convictions. What are you going to do? Who are you going to live in fear of, man or God? The greatest thing about living in the fear of the Lord is that He will help you. It is not your responsibility to change their decisions but to express your convictions. By choosing to fear Him, He will lead you, guide you and love you through the fearful times. Depend on Him; fear Him. For when all of life is said and done, we all stand before God one day. And there is no doubt whose side we will choose then.

January 22

Today's Reading: Exodus 4-6; Matthew 14:22-36
Today's Thoughts: All That You Need

Look at the birds of the air, for they neither sow nor reap nor gather into barns; yet your heavenly Father feeds them. Are you not of more value than they? **Matthew 6:26**

For some reason, my home has been chosen as the nesting place of several families of birds, particularly under the eaves of our house. I have never seen so many birds' nests lodged right above our windows and doors. As I walked out my back door the other day, I looked up at a bird nest just as the momma bird was flying away. The next thing I saw were three little bird mouths opened wide and waiting for their next meal. I was reminded of what Jesus said in Matthew 6:26 in that the birds do not have to worry about their next meal because our heavenly Father feeds them.

Jesus then says to us: "Are you not of more value than they?" I must admit I have spent many days worrying about earthly needs. I have wasted precious time trying to figure out how to provide for the things I need, or my family needs, or the needs of someone else in my life. I was even tempted to somehow try and provide for those baby birds. Even as I stood there looking at their little mouths crying out for food, I knew the Lord was cautioning me not to get involved. To interrupt their nest could be a death sentence for them, not a rescue. I realized that when I try so hard to intervene in what God is trying to do in my own life, I often end up causing more harm than good. I just need to trust Him and get out of His way. He will take care of all of my needs. We should act only when we know it is God moving us to act!

Take time and thank the Lord today for taking care of your needs. He already knows everything you want and need. Just as those baby birds are completely taken care of (even being sent to my welcoming house), our heavenly Father is taking care of us too. We are more valuable to the Lord than all of the birds of the air put together. If He cares for them, He will absolutely care for us!

January 23

Today's Reading: Exodus 7-8; Matthew 15:1-20
Today's Thoughts: Called to a Purpose

"I will make a distinction [a] between my people and your people. This miraculous sign will occur tomorrow." **Exodus 8:23**

Have you ever found yourself in a situation wondering how you got there? You feel as if you just woke up to find yourself in a strange place. Or maybe you feel like you are in the midst of a bad dream wishing you could wake up and start over. I wonder if Moses felt this way right after his first visit to Pharaoh.

Moses had already told the Lord more than once that he was not equipped for this job. Who is he to demand Pharaoh to let these people go? He is slow in speech, has a slow tongue, and by the way, he is 80 years old. Maybe God should choose someone else, someone younger and more qualified. So, God gives Moses a helper – his older brother Aaron. Now there are two old men set apart as the deliverers of Israel. Sometimes I really think the Lord has a sense of humor.

Moses did not need any help outside of God, nor do we. But God is gracious enough to give us those things that bring physical and emotional comfort. But even with support, at times we all feel a little, if not a lot, overwhelmed with our circumstances. Paul says in 2 Corinthians 12:9 that His grace is sufficient because His strength is made perfect in our weakness. God gets the glory when He works for us and through us in situations that are impossible to us.

As Moses and Aaron walked away from their first meeting with Pharaoh, they must have felt a host of emotions: fear, doubt and confusion to name a few. Even though Moses had spoken personally to the Lord God Almighty and knew that he was carrying out God's commands, how hard the situation must have been for him. We tend to read these accounts in the Bible and think that these people were more special or had some extra power from God. But in reality, these people are discussed in the Bible to demonstrate to us that the Lord uses people just as they are, not just back then, but today as well.

We need to understand that as believers we are all called for a purpose that was long ago determined by God. The Bible tells us that God knows all things, from the beginning to the end, and is in control over everything that happens. But we must also realize that God's ways are not our ways and that His thoughts are not our thoughts (Isaiah 55:8-9). Instead of saying to God, "Who, me? Why me, Lord?" remember Moses and pray that you can be as humble, faithful and obedient as he was. If He has called you to a task, He will give you the strength to endure to the end.

January 24

Today's Reading: Exodus 9-11; Matthew 15:21-39
Today's Thoughts: Listen to What You Hear

Give ear and hear my voice, Listen and hear my speech. Isaiah 28:23

Have you ever thought about the difference between *hearing* and *listening*? Did you know there was a difference? To "listen" means that we are paying attention to the sound, thoughtful and considerate attention to the sound. To "hear" refers to perceiving something with our ears, kind of like seeing with our ears. In life today there are literally thousands of noises a day that compete for our attention. There are so many different of noises and sounds that most of us do not pay attention to most of them. The television can be on, the phone ringing, and everyone in my house talking at once and I can still tune most of it out. It is not that I am not *hearing* the noise, but I am not *listening* to it.

God gave us ears to hear the beauty of sound and He gave us the ability to listen so that the sound can be understood. When both hearing and listening work together, we have an understanding of the purpose of the sound. You can hear music but not listen to the words. When both music and lyrics are understood, the beauty and meaning of the whole song is revealed to you. You can hear the words of a person speaking to you, but if you truly listen to the person, you will better understand their heart and the meaning behind their words. The same principle holds true with God's Word. We can hear the Word preached, taught and read. We can read the Word ourselves and hear our own voices speak it. But, when we really listen to God's Word, then our understanding begins to change. So many times, people leave church after having heard a sermon and not remember anything about it. When we are listening and giving consideration to what we are hearing, we will leave thinking about what we just heard. The next step is to start talking back to God, asking Him to open up more of our understanding.

Take time today and think about how much you hear versus how much you listen. How much are you missing of what God has for you? Is He talking to you but you are not paying attention? Take your Bible and read today's chapter in Isaiah. Instead of just reading the words and hearing them in your head, pray the Holy Spirit will give you the power to listen and to understand the message God has for you. When the Word of God truly becomes His voice in your life, then your hearing and listening will open up a new world of understanding for you. Your life will change. But it all begins with understanding the difference between hearing and listening—start listening to God today.

January 25

Today's Reading: Exodus 12-13; Matthew 16
Today's Thoughts: Who Do You Say That I Am?

He said to them, "But who do you say that I am?" Simon Peter answered and said, "You are the Christ, the Son of the living God." Jesus answered and said to him, "Blessed are you, Simon Bar-Jonah, for flesh and blood has not revealed this to you, but My Father who is in heaven. And I also say to you that you are Peter, and on this rock I will build My church, and the gates of Hades shall not prevail against it. **Matthew 16:15-18**

Who do you say Jesus is? Many of us would agree with Peter that He is the Christ, the Son of the living God. But do we know what that looks like in our daily lives?

Even though Jesus was Peter's friend and teacher, Peter was very aware that the religious leaders did not believe that Jesus was the promised Messiah. Despite all the miracles, Jesus did not come as the Jewish people envisioned. Jesus seemed to purposely perform miracles and teach things that bothered others and contradicted their lifestyle. People had conflicting views on who Jesus was. However, when Peter was asked who Jesus was, his answer was firm and Jesus rewarded Peter's words.

There are difficulties and conflicts in our lives today that do not reflect the Holy God, in whom we believe. Those things are placed in our lives as tests to examine what we will declare about Jesus. Can you say that despite the hardships, conflicts, struggles and stresses in your life, Jesus is your Christ, the Son of the living God? Does Jesus impact your life daily, making a difference in the way you think, feel and what you say? If your faith is wavering in belief, ask the Lord to reveal to you who Jesus is, just as He did with Peter. People and circumstances will change with seasons and times of life; but when all is said and done, we want our faith and convictions to be as solid as rocks so that the gates of Hades shall not prevail against us.

January 26

Today's Reading: Exodus 14-15; Matthew 17
Today's Thoughts: Move Mountains

So Jesus said to them, "Because of your unbelief; for assuredly, I say to you, if you have faith as a mustard seed, you will say to this mountain, 'Move from here to there,' and it will move; and nothing will be impossible for you. **Matthew 17:20**

The word "faith" has numerous definitions and interpretations. From the intellectual to the philosophical to the spiritual, I found myself back at square one trying to understand its meaning to my personal life. If "faith as a mustard seed" can move a mountain, then I want more faith. Can it be that easy? After reading pages of explanations, I seem to make it more complicated, which in turn, dampens my faith. What happens when we really search for true faith?

Paul says in Romans 10 verse 17 that "faith comes by hearing, and hearing by the word of God." It makes sense then that our search for faith must start in God's Word. It is interesting to note that Paul uses the word "hearing" instead of reading. When we read, study and meditate on God's Word, then we begin to hear it, as it penetrates our hearts and minds. To grow in faith, we must grow in the knowledge of God's Word. When we put the Word in our minds every day, we hear it in our thoughts all day long. Another component of faith is found in James Chapter One. James says that our faith will be tested. If we ask for things without having the faith to believe in the answer, then we will deal with doubt and unbelief. James Chapter 5 verse 15 says that "the prayer of faith will save the sick." So, once again we come back to the Word and prayer as being the key components to our growing in faith.

Do you want to move mountains in your life and see God do the impossible? All you need is faith. Sounds easy enough – but is it? For me personally, I have learned that being in God's Word every day puts me in a place of hearing Him more clearly about things on my heart and in my prayers. I find the strength to step out in faith, even when I have nothing tangible to cling to. I have a confidence that is not based on what I know or what has been proven to me. And, every time that I have acted or stepped out in faith, I have seen God do amazing works. Not once have I been disappointed in the Lord; He has always been faithful. I have been disappointed in myself, but not when I have the foundation of Scripture and prayer to guide me in faith. Make it a priority of your day to not only read the Word but also to hear it all day long. Let it guide your prayers and lead your steps. At that point, you just might see those mountains begin to move.

January 27

Today's Reading: Exodus 16-18; Matthew 18:1-20
Today's Thoughts: Be Set Apart

But know that the Lord has set apart for Himself him who is godly; The Lord will hear when I call to Him. **Psalm 4:3**

Do you know that you are set apart by God and for God? We are sanctified, set apart and made holy, because the Lord has set us apart from the ungodly. This sanctification starts to take place when we accept Jesus Christ as our Savior and are redeemed by His blood. It is a lifelong process and God's work within our life. We live *in* the world but are not to be *of* the world—we are to be set apart. Jesus prayed in John 17:17 that we would be sanctified by the truth, the Word of God. The Word of truth sets us apart from the world when we live by it and make it our daily bread. Even though as people we are all different and set apart from each other for various reasons, we are chosen by God to be set apart for Him. And sanctification is a process God uses to set us more and more apart from the world as we walk with Him daily.

Are you living a life set apart and made holy – sanctified – by the Word of God? Did you know that the Lord has set you apart for Himself? Take time today and look up verses on being set apart and sanctified. We need to take this calling very seriously and learn what a sanctified life really looks like. Pray that the Lord will guide you and that His Holy Spirit will teach you how to live a life set apart from this world, even though you live in this world.

January 28

Today's Reading: Exodus 19-20; Matthew 18:21-35
Today's Thoughts: Patience

Therefore be patient, brethren, until the coming of the Lord. See how the farmer waits for the precious fruit of the earth, waiting patiently for it until it receives the early and latter rain. James 5:7

I always cringe just a little (if not a lot) when I see the words "be patient." Not long ago we spoke at an event that had an interesting theme. The theme for this event embraced the honey "bee" and even the decorations had some kind of bee on it. There were stuffed animal-type bees, candy bees, and even flower corsages that a bee would love to buzz around. You get my drift. When we sat down at a table, there was a bee placed in front of each glass. The bee had a saying on it. Mine said "be kind." The one next to me said "be patient." I was so glad I got the "be kind" bee. Then, out of the blue, we were asked to move over one place and there it was—"be patient" staring back at me. I knew I was in trouble.

Since then I have had to wait on many things. Traffic seems extremely slow, far more than normal. This morning, we got stuck in a parking lot that was being paved and literally could not get out of it. We were running late for a meeting. And then, I went to the eye doctor. At five minutes past the hour, I heard him coming. Yea, he is right on schedule! Then, an emergency phone call came from nowhere. I waited 45 minutes. I sat in a semi-dark room and had a talk with God. He told me to be kind to everyone in that office and to keep waiting patiently. As I left, the doctor shook my hand, gave me a discount on the office visit, the receptionist gave me some free stuff, and another girl expressed to me that it had been a very long day for them. She sincerely thanked me for being "patient." I smiled, they smiled and I am pretty sure God smiled too.

When you get that little message and know in your spirit that you are being tested on something, like patience, stop and pray. The Lord had to remind me of the fruit that He wanted to bear through me – not about me, but about Him. Be alert to those things that God wants to do through you today. Stop and pray and let the Lord be your guide. You never know how it will turn out to bless someone else.

January 29

Today's Reading: Exodus 21-22; Matthew 19
Today's Thoughts: Faith or seeking a sign?

But He sighed deeply in His spirit, and said, "Why does this generation seek a sign? Assuredly, I say to you, no sign shall be given to this generation." Mark 8:12

There are four gospels that describe the life of Jesus, from His birth to His death. Even though these accounts were written by different men at different times, many of their stories are literally identical in the details. Even though each gospel is unique in its own way, many of the miracles and events are shared throughout them. How is this possible? Have any of us ever told the same story the exact same way the second time? And anyone who tells the story next would no doubt make changes to it. The only way this is possible is through the Holy Spirit of God. Paul tells us in 2 Timothy 3:16 that "All Scripture is given by inspiration of God." Yet many people still refuse to believe that the Bible is completely inerrant and perfect in its every detail.

One of the miracles told in all four gospels is how Jesus fed the people with a few fish and some loaves of bread. On one occasion He fed five thousand people and on another, He fed four thousand, having leftovers each time. In Mark chapter eight, the religious people are the ones questioning Jesus, asking Him for a "sign." After all that Jesus had done in this miracle alone, the people still did not believe in Him. What other sign could He possibly give them? Jesus performed miracle after miracle for the three years of His ministry, in full view of everybody looking. He even foretold of His death, which happened just as was prophesied, but many would still not believe. I am sure the Lord sighs deeply even now, in those times when He shows Himself so clearly to someone, yet they refuse to believe Him. They want more proof.

There is a reason that Hebrews 11:6 states that *faith* is the only way to please God. The Lord knows how hard it is for us to really believe in Him and His promises. Look up and see all that Jesus is doing in your life right now. Do you trust Him or are you still seeking a sign from Him for more proof? If you are struggling in your faith, ask the Lord to increase your faith. So often in life, we get so far down the road before we realize what God did for us a long time before. If only we could realize His signs when He is working in our lives and praise Him for His faithfulness to us. Are you seeking a sign today? Seek Jesus first and you may just notice all kinds of signs that you have been missing.

January 30

Today's Reading: Exodus 23-24; Matthew 20:1-16
Today's Thoughts: Little by Little

Little by little I will drive them out from before you, until you have increased, and you inherit the land. **Exodus 23:30**

In Chapter 23, God grants the Israelites various promises, including utterly overthrowing and completely cutting off their enemies. The Lord further states that He will bless their bread and water and take their sicknesses away. With the Lord on their side, the Israelites cannot lose. But in verse 30, the Lord says that He will fulfill His promises by driving out their enemies little by little. Why? Why doesn't God just wipe their enemies out like He did the Egyptians as they entered the Red Sea? What benefit is there to do it little by little?

Exodus 23:29 says "I will not drive them out from before you in one year, lest the land become desolate and the beast of the field become too numerous for you." In other words, a place of instant abundance and blessings would be harmful for the Israelites. Blessings are not true blessings unless they come when we are able to handle them. Think of areas in your life that you would love to see the Lord work in you today. Maybe you need deliverance from sin that seems to have trapped you or maybe it is your finances. The list can go on and on. God wants to deliver you and bless you, but He will only give you what you can handle. God knows what is best for us. Even the blessings can become curses if we do not have the wisdom, maturity and honesty to handle them.

The Lord's timing is never too late nor too slow. He is patient and kind, assessing the whole situation, so that when He fulfills His promises in you, you will be able to receive them in peace. We want the quick fix or the instant gratification. We want the immediate and the miraculous. However, God's ways are not our ways and He desires to give us a future life in His promises. Little by little, complete victory will be yours by having faith and patience day by day.

January 31

Today's Reading: Exodus 25-26; Matthew 20:17-34
Your Thoughts on Today's Passage:

February 1

Today's Reading: Exodus 27-28; Matthew 21:1-22
Today's Thoughts: Just Ask

So Jesus answered and said to them, "Assuredly, I say to you, if you have faith and do not doubt, you will not only do what was done to the fig tree, but also if you say to this mountain, 'Be removed and be cast into the sea,' it will be done. And whatever things you ask in prayer, believing, you will receive." **Matthew 21:21-22**

My sister called the other day complaining about her boss at work. She began telling me about the problems she was having on her job and how they were beginning to disrupt her home life. How many of us can relate to this? After listening to her stories, I realized that the root of the problem stemmed from a breakdown in communication between my sister and her boss. As the picture came into focus, I realized that my sister had issues (wants, needs, desires) that needed addressing. However, she was not telling all of this to her boss. In other words, she somehow expected her boss to know and understand what she needed. I told her that maybe she should *tell* her boss what her issues were and *ask* for help. Sounds easy enough, but for some reason we just do not know how to ask for the things we want or how to discuss the problems we are having in certain situations.

What about our prayer lives? Do we ask the Lord for the desires of our hearts? Do we honestly tell the Lord our problems, even when they involve Him? Today's verses clearly instruct us on how to pray. We are to have faith and not doubt. We must come to the Lord in prayer and truly believe in our hearts that He hears us and will answer us. The faith to move mountains is amazing faith and is available to all of us, if we believe. Try not to get hung up on how to pray, what to pray and praying God's will. Just go to the Lord and ask, believe, and receive. Even though God already knows our hearts' desires, we need to ask Him for them because it increases our faith and blesses us in the process. Remember this: the Lord will answer as He sees best. That is why faith is so important in our prayers. We must trust in God's final answer, regardless of what we think. If our prayers are continuing without ceasing, alongside His Word, and from our honest hearts, then His answers will be received in our hearts of faith.

February 2

Today's Reading: Exodus 29-30; Matthew 21:23-46
Today's Thoughts: Growing Up

When I was a child, I spoke as a child, I understood as a child, I thought as a child; but when I became a man, I put away childish things. **1 Corinthians 13:11**

When we first come to Christ, we are like babies. We crawl, then we wobble and stumble until we finally start to walk. Once we learn to walk, we want to walk faster, farther and without hindrances. Then we realize that the only way to go is upward, aiming for heaven. We do not want to move sideways any longer, but straight up. So we look up and try to figure out how to move up with the same pace... farther, faster, and without hindrances.

But God does not desire for us to move up as of yet. In John 17:15, Jesus prayed, "I do not pray that You should take them out of the world, but that You should keep them from the evil one." Being taken out of the world would be to move up and to get out of here. That is the natural desire of our heart when we really come to know Jesus. We just want to be with Him and we just want to see Him and please Him. Wouldn't it be great if the moment we met the Lord on earth, He brought us up to heaven? We would accept Him in faith on earth but immediately see Him face-to-face in heaven. But...that was not His will nor His prayer for us. His desire is that we learn to advance from crawling to walking. His desire is that we walk by faith. Walking by faith does not come naturally. The believer's stages of development are not easy but necessary. The Lord wants us to become mature, healthy, spirit-filled believers able to continue the work He began. Growing up is difficult because it always involves discipline. We are not born physically or spiritually knowing what is right and wrong. Many times we need to learn through trial and error. God is patient and loving to guide us and love us through those hard times of discipline. So, the Lord's plan for us is not to *move up* but to *grow up*.

Pray about it: *Oh Lord, thank You for being my Lord. You love me so much to help me through every stage of growth and maturity. Remind me that You are always on my side, guiding me and directing me to the next stage for my own good. Mature me so that You may use me. I love You Lord. Let my life reflect my love for You.*

February 3

Today's Reading: Exodus 31-33; Matthew 22:1-22
Today's Thoughts: You Are Called by Name

Then the Lord spoke to Moses, saying: "See, I have called by name Bezalel... And I have filled him with the Spirit of God, in wisdom, in understanding, in knowledge, and in all manner of workmanship. And I, indeed I, have appointed with him Aholiab... and I have put wisdom in the hearts of all who are gifted artisans, that they may make all that I have commanded you" **Exodus 31:1-6**

Do you know that the Lord has called you by name and that He has a specific purpose for your life? Do you know that not only has He called you but also that He has filled you with His Holy Spirit who will teach you all things? (1 John 2:27) Even if you answer "yes" to these questions, you may not fully grasp the meaning, most of us probably do not. The two men named in the above verses were appointed and anointed by God to complete a specific task to design and construct the tabernacle. Ordinary men performed an extraordinary and supernatural task. We also have the indwelling Spirit of God who wants to guide us in performing extraordinary and supernatural tasks. But, *how*?

We first must believe that God's Word is true, without exception or compromise. He has promised us His Holy Spirit to be our Helper and Teacher (John 14:26) and further says that we are sealed with this promise (Ephesians 1:13). Throughout the Book of Acts several accounts are mentioned regarding the infilling of the Holy Spirit. When we accept Jesus into our hearts, He enters in and lives within us through His Spirit. But the first step is to *believe*.

As we claim this promise as truth, we need to start asking for the same kinds of gifts as given to the two men in Exodus chapter 31. We should pray for wisdom, knowledge and understanding so that in all things we are seeking to do God's will and to fulfill His purpose. Paul says in Ephesians 4:1 "to walk worthy of the calling in which you were called" and the only way we can truly know our calling is to seek a spiritual wisdom and understanding that only comes from God. Through His Holy Spirit, He will teach us and lead us to perform and complete the tasks He has chosen for us.

To summarize, if you have received Jesus Christ as your Savior, then you are filled with His Holy Spirit. But to truly live out your calling, you must believe in faith in what His word teaches and pray for the spiritual wisdom, knowledge and understanding to apply His guidance to your life. Bezalel and Aholiab were ordinary guys called for a supernatural purpose. They received, they believed and they performed. Start praying today for the Lord to train you to yield more to His Spirit and to teach you with His wisdom how to walk worthy of your calling.

February 4

Today's Reading: Exodus 34-35; Matthew 22:23-46
Today's Thoughts: Go to the Lord First

He also hired a hundred thousand fighting men from Israel for a hundred talents of silver. But a man of God came to him and said, "O king, these troops from Israel must not march with you, for the LORD is not with Israel-not with any of the people of Ephraim. Even if you go and fight courageously in battle, God will overthrow you before the enemy, for God has the power to help or to overthrow." Amaziah asked the man of God, "But what about the hundred talents I paid for these Israelite troops?" The man of God replied, "The LORD can give you much more than that."
2 Chronicles 25:6-9

Amaziah was king of Judah and though he did what was right in the eyes of the Lord, he did not follow the Lord wholeheartedly. One day Amaziah decided to muster his army, so he took count all of his able-bodied fighting men. Thinking he needed more troops, he hired another hundred thousand fighting men from Israel. Now at this time, the kingdoms of Judah and Israel were divided, each with their own kings. The Lord was with Judah because of His covenant with David but the Lord was not with Israel, especially as their kings were evil, wicked and against the Lord. So, the Lord sent a prophet, "a man of God," to speak with Amaziah about sending the men from Israel back home. Even though Amaziah did as the man of God instructed, his first concern was about the money that he had already spent. The response from the man of God should speak to our hearts even today; "The LORD can give you much more than that."

Sometimes we fail (or forget) to realize that our God is in charge and has ownership of all things, even all the money of the world. We can get so caught up in making things balance out that we think we do not need to depend on the Lord. Many of us hate to waste money or waste things of value. We try to be good stewards. We try to make the right decisions. But, sometimes we act before we pray and we make decisions that may not be God's will for us.

The best course in making decisions is to take them all to the Lord first. But if you find yourself in a place of looking backwards (as we all do at times), then ask the Lord to work it out and even to replace what was lost. You will be amazed to watch how God works through our situations, even our mistakes, and He can replace what was lost in the process. Do not bear the burden alone—take it to the Lord. He loves to rescue. The rest of the story of Amaziah? He went into battle after sending the hundred thousand troops home and the Lord delivered him the victory. He will give us the victory when we seek His help in all things.

February 5

Today's Reading: Exodus 36-38; Matthew 23:1-22
Today's Thoughts: Know, Believe and Understand

***For God so loved the world that He gave His only begotten Son that whoever believes in Him should not perish but have everlasting life.* John 3:16**

I remember the time that I stood up in Sunday School class and quoted John 3:16 by memory. My voice cracked and my knees knocked as I nervously recited the words of this verse. To this day it remains one of my life verses, not only for this life, but also for my eternal one as well. The love of God has given me the assurance of everlasting life because I believe in His Son and what He did on the cross. I know He loves me.

More than once we have had people come up to us after one of our teachings and share with us that they have finally believed that God truly loves them. Many people who have known the Lord for years and have served Him in faithful works do not fully understand or believe that God loves them unconditionally. The world is seeking to find true love, but in all the wrong places. True love can only be found in the One who created us, our Heavenly Father. For God so loved the world...and yet so many refuse to believe.

I pray today that you will know the love of God. It is His will for us to "know, believe and understand" who He is. (Isaiah 43:10) If you are struggling with knowing the love of God, ask Him to show you and to help you believe and understand who He is and who He wants to be in your life. Do not waste another day wondering...seek out the truth.

February 6

Today's Reading: Exodus 39-40; Matthew 23:23-39
Today's Thoughts: Working Weary or Working With the Lord?

Thus Moses did; according to all that the Lord had commanded him, so he did. And he raised up the court all around the tabernacle and the altar, and hung up the screen of the court gate. So Moses finished the work. **Exodus 40:16, 33**

Have you felt overwhelmed by too much work? Many times, we are overloaded with overbooked calendars and too little time to do the things we have to do, not to mention the things we would like to do. Most of us are always busy trying to complete the tasks or jobs for the day. Our society is works-based in every way imaginable. The eighties were a decade that made longer workdays the in-thing. Forty-hour weeks were considered short in comparison to the more popular sixty to eighty-hour weeks. Stay-at-home moms had to get busier too. Today, with advances in technology and a "faster is better" mentality, doing more is a must. Even our kids are obliged to participate in extra-extracurricular activities, thinking they are necessary for higher education opportunities. As a result, many of us are worn out, stressed out, and burned out from so much work.

In studying the construction of the tabernacle and its articles, we do not read of any overwhelming or stressful problems. The Lord commanded Moses who commanded the people to do all that the Lord had spoken. Simply put, Moses did what the Lord commanded and finished the work. What are we to learn from this story? The work that Moses did was given to him by the Lord. Moses followed the instructions to the exact detail, nothing added, nothing omitted. The Lord gave the gifts that were needed to do all of the designs and the Lord had already provided the required materials. The building of the tabernacle is a beautiful example of a perfect union between God and man. And the results were nothing short of miraculous.

Think for a moment about the work you are doing today. Did the Lord instruct you to do these things? Has God given you the tools and talents to carry out the work? Are you working in unison with God, following His instructions and doing all that He says? The things we choose to do outside of the Lord's plans for us will lead to fatigue, frustration and futility in our efforts. We get so busy just being busy that many of us do not stop to include the Lord in our activities.

Have you spent time with the Lord today? Is He part of your "work" day? Maybe there is a tabernacle project in your life and God wants to do something miraculous through you. Pray that today is the day you let God take over all of your work. Get together with Him and see what happens.

February 7

Today's Reading: Leviticus 1-3; Matthew 24:1-28
Today's Thoughts: In Order

The sons of Aaron the priest shall put fire on the altar, and lay the wood in order on the fire. **Leviticus 1:7**

I was reading a few weeks ago about Abraham preparing the altar for Isaac. Abraham had laid the wood *in order* on the altar before placing Isaac on it. The more I looked at the words "in order" the more I wondered why the wood had to be laid in a specific manner. How could wood be distinguished in such a way as to know if it was "in order" or not? The Lord brought to my mind the other ways that He established His order for things. In Exodus 28:10, the names of the tribes of Israel were placed on the breastplate of the ephod in order of their birth. In Exodus chapter 40, Moses places the articles of the tabernacle in the order as God had instructed him. Even the showbread was placed "in order" on the table "as the Lord had commanded Moses." (Exodus 40:23) We see God's purpose for order in many places in Scripture.

Our God is a God of order and He takes it very seriously. Isaiah, the prophet, told King Hezekiah "Thus says the Lord: 'Set your house *in order*, for you shall die, and not live.' (2 Kings 20:1) Hezekiah was not going to die because his house was not in order, but because he had a sickness unto death. The Lord wanted his house in order before he died. This priority of order is also seen in the New Testament as Paul would tell the Corinthians to "let all things be done decently and in order." (1 Corinthians 14:40) He was referring to how the gifts were being used in the church as relating to the conduct of the people.

Is your house in order today? Did you know that is an important part of our walk with the Lord? When our houses are in order, we have more peace and contentment than when they are in disarray. As women, we set the tone for the home and are accountable before the Lord for how we honor and obey Him in our responsibilities. Are you struggling with disorder? Ask the Lord to reveal ways you can improve things that are out of order in your life. If God wants wood to be put in order, how much more does He want our homes and days in order? We just need to start asking Him to show us how. We need to be willing to give Him the first fruits of our days and let Him lead us in the rest of it.

February 8

Today's Reading: Leviticus 4-5; Matthew 24:29-51
Today's Thoughts: The Power of Unity

1 Now the whole earth had one language and one speech. 4 And they said, "Come, let us build ourselves a city, and a tower whose top is in the heavens; let us make a name for ourselves, lest we be scattered abroad over the face of the whole earth." 5 But the Lord came down to see the city and the tower which the sons of men had built. 6 And the Lord said, "Indeed the people are one and they all have one language, and this is what they begin to do; now nothing that they propose to do will be withheld from them. **Genesis 11:1, 4-6**

The tower of Babel is a familiar story in the Old Testament. After the flood, Noah and his sons were fruitful and multiplied, populating the "whole earth." The people had one major commonality: they all spoke the same language. In the course of time, many of them moved into the same region and decided to build a city together. This city would have a tall tower, one so spectacular that they would be well-known for creating such an awesome sight. Their intentions concerned the Lord as He looked upon what they were doing. Once again, man was heading in a direction of self-serving, self-loving ambition that would only lead them farther away from God. God intervened and confused their speech. No one understood each other anymore, hence Babel means confusion. The tower of Babel represented the place where the people's language was no longer of one accord. God scattered the people abroad, with only those who understood each other staying together and multiplying.

Though this story is quite familiar, as I read through it again, I noticed something very interesting. In verse 6, the Lord recognizes that the people have a unity of such power that "nothing that they propose to do will be withheld from them." The NIV Bible uses the phrase "nothing will be impossible" for them. Even though these people were in sin, their unified focus gave them an extraordinary ability to achieve the impossible. How much more can we achieve if we work together as the body of Christ? The apostle Paul tells us that gifts were given individually to those in the body of Christ to work together as one for the whole body of Christ.

As Christians, we need to recognize our power in being unified together. When we stand together for a common purpose, surely nothing can stop us. Unfortunately, we too often stand alone. Pray about getting involved in your church ministries. How can you join forces with the body of Christ to help achieve certain goals? The only thing that usually stops us is fear. But don't worry, instead of God coming down to confuse our language, He will come down to help us better understand each other through His Word.

February 9

Today's Reading: Leviticus 6-7; Matthew 25:1-30
Today's Thoughts: To Seek Him Face to Face

So I said: "Woe is me, for I am undone! Because I am a man of unclean lips, And I dwell in the midst of a people of unclean lips; For my eyes have seen the King, The LORD of hosts."
Isaiah 6:5

As Christians, our desire is to be with Jesus. Personally, I would love to see Him face-to-face. I would talk to Him about my heart's desires and goals, and even if He didn't respond, I would love to read His facial expressions. To be in the presence of the Lord seems like the best thing imaginable. Isaiah did not have to imagine it; however, because he experienced it. Of course, his experience is more accurate than my imagination. Isaiah did see the Lord, but Isaiah did not talk about himself or go over his prayer requests in God's presence. Isaiah was confronted with his sin. Isaiah was also aware of the sins of the people and the inadequacy all of us face in the presence of the Lord. Isaiah realized that we are all undone, unworthy and unclean. We have no way of changing these things for we live in a midst of people and within a world of sinfulness. Seeing the Lord brings forth a holy fear that we are nothing and we deserve nothing.

It is amazing to me that Isaiah brings attention to his unclean lips (instead of his eyes having seen the Lord). The angel had a treatment for his unclean lips, touched by a coal from God's altar. With that process, Isaiah's iniquity was taken away and his sin purged. Why his lips and not his heart or his eyes? If you think about what Jesus taught, the lips make sense in the spiritual realm. Jesus said that out of the abundance of the heart, the mouth speaks (Luke 6:45). Isaiah's lips represent his heart. Isaiah would respond to the call of God, but in order to go and speak for God, his heart needed to be touched.

This same process starts with our eyes being opened to our unworthiness as we come before the presence of God. Then, we will realize that our heart needs to be cleansed so we can represent Him with the words of our lips. This is a continual process. It is easy to be swayed away and think we are being used by God because of our works of righteousness. It is not about us. We need to be about Jesus. Today, ask Him to send you. Respond to His call. But first, set your eyes on the Lord's majesty, awesomeness and purity. Understand that you are not worthy to be used but He has made a way to wash you through the blood of Christ so that you can speak for Him from a cleansed heart. All you have to say is, "Here am I! Send me."

February 10

Today's Reading: Leviticus 8-10; Matthew 25:31-46
Today's Thoughts: Not Alone

Then the rib which the Lord God had taken from man He made into a woman, and He brought her to the man. And Adam said: "This is now bone of my bones And flesh of my flesh; She shall be called Woman, Because she was taken out of Man." Therefore a man shall leave his father and mother and be joined to his wife, and they shall become one flesh.
Genesis 2:22-24

When God put Adam in the Garden of Eden, He gave him everything he needed. Adam was in charge of the plants and the animals. He would never get hungry or thirsty. He never had to worry about bad weather or shelter from the elements. Best of all, Adam had a personal relationship with God, one can only imagine how wonderful that must have been. But even with all of his needs being met, God knew that Adam was alone. None of the animals served as a comparable mate, even though Adam knew them and named them. God loved Adam so much that He made a mate just for him – a woman.

In society today, there is much debate about the roles of man and woman, especially related to marriage issues. There are those who disagree with the biblical account of what God intended in creating man and woman and why He put them together in the first place. We get hung up on the concepts, duties and rights of marriage to the point that the true meaning of it is lost. The most basic element of marriage is companionship, to never have to be alone. God's heart for us is that we never feel alone. Whether married or not, we need friends and relationships that encourage and edify us. Loneliness can be debilitating and depressing to the point of death.

Where are you today? Most of us are living in the Garden, where our basic needs are met, but what about true companionship? If you are lonely and feeling isolated, start praying to the Lord for His help. The enemy will keep you isolated as long as he can. Take the steps to get involved in a Bible study with small groups and fellowship opportunities. Volunteer and get involved to help others. In helping others, you also help yourself. You never know what God has in store for you. One thing you can be sure of – He does not want you to be alone.

February 11

Today's Reading: Leviticus 11-12; Matthew 26:1-25
Today's Thoughts: Extraordinary Prayer

"God heeded the voice of man" Joshua 10:14

There are numerous scriptures and stories throughout the Bible that encourage us to pray. Prayer is so important in our lives as Christians, yet it often gets reduced to either 911-type cries for help or routine words that we are accustomed to praying. In the case of Joshua and the battle with the Amorites, he asked the Lord for a miracle of major proportions. Joshua dared to ask God that the sun would not set until their enemies were defeated. Not only did God answer Joshua's prayer but verse 14 further states that "there has been no day like that, before it or after it, that the Lord heeded the voice of a man; for the Lord fought for Israel." The Lord listened and responded to the voice of man.

Does God heed the voice of man today? James 5:16 says that the effective fervent prayers of a righteous man avails much. Jesus tells us to ask, seek and knock and that whatever we ask in His name will be given. The key to answered prayers, however, is in praying God's will. Because we have the Holy Spirit and the Word of God, we can know how to pray God's will for our lives (I should add "better know" because there are still times when we all struggle to truly understand how to pray the will of God.) But regardless of what we know or how we pray, the point is that we need to pray to the Lord for everything. We should not hold back from asking the Lord for anything we want. It is up to Him to decide if or how He answers.

This day in Joshua's life was extraordinary to say the least. God gave him extra daylight time to completely destroy the five kings who had fled from him. God not only answered Joshua's prayer, but He divinely intervened in this battle Himself. When God has a plan, no one can thwart or hinder what He is going to do. Sometimes we need to pray just for peace in the situation we are in, knowing that the Lord will have His way no matter what we do or do not do. God kept His promises to Joshua and the Israelites and He demonstrated His faithfulness to them in amazing ways.

Is this the same God in our lives today? Absolutely! Do not stop praying and do not allow the enemy to hinder your faith in your prayers. The Lord may not extend our days by keeping the sun from setting, but there are miracles that He does perform in our lives that hold just as much impact. Has God given you promises? Believe in faith that He will fulfill them in your life, maybe not in your timing or by your methods, but in His. He is the same God today as He was to Joshua. Are you expecting the same of Him?

February 12

Today's Reading: Leviticus 13; Matthew 26:26-50
Today's Thoughts: Seek His Promises

**"And blessed is he who keeps from stumbling over Me."
Matthew 11:6**

While teaching on Joshua 1, I realized that God clearly told Joshua that He fulfills His promises. God even clearly restates what those promises entailed. However, today, many people do not know how to get promises from God. And even if God is speaking to them, many Christians do not know how to wait it out, and so they fail to watch God work it out to completion. As a result, many Christians stumble over Jesus and the way He works in our lives. But that is not God's intention as He desires to speak to us and reveal to us things that are to come.

My mom came over today and I asked her how she would teach someone to get promises from God. Immediately, she started giving me illustrations of when God gave her promises. So, I asked her again, "But how did you learn to get a promise from God?" She used another illustration of a promise God gave her to try to speak in the generalities of teaching someone else. I was thankful for the discussion because I had also realized that it is very difficult to come up with a "basic formula" in receiving a promise from God.

God's Word is filled with promises He gave in the past which can be applied to my present circumstances. I can give a very simple way of how I get a promise from God. First, I start by praying to Him about a burden while asking for His will to be done in it. It is not God's will for me to be stressed out and unsettled, so I start reading the Word, looking for a passage that addresses what I am praying about. If a verse pops out, I will start meditating over it, asking God about it and looking up the cross references for it. I know that it is a promise from God if it addresses the burden on my heart in future terms while giving me a present day peace. I also know that it is a promise and not a command if I can't do anything to fulfill it by myself. God wants me to trust and have faith and He will accomplish the rest.

Today, try to read segments in the book of Isaiah. God has given me so many promises in that book. Keep seeking the Lord's will until He gives you a verse. Then trust Him. He is more than able to address and accomplish all that concerns you today and every day!

February 13

Today's Reading: Leviticus 14; Matthew 26:51-75
Today's Thoughts: You are Not Alone

For we do not have a High Priest who cannot sympathize with our weaknesses, but was in all points tempted as we are, yet without sin. Let us therefore come boldly to the throne of grace, that we may obtain mercy and find grace to help in time of need. **Hebrews 4:15-16**

Have you ever poured your heart out to someone but felt worse afterwards because you knew that no matter how hard you tried to express your feelings, they just did not understand? True understanding and empathy come when the person you are sharing a personal problem with has lived through the same type of situation. Several years ago my dad died of a heart attack, suddenly and without warning, and the shock was painfully tough for our family. But comfort came from those who experienced similar traumas, including my two half-sisters whose dad had died a few months earlier. Their dad died the same way, and they were dealing with the same type of pain. Through the pain and suffering a close bond was formed between us as our hearts were drawn to each other, and even to this day, we share a relationship like sisters who have suffered together.

When I think of the sufferings that Christ endured here on earth, I am comforted that He understands my sufferings. No one was more rejected, ridiculed, misunderstood, and alone than Jesus. Even though He was fully God, He was also fully man. Jesus dealt with the same emotions and temptations in his flesh just as we do, yet He was without sin. But Jesus never focused on His own pain; His concern and compassion was for the people. Even when His friend Lazarus died, Jesus cried. He understood the pain that Lazarus' sisters were feeling and He knows today the pain that we feel. For me, I try to remember these things in my weakest moments, and I know that I am not alone.

Are you hurting today? Maybe you feel as though no one really understands or even cares. Jesus says to "come boldly to the throne of grace" and there you will find help in your time of need. Come boldly today and seize the mercy and grace that is freely given by Jesus. His sufferings were for just this reason, so that we can all pour out our hearts to the One who truly understands and the only One who can truly help us. Go to Jesus today. He loves you, He understands you, and He will help you.

February 14

Today's Reading: Leviticus 15-16; Matthew 27:1-26
Today's Thoughts: As Christ Loves Us

And now abide faith, hope, love, these three; but the greatest of these is love. **1 Corinthians 13:13**

Valentine's Day is celebrated as a day to express love and most notably, romantic love. The history of Valentine's Day, along with the saint for whom it is named, is shrouded in mystery. Both Christian and ancient Roman traditions make up its origins, but the exact details of what happened to St. Valentine are uncertain. What is certain is that through a series of events dating back to the middle ages, February 14 became the date to commemorate *Valentine's Day*. Roses and chocolates will be sold at exorbitant prices and in massive quantities, all in hopes to express our love in that special way. What is the most special way to express our love?

The thirteenth chapter of First Corinthians is known as the *love* chapter of the Bible. The apostle Paul writes about true love, what it is and what it is not. It "suffers long and is kind; love does not envy; love does not parade itself, is not puffed up;" (verse 4). It "does not behave rudely, does not seek its own, is not provoked, thinks no evil;" (verse 5). Paul puts it very simply in verse 8, "Love never fails." Flowers fade and candies melt, but true love never fades, never dies, never fails. The most special way to express our love is to practice these qualities Paul speaks of as we learn to love beyond ourselves.

Romantic love is a wonderful feeling and being in love is truly a gift from God. Those feelings of heightened joy and excitement seem to make everything else less significant. But today, let's ask the Lord to help us love beyond the feelings of the gifts and the favors. Let's pray that we can love as Christ loves us and that we can express that love in ways that bless those around us more than anything else we can give. Why not make today a day to take love more seriously than you ever have before? Pray that love becomes the motivation for all that you do. Enjoy the flowers and candies and candlelight dinners, but remember to keep love in your heart, not just in the festivities.

February 15

Today's Reading: Leviticus 17-18; Matthew 27:27-50
Today's Thoughts: How to Have a quiet time with God.

The following is an excerpt taken out of the Daily Disciples Prayer Journal: Building Divine Prayer Time.

Here are some of our ideas to help you make your "quiet time" a habit…

Choose a time and stick with it…whatever time you choose, make it a priority. You can put up sticky notes to remind yourself on the bathroom mirror, car dash board, by the telephone or even on the refrigerator. It takes 21 days of doing something consistently to make a habit.

Choose a place…some place where you can be alone and not distracted. Choose a consistent place where you can keep all your materials. And you want a place that you go often.

Open with a prayer …start talking to God and asking Him to hear your heart.

- Refer to "How to Pray" for tips on praying on www.dailydisciples.org website
- Be honest with the Lord and tell Him how you are doing
- Focus on Him first, then on your needs
- Keep your time frame in mind, if you are limited for time

Open your Bible…depending on your available (try not to short cut this time). Pick a reading plan that you will follow everyday.

Keep a Prayer Journal…write down any notes on the verses and your prayers.

We encourage you to have a quiet time with God. That is how your relationship with God will grow. Be sure that you listen with your eyes as you read the Word and then be quiet before Him as your ears will hear His voice. If you are looking for more instruction, we have Bible studies, CDs, DVDs and we wrote a book called, "Building Divine Intimacy with God." They are on our e-store at www.dailydisciples.org

February 16

Today's Reading: Leviticus 19-20; Matthew 27:51-66
Today's Thoughts: An Idol in Your Life?

Do not turn to idols, nor make for yourselves molded gods: I am the Lord your God. **Leviticus 19:4**

When we think of idols, most of us do not think of worshiping some molded statue. Today, some religions do bow down to gods with the image of a man, woman or creature, but certainly not Christians. We know that an idol is anything that we prioritize over God, not necessarily a literal man-made god that we worship. But then again, we do have the "Oscars." Each year Hollywood gives out these little gold statues to its most honored artists. The evening gowns, the jewelry and the parties are also a big part of this night of awards and acclamations. However, it is not just for those who attend the event and are nominated for the golden honor…it seems that many of us get caught up in the excitement as well.

As Christians, what are our responsibilities in supporting or not supporting this industry? How much should we care about the movies and about these little golden statues? For some, it is harmless entertainment. For others, it is a continuing sign of moral decay in our society. Where are the lines for us as Christians? The answers to these questions are based upon a personal conviction for each one of us. If we somehow are choosing these idols over our Lord, then maybe we should really pray about our level of interest.

The bigger issue is that God warned His people to turn from their idols and false gods because He knew where it would lead them…to destruction. The same is true for us today. When we focus on things outside of Jesus, our focus is marred with images that are not healthy. No greater place breeds false images than Hollywood. Think about your level of interest. If you feel a nudge in your spirit, ask the Lord to show you areas that need refocusing. We must also remember that our minds tend to replay images over and over in our minds, so be careful of the content in those images. If you find yourself thinking on things of the world too much, ask the Lord to change your focus.

February 17

Today's Reading: Leviticus 21-22; Matthew 28
Today's Thoughts: What Spiritual Temperature are You?

I know your works, that you are neither cold nor hot. I could wish you were cold or hot. So then, because you are lukewarm, and neither cold nor hot, I will vomit you out of My mouth. **Revelation 3:15-16**

In my house, temperature can be a "hot" topic. In the winter, the thermostat can never be too hot for me, but my husband loves to sleep in the cold. I like foods taken straight out of the fridge but he likes them steaming hot out of the oven. One extreme or the other tends to dominate in several areas in our home, but how extreme am I when it comes to the things of the Lord? What is my spiritual temperature? How many areas of my life am I content with just being lukewarm, especially when it comes to living for Jesus?

We as Christians are often afraid to live at the extremes. We are uncomfortable standing out and speaking out too much. We rationalize our positions by saying that we do not want to offend anyone or come across as too zealous in our faith. We even justify sinful behaviors by telling ourselves that if we blend in with the crowd, then we can be a more effective witness. Where did these ideas really come from? Are they from the Lord? Not according to Revelation 3:16. Jesus says He will "vomit" us out of His mouth if we are lukewarm. He wants us to pick a position; we are either cold or hot, for Him or against Him. We either take up our cross and follow Him with our whole hearts or we follow our own desires. To live in the gray areas is unacceptable, regardless of our human rationalizations.

Where are you today? Do you have a lukewarm relationship with Jesus? I fear greatly for many people who sit in church every Sunday proclaiming to know Jesus but having no evidence of Him in their lives. Many of us look righteous on the outside but are numb on the inside. Our churches are filled with complacent Christians who are quite content to live in the gray areas, not wanting to get too uncomfortable. Will Jesus say "well done good and faithful servant" (Matthew 25:23)? Or will He say "I never knew you; depart from Me" (Matthew 7:23)? Ask the Lord to light a fire in your heart today that will set a blaze in any area of your life that has become lukewarm and complacent. Life here on earth is but for a moment, but eternity is forever.

February 18

Today's Reading: Leviticus 23-24; Mark 1:1-22
Today's Thoughts: What God Calls You

But you are a chosen generation, a royal priesthood, a holy nation, His own special people, that you may proclaim the praises of Him who called you out of darkness into His marvelous light; 1 Peter 2:9

Have you ever thought of yourself as royalty? Most likely, you have not. We think of kings and princes as royalty, with luxuries and extravagances that far exceed our ways of life. But what is "holy" royalty? Have you ever thought of yourself as holy royalty? Did you know that you are considered *holy – special – chosen* by God? Look closely at the verse and you will see that the second word is "you." The Lord is speaking to *you*, personally. So often we tend to see ourselves in the negative. We may believe that Jesus saved us from our sins, but to believe that He sees us as royalty is much harder to accept. Today, Jesus wants you to know and believe the truth of this verse. You are His own special person. He chose you; He loves you; and He calls you His royal priest. Jesus called you out of darkness so you can live in His marvelous light.

If we could only truly grasp how much Jesus loves us! While we were yet sinners, Christ died for us (Romans 5:8). When God looks at us, He sees His Son, Jesus. Because we are covered in the blood of Christ, we are holy and blameless (Philippians 2:15) before God. What are we to do in return? We are to proclaim the praises of God. We should be singing God's praises from the rooftops, giving glory and honor to His unfailing love and unending mercies. Despite being in the midst of a crooked world, He has set us apart to glorify Him.

Take time today to thank the Lord for choosing you as His own and for calling you out of the darkness. Spend time worshiping and praising Him for His goodness. The Lord has clothed you in royalty and righteousness. Look up, see His glory and believe with all of your heart that He has a plan and a purpose for your life. (Jeremiah 29:11) Proclaim your praises of Him to others and let your light shine so that others may see.

February 19

Today's Reading: Leviticus 25; Mark 1:23-45
Today's Thoughts: Relationship Wins Over Religion

***Therefore, as through one man's offense judgment came to all men, resulting in condemnation, even so through one Man's righteous act the free gift came to all men, resulting in justification of life. For as by one man's disobedience many were made sinners, so also by one Man's obedience many will be made righteous.*
Romans 7:18-19**

When we begin a new Bible study, regardless of the book or topic being taught, we always do an overview that explains why we are sinners and why we desperately need a Savior. We usually start in Genesis Chapters one through three and discuss what happened in the Garden when man fell and sin entered the human race. After Genesis Chapter three, the rest of the Bible tells us about Jesus, our salvation through Him, and our reconciliation with our Heavenly Father. Our only hope for eternal life with our Lord in heaven rests solely on our faith in His Son Jesus Christ.

We do not teach about a religion but about a relationship with Jesus. Jesus came and died for us so that we could have a relationship with Him, regardless of the sin that exists within us. One of the hardest struggles for Christians is resting in the work that Jesus has already done for us. Once we believe in Jesus, truly receive Him into our hearts, then, we are filled with His Holy Spirit and sealed for the day of redemption. (Ephesians 1:13; 4:30) The seal cannot be broken. But so often, we live as if we still have something to prove to God. So many of us are still trying to work out our salvation in our own strength. We are trying so hard to please God that we are not letting His Holy Spirit truly fill us each day. Tension and stress hurt us in more ways than just physically or emotionally. Any relationship filled with stress and strife will ultimately be in danger of falling apart. The same is true of our relationship with Jesus. He will never leave nor forsake us (Hebrews 13:5), but we pull away from Him, by leaving and forsaking His presence.

February 20

Today's Reading: Leviticus 26-27; Mark 2
Today's Thoughts: Strength in Jesus

Therefore I take pleasure in infirmities, in reproaches, in needs, in persecutions, in distresses, for Christ's sake. For when I am weak, then I am strong. **2 Corinthians 12:10**

The Bible is filled with stories and teachings that go completely contrary to the views of the world. In Matthew 6:24, Jesus warned us that we could not serve both God and mammon (money). Later, the apostle John would write: "Do not love the world or the things in the world. If anyone loves the world, the love of the Father is not in him." (1 John 2:15) But today's Scripture verse seems to go even a step farther. Paul tells us to "take pleasure" in the things in our lives that make us weak, such as sickness and all other types of distresses. How many of us can truly say that we find pleasure in being sick, having financial problems, family struggles or even persecution from various sources?

As Christians, we are told not only to believe in Jesus, but to follow Him. In Luke 14:27-28, Jesus speaks of those who choose to follow Him, that they must leave everything else behind. He tells them to count the cost before they make their decision. We live in a world today that promotes "self" and all of the accolades that go along with self gratification and self satisfaction. We want to be comfortable, we do not want to suffer. Even the weather can dampen our spirits. We have our days so well-planned, so tightly-scheduled, that even one deviation from our plan can send us into a frenzy. Add sickness or some other types of problems and we soon find out where our strength is based. If our strength is not in Christ, then we are headed for a fall. Some of us fall several times a day. Jesus says that when we are weak, He is strong. When we truly follow Jesus, we discover just how weak we really are. It is in the weakness of our lives that He is able to show His power and might. Paul learned to glory in his weaknesses because that is when he saw the Lord work in a more powerful way through him.

We will have troubles in this world, despite our efforts to avoid them. We will have sickness, needs, and distresses, but contrary to the world's teachings, we can take pleasure in these times of weakness. Why? Because Jesus will show us His strength and power in glorious ways. Sin brings these bad things into our world, but Jesus has made a way for us to live through them with peace and joy. Today, instead of complaining about your circumstances, regardless of how painful they may be, ask the Lord to give you His strength through them. Allow yourself to be weak and dependent on Jesus. And, finally, give Him thanks for all that is happening in your life. Look up and see the glory of the Lord, for He is above and controls the activities of this world.

February 21

Today's Readings: Numbers 1-2; Mark 3:1-19
Today's Thoughts: Just the Truth

If we say that we have no sin, we deceive ourselves, and the truth is not in us. If we confess our sins, He is faithful and just to forgive us our sins and to cleanse us from all unrighteousness. If we say that we have not sinned, we make Him a liar, and His word is not in us. 1 John 1:8-10

Many of us equate truth with whatever is in print. I have heard people say, "I read it in the paper. The writers wouldn't write it if it wasn't true." However, I am reminded of two different news articles by staff reporters who lied about their facts. The first one involved a long time employee of the *New York Times*, one of the most prestigious newspapers in the world. The reporter confessed that he lied frequently and deliberately as a means of generating a good story. The second instance that made a lasting impression on me is about a reporter from *USA Today* who committed the same types of deception. Both men covered stories involving the tragedies of 911 and the war in Iraq. Much of their information was falsified to the extreme, using photos of individuals supposedly killed; later, these people were discovered to be very much alive. Not only do these revelations speak volumes as to what goes on behind the scenes of the media world, but also it leaves the public gullible and ignorant, with no way of knowing what the truth is.

Where does "truth" really exist? Do we actually believe that anything written in print is somehow more believable? The Bible says that "if we say that we have no sin, we deceive ourselves, and the truth is not in us." So, in essence, we are all sinners and we are all capable of lying or falsifying information. I think we expect more integrity from those who make a living out of reporting the news. But the lesson for all of us is: *don't believe everything you read.* There is only one source that is truth and it is the Holy Bible. Anything else written by man has the opportunity for errancy and deception. Why? Because we as human beings are sinners, capable of lying and deceiving at every turn. We even justify our exaggerations by calling them "little white lies" or "just stretching the truth." The truth does not "stretch!"

Our only hope is in confessing our sins to "He" who "is faithful and just to forgive us our sins" and to save us from perishing in them. He is Jesus Christ. He is Truth. Today, we must plant His Word in our hearts and minds so that His truth will guide us in our daily lives. Why do we waste so much time reading and listening to the world's news of death and destruction when we have the Good News that brings hope eternal? Think about it and make the switch today.

February 22

Today's Readings: Numbers 3-4; Mark 3:20-35
Today's Thoughts: Listen and Obey

Then the word of the LORD came to Elijah: "Leave here, turn eastward and hide in the Kerith Ravine, east of the Jordan. You will drink from the brook, and I have ordered the ravens to feed you there." So he did what the LORD had told him. He went to the Kerith Ravine, east of the Jordan, and stayed there. The ravens brought him bread and meat in the morning and bread and meat in the evening, and he drank from the brook." **1 Kings 17:1-6**

The message of the Lord to His prophet, Elijah, is readily accepted by the reader. Because the Lord's commands are past tense, we have hindsight to say, "Isn't it great that Elijah obeyed and did what God told him to do?" But if this was a command that God gave you, would you be as obedient? Let's take the black and white words of the Word and look at this in living color. God told Elijah *to hide, to drink from a brook and to eat road kill from ravens*. These commands would be hard for me to accept. Personally, I would doubt if I was really hearing God's voice speaking to me and I would wonder why I was the one running and hiding. But the Bible says in verse 5 that "he did what the LORD had told him."

God will only give us as much as we can handle. As we become faithful to do what the Lord tells us in the small things, He will entrust us with bigger commands. Elijah knew what it meant to listen and obey. He had a mighty ministry but the works of the ministry meant nothing in comparison to the obedience of his heart to do what the Lord had said. God can entrust a mighty ministry to an obedient heart. And an obedient heart just wants to please God regardless of the extent of the ministry or the commands.

We have the Bible; the Bible is the book of what God wants for us. The Lord will speak to you through the Bible if you are really willing to listen. The Lord will lead you if you really want His direction. The difficulty comes when God's word doesn't seem to make sense to our situation. So instead of doing what He says, we choose our own ways. Faith goes against the flesh. Most things God says to me make no sense to my intellect or feelings or circumstances. But, when I listen and obey the smallest of commands, I have seen His faithfulness develop my faith to handle the next command like Elijah. I want to do all the Lord tells me to do. I want to run the race He has set before me, even if it means hiding out and eating ravens' food.

Pray about it: *Lord, help me to be faithful with the little things so that I may be faithful with all You have for me.*

February 23

Today's Reading: Numbers 5-6; Mark 4:1-20
Today's Thoughts: God's Blessings

And the Lord spoke to Moses, saying: "Speak to Aaron and his sons, saying, 'This is the way you shall bless the children of Israel. Say to them: "The Lord bless you and keep you; The Lord make His face shine upon you, and be gracious to you; The Lord lift up His countenance upon you, and give you peace." ' "So they shall put My name on the children of Israel, and I will bless them."
Numbers 6:22-27

When I was in high school, I attended a church in which the presence of the Lord and His peace were evident from the moment I walked through the door. The pastor exposited the Word of God, chapter by chapter and verse by verse in a simplistic, but powerful and compassionate way. At the end of each service, he would close his sermon with these words of blessing from Numbers 6.

If we meditate on these verses, we can discern the heart of God. Three times the Lord uses the word "bless." As the Lord fills us with His countenance and peace, we are able to represent Him to others. We cannot be blessed apart from God because blessing is two-fold. His face shines upon us so that we in turn bless Him. His desire for us is to bless us and keep us in His care. He wants His face to radiate through our faces so that others may know we serve a God who makes us shine with His favor and His countenance. The Lord is gracious and desires for us to live with peace. These spiritual characteristics take on physical attributes displayed in our lives.

As you look at your life today, do you reflect the countenance of the Lord who loves you and knows you by name? Do you live in fear or in peace? Not only does He know us by name, but He places His name on us, so that our words, attitudes and actions represent Him. Do you have peace, even in the midst of trying circumstances? By understanding the heart of God, our desire should be to draw closer to Him. The Lord is consistent. He is the same yesterday, today and forever. His heart's desire is to bless you and to give you peace. We realize that we are not blessed because of anything we have done but because of His mercy through the blood of Christ. Ask for the Lord's blessing on your life today. If you have come to know Jesus Christ, you now bear His name and are called a child of God. And, we know from Numbers 6:27 that by receiving His name as His child, He will bless us.

February 24

Today's Reading: Numbers 7-8; Mark 4:21-41
Today's Thoughts: Money Issues

Command those who are rich in this present world not to be arrogant nor to put their hope in wealth, which is so uncertain, but to put their hope in God, who richly provides us with everything for our enjoyment. Command them to do good, to be rich in good deeds, and to be generous and willing to share. In this way they will lay up treasure for themselves as a firm foundation for the coming age, so that they may take hold of the life that is truly life. **1 Timothy 6:17-19**

How do we keep a spiritual perspective about money while living in our materialistic society? Money seems like a good thing, even spiritually speaking. How can Paul say that wealth is so uncertain but the after-life is such a thing?

Often Jesus would be speaking to someone about the eternal while they would interpret His words in the temporal. For example, when Jesus talked about the living water and the bread of life, He was speaking of spiritual elements that affect our earthly lives. But the Samaritan woman asked for the living water so she wouldn't have to return to the well, and the people asked for the bread Jesus promised because they wanted to be fed continually on earth.

Like Jesus, Paul is saying the same things in these verses above. He is trying to help us focus on eternal values by using temporal tools. Paul is saying: "Do not trust in your money to provide for your needs, but depend on God." He is explaining that we should give our earthly materials generously and willingly for God's kingdom. By keeping a spiritual perspective on earth, we are laying for ourselves a firm foundation in the life that is to come. Paul is saying to use your money on earth for the things of God. This way, God's work will continue here, while laying a firm foundation for you in the coming age.

"No one can serve two masters; for either he will hate the one and love the other, or he will hold to one and despise the other. You cannot serve God and mammon." Matthew 6:24

Pray about it: *Lord, help me to have a spiritual perspective towards money. Help me to be generous and willing to share. I want to lay up treasures in heaven but I need help to trust in you to provide for my needs on earth. Please increase my faith.*

February 25

Today's Reading: Numbers 9-11; Mark 5:1-20
Today's Thoughts: Be a Doer of the Word

***Now when the people complained, it displeased the Lord; for the Lord heard it, and His anger was aroused. So the fire of the Lord burned among them, and consumed some in the outskirts of the camp.* Numbers 11:1**

God knows our needs and hears our words. God opens His hand and satisfies the desires of every living thing (Psalm 145:16). He knows our wants before they are even formed on our tongue. He is faithful in all things and will not let the righteous beg for bread. So why are we so discontent? Why do we complain to each other? Why don't we pray more and trust Him more?

There is a consistent theme in the lives of the Israelites that left Egypt. They complained. Over and over, they saw the miracles of the Lord to provide for them and protect them. They were witnesses of His ever constant presence and received daily manna from heaven. But it wasn't enough. They wanted more and bigger and better and newer. They even complained that they wanted what they had when they were slaves. How quickly they forgot. It was because of their oppression in Egypt that God intervened to save them from slavery. God answered their prayer in unbelievable ways, but it wasn't enough. And over and over, they were punished for this behavior but they didn't change. Why?

I see patterns like this in my own life too. The Lord will reveal to me a message in the morning like "be thankful today" or "be quiet today." For some reason though, I remember His message after I disobeyed. I have frequently asked the Lord how the Holy Spirit was able to impress upon me my sin immediately after I did it instead of helping me before. The answer is always the same…my lack of self control. We speak without thinking, complain to whoever will listen, and become forgetful in the busyness of the day. I have learned, slowly but surely, to write down those things He told me in the morning on a 3x5 card. I then place the card in my pocket or on my dash board to help me to remember. The responsibility isn't all on the Lord. We need to be active in our faith and fight for victory. Let's stop being hearers and complainers and become doers of the Word. Join me in entering the promised land of God's rest and peace today, for we have a God who parts Red Seas and pours water from rocks.

February 26

Today's Reading: Numbers 12-14; Mark 5:21-43
Today's Thoughts: For the Lord

Whatever you do, work at it with all your heart, as working for the Lord, not for men.   **Colossians 3:23**

We are alive to bless the Lord, not for the Lord to bless us. Although God desires to bless us, our desire needs to be that we just want to please the Lord. That means that everything we do should be done to please the Lord and not man. We can do dishes for the Lord, we can wipe noses for the Lord, and we are to help others for the Lord. It is not about being appreciated and complimented by others but by Him. A good test to check your motives is to examine what you do when no one is looking. Do you realize that God is watching? The Lord is with you all the time. When we become aware of His continual presence, we change. And this change is good because as we align our hearts to please Him, we receive His peace.

God has given us internal monitors that let us know if what we are doing is right or wrong before Him. When we make choices that please Him, we have peace. If we are making wrong choices, the Lord convicts us. We become uncomfortable with a lack of peace. We realize quickly that we are not pleasing God when gossiping or speaking negatively. When this happens to me, I sense the quenching of the Holy Spirit and I want to get out of that conversation as fast as I can. Praise God for His conviction. I don't want to do anything that quenches the Spirit because I have learned that whatever quenches His Spirit hurts Him because it is bad for me.

Let's live to bless the Lord. Ask Him to convict your heart today when you are doing anything that isn't right before Him. Some day we will each give an account of our actions before Him. If we become aware of His presence now, we will live a life that brings Him glory, which in turn blesses us.

February 27

Today's Reading: Numbers 15-16; Mark 6:1-29
Today's Thoughts: God Anoints and Appoints

Now Korah..., with Dathan and Abiram the sons of Eliab, and On..., took men and they rose up before Moses with some of the children of Israel, two hundred and fifty leaders of the congregation, representatives of the congregation, men of renown. They gathered together against Moses and Aaron, and said to them, "You take too much upon yourselves, for all the congregation is holy, every one of them, and the Lord is among them. Why then do you exalt yourselves above the assembly of the Lord?" **Numbers 16:1-3** *... Now it came to pass, as he [Moses] finished speaking all these words, that the ground split apart under them, and the earth opened its mouth and swallowed them up, with their households and all the men with Korah, with all their goods.* **Numbers 16:31-32**

Isn't it amazing what envy can do? Korah was extremely jealous of Moses and Aaron. Even though Korah complained and judged Moses and Aaron, he was really criticizing God. God was the One who anointed and appointed Moses and Aaron. God was the one to also defend them. Korah's jealousy planted seeds of dissension among brothers, causing an entire community to suffer consequences because of the rebellion.

We need to remember today that it is God who anoints and appoints certain people to accomplish certain tasks. Some of us are anointed with gifts that stand out while others receive gifts that are more private (like intercessory prayer). But the Spirit distributes different gifts for different callings to all types of people. For a variety of reasons, we become critical and jealous, leading to an attitude of arrogance and rebellion. Those bad seeds sow destructive fruit. People who eat that rotten fruit get sick and spread those destructive germs to many others.

I remember when the Lord first put on my heart the desire to teach. I would think thoughts like, "I would have taught that passage differently" and "I wonder why he/she didn't teach this part of the passage." I had a choice to sin with those thoughts or not. Those thoughts could have become critical, but I had to take those thoughts captive or my mind could have led my mouth to sin against them. God was just giving me a new desire and a new gift, stirring my heart to see and hear things differently. I didn't know that at the time. My heart could have become like Korah's during that transition. We need to be careful. God knows what He is doing. Be sure that you are found blameless before Him when it comes to your opinions about Christian leaders. God has placed them in those positions. If you are becoming critical or disrespectful in any way, you might miss out on a new and very special anointing. The test of time will tell...

February 28

Today's Reading: Numbers 17-19; Mark 6:30-56
Your Thoughts on Today's Passage:

March 1

Today's Reading: Numbers 20-22; Mark 7:1-13
Today's Thoughts: Match Outside to Inside

Then the Pharisees and Scribes asked Him, "Why do Your disciples not walk according to the tradition of the elders, but eat bread with unwashed hands?" He answered and said to them, "Well did Isaiah prophesy of you hypocrites, as it is written: 'This people honors Me with their lips, but their heart is far from Me. and in vain they worship Me, teaching as doctrines the commandments of men.' "For laying aside the commandment of God, you hold the tradition of men--the washing of pitchers and cups, and many other such things you do." Mark 7:5-8

I was speaking with someone the other day who *used* to take part in Bible studies. Although she did not tell me exactly why she is not presently involved in a Bible study, she did share with me an observation that I knew was at the root of her issue. She commented that people act a certain way in church but a different way outside of it. Obviously, she was bothered by the Christian's hypocritical behavior. Many of us look really good on the outside but are falling apart on the inside. In other words, we put on a good show. The problem is when we try to put on an act before the Lord. We honor Him with our lips but our hearts are far from Him. Jesus wants the truth of what is in our hearts, not how we portray ourselves on the outside. Clearly He sees the thoughts and intents of our hearts.

The Pharisees kept the law and man's traditions to the strictest of standards. Their hands were ceremonially cleansed before they touched any bread to eat. Even their utensils were cleansed with the same ceremony. The outside looked spotless to the point that they believed themselves to be more spiritual and pure than those who did not perform these rituals. The same is true for people in the church today. Many Christians work hard at being righteous and look the part to those around them, but God knows what is in their hearts. If our activities are whitewashed by man's traditions, then the fruit of that will be seen in our behaviors. There will be a lack of sincerity and realness with people. Not only does the Lord see through it, but so do others.

Where is your heart today? Are you worshiping with your lips or with sincerity in your heart? Do not just keep going through the motions, whether in your ministry services, church activities or daily checklist of religious do's and don'ts. Get real with the Lord today. Be careful in following the traditions of man to the extent that the Lord is far from your heart. Pray that your outside will match your inside and ask the Lord to remove your fears about how real that might look. If we all would commit to this process, we might just see a lot more people in Bible studies and church.

March 2

Today's Reading: Numbers 23-25; Mark 7:14-37
Today's Thoughts: The Fast I Have Chosen

"Is this not the fast that I have chosen: To loose the bonds of wickedness, To undo the heavy burdens, To let the oppressed go free, And that you break every yoke? Isaiah 58:6

I remember the first time I ever fasted. I spent most of the time looking at the clock and thinking about my stomach. But I have kept the practice of fasting regularly, despite the discomforts in my body. Throughout the years, I have committed to several different types of fasts and for different, yet specific, reasons. I have experienced mountain-top highs and dark-valley lows during times of fasting. I have learned that fasting in its essence is a time of cleansing and purifying, which means I spend a lot of the time confessing and repenting of my sins. When all is said and done; however, fasting has been one of the most powerful tools in my walk with Christ.

The main thing I had to understand very early in my experiences with fasting was that if I tried to do it on my own, then failure seemed to be the result. I have proclaimed fasts for reasons that were not from God. I have determined the type and the length of a fast, then seen it completely fall apart in my weakness. Now when I fast, I try to make certain that it is the fast God has chosen, otherwise, I end up feeling worse.

Today's verse gives us an example of the type of fast God chooses for us. Fasting is an excellent weapon against the enemy's strongholds and bondage in our lives. God appoints certain times of fasting for us to find freedom from wicked bonds and heavy yokes that burden us down. The question for us: are we willing to make the sacrifice for spiritual freedom? It is not easy, but it is so worth it. Are you struggling with an area of bondage? Do you feel burdened and cannot seem to overcome? Pray for clear direction as you ask the Lord to lead you in to a specific fast for your situation. He may impress upon you to fast from certain foods or television shows or daily activities. I believe God honors any attempt we make at fasting, whether it is a food fast or something else. I also believe that He needs to be involved from the very beginning; otherwise, we may ditch it earlier than planned. Pray about fasting and see what God shows you. This may be just the break you have been waiting for.

March 3

Today's Reading: Numbers 26-28; Mark 8
Today's Thoughts: Let the Lord Decide

"Why should the name of our father be removed from among his family because he had no son? Give us a possession among our father's brothers." So Moses brought their case before the Lord. And the Lord spoke to Moses, saying: "The daughters of Zelophehad speak what is right; you shall surely give them a possession of inheritance among their father's brothers, and cause the inheritance of their father to pass to them. **Numbers 27:4-7**

In front of one of our public schools, there is a banner that says, "We celebrate *fairness*." My son commented on it the other day and said, "But life isn't fair." There is a time in each of our lives that we realize that the ways of the world are not always right, good and fair. And each of us does have to come to terms with the ways of the world. However, because we are believers, we do not have to accept the things at hand but we can go to God for everything that troubles us, especially when life does not seem fair.

These daughters in Numbers 27 realized that it would be unfair to not receive an inheritance from their father just because they are girls. The possessions went to the men in those days and not women. No one considered it a dilemma because when the girls got married, they received land under their husbands. But these ladies didn't want to settle for their husband's land. God had given land to families and their father's land would now go to another family. This was not fair. They wanted their father to receive from the Lord as everyone else. God agreed.

When we face issues in our lives that do not seem fair, we need to go to God. Even though the world is not fair, Jesus has overcome the world. We should not just accept the ways of the world or the laws of the land when they vary from God's Word or His Promises. These girls knew the laws of God and asked for specific interventions on their behalf. Often, laws are written for the majority but our God also listens to the minority. Come to Him if you are struggling and let the Lord decide what is "fair" for you.

March 4

Today's Reading: Numbers 29-31; Mark 9:1-29
Today's Thoughts: Dealing with Our Sin

For what I am doing, I do not understand. For what I will to do, that I do not practice; but what I hate, that I do... For I know that in me (that is, in my flesh) nothing good dwells; for to will is present with me, but how to perform what is good I do not find. For the good that I will to do, I do not do; but the evil I will not to do, that I practice. **Romans 7:15, 18-19**

I read an article about a family's fight against pornography. The husband (and father) had become addicted to pornographic websites, which led him into an assortment of related sinful activities. After repeated confrontations and family interventions, he chose to continue this path of destruction. The end result was a family torn apart amidst tremendous pain and confusion. One of the saddest facts of this story is that he was active in his church, served as a deacon, and had a great reputation in the community. Is this story an anomaly, a rare case that happens infrequently in the Christian church? Or is it more common but hidden? This article quoted statistics suggesting that almost half of all Christian homes are affected by pornography. Unfortunately, the Internet provides easy and secret access to these types of websites.

This story is tough to reconcile, especially when we think about how many families are impacted by this sin. It is also tough to realize that we, as Christians, are dealing with all types of sins. Pornography is only one example. Regardless of who we are, we struggle with sin and its consequences. Even the apostle Paul said that he knew that "nothing good" dwelt within him. He struggled with his flesh, just like we do. But unlike the man in the story, Paul dealt with his sins differently. He confessed them and he repented to God. Paul knew that his only hope was in Jesus. He knew that by the blood of Christ his sins past, present and future—had been washed away. Because he kept his struggles against his flesh before the Lord, he was able to walk in the light of Jesus. It is when we hide our sins, try to justify them, make excuses for them, or completely deny them that we are mastered by them.

Regardless of what you struggle with today in your flesh, please do not hide it from God. He already knows everything about you, even your deepest thoughts and desires. All He asks is that you come to Him and confess. Tell the Lord your thoughts and feelings, how you honestly feel, and then ask forgiveness. You may have to do this several times a day. But it is through this process of continually crying out to God, consistently bringing these hidden things to the Light, that healing comes. Do not give up. Jesus did not come for the righteous; He came for sinners. Jesus does not ask us to heal or fix ourselves. He does not ask us to do anything but come to Him. He will do the rest.

March 5

Today's Reading: Numbers 32-34; Mark 9:30-50
Today's Thoughts: You Can Be Sure

And so we have the prophetic word confirmed, which you do well to heed as a light that shines in a dark place, until the day dawns and the morning star rises in your hearts; knowing this first, that no prophecy of Scripture is of any private interpretation, for prophecy never came by the will of man, but holy men of God spoke as they were moved by the Holy Spirit.
2 Peter 1:19-21

My son went through a season of time when he would ask me the same question. "Mom, how do we know that what we believe about God is true?" I encourage those questions because I am glad that my son is honest with me and can openly question these things. I pray that the answer I give him always comes from the Lord. I am so thankful that I worship a God who can handle any controversy, doubt or question. In 1 Peter 3:15, the Lord tells us to "always *be* ready to *give* a defense to everyone who asks you a reason for the hope that is in you, with meekness and fear." So, if we are not asking those questions ourselves, we may not always be ready to have an answer for someone else.

How do we know that our belief in Jesus is true? Some people base their faith on a feeling. Others would say that it is based on all the years of tradition that have withstood the test of time. And some would say, "Because it works for me." But the reason why we know that Christianity is true is because of the *sure word of prophecy* spoken of in 2 Peter 1:19-21.

Prophecy (or the prediction of future events) proves that God holds time and events in His hand. Fulfilled prophecy clearly shows us that God knows the end from the beginning. The Old Testament prophets predicted the coming of Jesus Christ. When Jesus came to earth, He fulfilled 333 prophecies, which is why one of God's rules in the Law of Moses was to stone someone who was a false prophet. God does not make mistakes. He would not speak through someone and give false words for the future. God is completely in control and nothing surprises Him. If God said it, we can believe it. When we understand that we can trust Him with our future, then we will know that we can and do trust and believe.

March 6

Today's Reading: Numbers 35-36; Mark 10:1-31
Today's Thoughts: Child-Like Faith

"Assuredly, I say to you, whoever does not receive the kingdom of God as a little child will by no means enter it." **Mark 10:15**

My sister and I had a good laugh the other day as we shared a conversation she had with her granddaughter. There was a minor problem with her car engine so my sister reached in the glove compartment for the owner's manual. As she was looking through the pages, her granddaughter said, "Mamaw, why are you looking in that book?" To which my sister replied, "Because I want to fix this problem with my car." At that point, little 4 year-old Emma ran into the house and came out with her own book. "Here, Mamaw, this book will tell you how to fix your car." My sister looked at the book and said, "It is your Bible, Emma." "Yes, Mamaw, everything you need to know is in this book, even how to fix your car." What great faith!

I wonder how much more God would fix in our lives if we truly lived out what the Bible teaches us. Little Emma is right when she says that her book has all of the answers. God's Word has the answer for every question we have and for every step we need to take. The only thing we need to do is to have the faith to believe it and obey.

Pray for that child-like faith today and ask the Lord to plant His Word on your heart and mind. *"Study this Book of the Law continually. Meditate on it day and night so you may be sure to obey all that is written in it. Only then will you succeed."* (Joshua 1:8)

March 7

Today's Readings: Deuteronomy 1-3; Mark 10:32-52
Today's Thoughts: A Blind Man's Prayer Request

And when he heard that it was Jesus of Nazareth, he began to cry out and say, "Jesus, Son of David, have mercy on me!" **Mark 10:47**

In Mark 10:46-53, Jesus left the city with a large crowd following close behind. Most likely this crowd was creating a great deal of noise and excitement. A blind man begging for help would be sensitive to hear the activities as the people scuffled by. It is amazing that a blind beggar would know about "Jesus of Nazareth" and somehow would know that Jesus could help him. The blind man cried out, "Jesus, Son of David, have mercy on me!" While Jesus walked on earth, He answered that kind of prayer frequently. It must have been music in our Savior's ears, because that is why Jesus came, to grant mercy to blind beggars like you and me. Jesus stopped, called to him and asked, "What do you want me to do for you?" This man could have asked for money or help with some other desperate need, but instead he asked specifically for his sight. Jesus said that his faith healed him as the beggar immediately received his sight.

There are many situations and circumstances in our lives where we are blind to the will of God. But how many of us cry out to God for mercy? How many of us really bring our specific need to the Savior to ask for His touch? Too many times, we are *faithless* to ask for what we really want. Too many times, we hear that Jesus is near but are apprehensive to call out to Him. This blind beggar received what he asked for. This blind beggar knew that Jesus could touch him and change him. He did not hold back but yelled into a crowd full of people. Jesus says in Matthew 7:7 "Ask, and it will be given to you; seek, and you will find; knock, and it will be opened to you." We need to take Jesus at His word. We need to understand that God desires to grant mercy and He wants us (the spiritually blind) to see. Are you in need of His mercy today? Tell Jesus what You want Him to do for you...tell Him that you need Him today. Just think, it is not necessary to yell or push through a crowd – He is just a prayer away.

March 8

Today's Reading: Deuteronomy 4-6; Mark 11:1-18
Today's Thoughts: Stop to Hear His Voice

Whether you turn to the right or to the left, your ears will hear a voice behind you, saying, "This is the way; walk in it." Isaiah 30:21

Which way? This way or that way? Today, there are so many paths we can follow. We have the advice of our family, friends, the media and then there is our own instinct. But how do we know the right way? How do we discern which way is really best for us?

The other day, I counseled a woman who was convinced that she was doing God's will by serving the church. She gave of herself, her finances, and her time. She experienced the Lord's help as He gave her great ideas and even multiplied her time to keep going. But, soon, she came to me full of bitterness, disillusioned by the people and the staff. She felt confused and angry. She asked questions like, "How could this be God's way? How could they so disregard her following the Lord? And where was He in their actions?"

We ask the same questions, "Which way Lord? What is Your will, Lord?" and then we follow the direction He gives. But sometimes, we find ourselves angry and resentful. How come? What happens? Frequently, God answers our prayer and uses us and guides us. However, once He has shown us what He wants, frequently we stop asking for further wisdom and counsel and just go on our own way in the name of the Lord. This verse talks about God speaking from behind us. And why would God need to speak from behind us unless we have run ahead of Him? So often, He leads us to a certain place, position, or calling, then we run ahead. When we run ahead, we lose His way and make it our way. That is what happened to my friend who was seeking counsel. She knew that the Lord had one job for her to do. When that job was completed and very blessed, she determined in her heart to do more. More wasn't asked of her by the church or by the Lord. And it is because of the "more" that she grew angry. God now called to her from behind to lead her forward in a new direction or way to walk.

God is faithful. He will lead you and answer you when you call out to Him. Just remember to keep asking continually as you proceed. Always follow His leading. If you begin to experience anger, bitterness and judgmental thoughts, you have probably run ahead of God. Stop until you can hear His voice from behind you. He will lead you back into His way and help you to walk in it.

March 9

Today's Reading: Deuteronomy 7-9; Mark 11:19-33
Today's Thoughts: Because of Who He is

The Lord did not set His love on you nor choose you because you were more in number than any other people, for you were the least of all peoples; but because the Lord loves you, and because He would keep the oath which He swore to your fathers, the Lord has brought you out with a mighty hand, and redeemed you from the house of bondage, from the hand of Pharaoh king of Egypt.
Deuteronomy 7:7-8

Moses reminds the children of Israel that it was not because of anything they had done to be chosen as His people. God had made a covenant with Abraham and promised him that his descendants would be more numerous than the stars. God showed Abraham the land in which they would dwell. God also told Abraham that his descendents would have dark years of bondage. It would be several hundred years later before the Israelites would actually enter the Promised Land. But by the end of this appointed time, they consisted of twelve tribes and millions of people. God knew the end from the beginning. He knew how vast a nation they would be. He set His love on them because He is love, not because of who they were or what they did.

By God's great mercy and grace, He chooses us today. He loves us beyond our understanding. He has plans for us beyond our imaginations, not because of who *we* are but because of Who *He* is. When we begin to appreciate God's character, then we can begin to rest in Him. We cannot earn rewards by impressing God with our works. We can never be good enough or do enough good deeds. In reality, it is not about us at all. It is all about God. Everything God does, He does because He is God. God is love. God is righteous. God is faithful. God is perfect in every way. God is all-knowing, all-powerful and ever-present. God cannot be anything but who He is. Therefore, God will *never* not love us. God will *never* let us down. God can never go against His own character.

The Bible clearly teaches us God's character. When we spend time in His Word, we learn more and more about Who He is. We learn of His activities throughout history and how He dealt with the people. We are given amazing insight into His plans and purposes through His prophets. We learn of His never changing love and faithfulness in ways that pierce our hearts. And most of all, we learn of how He transacted His covenant of love to us by sending His Son to die on a cross so we could inherit the land He has for us. As the song goes, "Not because of who I am, but because of what You've done, not because of what I've done, but because of Who You are." We need to get to know who the Lord is and who He wants to be in your life today.

March 10

Today's Reading: Deuteronomy 10-12; Mark 12:1-27
Today's Thoughts: Can We Know His Thoughts?

Surely the Sovereign Lord does nothing without revealing His plan to His servants the prophets. **Amos 3:7**

Truly we, as believers, can know His plans before they happen. Many times, we hear others say, "I knew that was going to happen." But how easy it is to say that after the fact! God is saying that we can know before.

I learned early in my walk with the Lord to be quiet before Him. I would read a Scripture passage until I felt God was speaking to me. I would determine that the Lord was speaking because the verse would seem special in some way, applicable in some way or I considered the verse to be a new thought. I would just stop and ponder and pray it back to God, asking Him to make His message clear to me for the day. In the quietness, the Lord would speak to my heart. It would start as an impression at first. I would pray about it and ask God questions about the impression until His wisdom became "louder or firmer" than my thoughts, questions and distractions. Next, I would step out in faith and test the message. If it is the Lord, He would confirm it through bringing it to pass or giving me similar messages through His Word or others.

Today, I pray this verse to the Lord, knowing that He will reveal His plan to me because I long to be His servant and want to live in His will. God is faithful and **if** you ask, He will answer. **If** you learn to be still, He will speak. **If** you listen, He will reveal His plan. After years of relating to the Lord like this, you will then know that you know His will before the fact, for He promises that He does nothing without revealing His plan to His servants the prophets. So the first step is to find out **if** you are His servant.

March 11

Today's Reading: Deuteronomy 13-15; Mark 12:28-44
Today's Thoughts: The Perfect One

"How long will you keep us in suspense? If you are the Christ, tell us plainly." **John 10:22-24**

Do you really believe that the Jewish religious leaders wanted clarity and confirmation that He is the Christ? No. Their mission was consistent in wanting Him to publicly declare that He was Messiah in order to justify attacking and eventually killing Him. Their motive doesn't surprise me, but Jesus' consistent patience, sincerity and honesty absolutely surprises me. He continually answers their questions and exposes their hearts. These religious men think that they are so perfect—perfect in their knowledge of the scriptures, perfect in judging and criticizing. And now they think that they are perfectly justified in killing the only Perfect person who ever lived. Jesus once again repeats over and over that they will not understand because they are not His sheep. Jesus' sheep hear His voice, know Him and follow Him. Jesus tells us that His Father fully supports Him and loves Him. And so we know that no one will snatch us out of the Father's hand either. The Father, Son and Holy Spirit are united in their plans and purpose for sheep preservation. They are committed to the security and safety of the sheep. As His sheep, we have the responsibility to hear, obey and to follow where He leads. But if you know Jesus, regardless of how far you stray, He will eventually lead you back for nothing can separate you from the love of God.

Pray about it: *Oh Lord, thank You for being my Good Shepherd. You know me, love me, call me by name, and lead me. You also tell me that nothing will snatch me away from You. I am so thankful that You will complete the work You started in me and that You will get me to where I need to be. Please help me to live by fully trusting You. I want the attitude of my heart and the meditations of my mind to be without fear, doubt, worry and insecurity. Thanks for taking care of me.*

March 12

Today's Reading: Deuteronomy 16-18; Mark 13:1-20
Today's Thoughts: What Do You Boast In?

Some boast in chariots, and some in horses; But we will boast in the name of the Lord, our God. **Psalm 20:7**

My friend had a blood test that came back questionable. I asked her mother about the doctor and the following steps that would be taken. Her mom's answer ended in explaining to me the credibility of the doctor.

Intelligence, money and strength are three attributes that we love to measure in ourselves and others. Our intelligence can be measured by how hard one studied and the grades one received. We form an instant opinion when we hear certain colleges were attended with the credentials that were earned. Money tends to reflect our success. If someone is wealthy and manages their money well, we give credence as well as our respect. And strength is the last characteristic that helps us to form an opinion about someone. Are they athletic? Do they have high energy to get important things done? Do they have a disciplined body to maintain stressful positions and self control? These three things we assume we can boast about. These three things others boast about for someone else.

God does not value our intelligence, wealth or strength. God values a person who trusts and boasts in Him. All three of those characteristics are gifts—God given gifts. In light of eternity, they mean nothing and each one could be taken away in an instant through illness alone. It is great that we have good doctors to attend to us, but God is the Author and Sustainer of life. I would rather boast in the name of the Lord, my God than to put my trust in any person, despite their credentials. Turn to the Lord and seek His face. Know that you serve a God who gives gifts to men and the greatest gift given was Jesus. Do we really trust in Jesus each day? And do we boast in His name?

Pray about it: Lord, let me not boast in my wisdom, nor in my strength, nor in my riches. I want to boast in knowing You, believing in You and understanding Your ways. (Jeremiah 9:23-24)

March 13

Today's Reading: Deuteronomy 19-21; Mark 13:21-37
Today's Thoughts: The Christian life is like climbing stairs.

Therefore we also, since we are surrounded by so great a cloud of witnesses, let us lay aside every weight, and the sin which so easily ensnares us, and let us run with endurance the race that is set before us, looking unto Jesus, the author and finisher of our faith, who for the joy that was set before Him endured the cross, despising the shame, and has sat down at the right hand of the throne of God. **Hebrews 12:1-2**

We realize early in our Christian life that we start on the bottom floor of the staircase. Our main goal is to get up to the top. While first beginning to ascend, the climb seems to go easy and fast. After all, the need for change was evident and our goals for coming to Christ were clear. Who wants to stay on the bottom step? As we keep climbing, sometimes we wonder if we will ever make it to the top. Our muscles ache and our heart pounds at times. We pray for an elevator or escalator to make the climb easier. But God makes sure that we consciously have to make a choice to go each step, and God is not in a rush. He waits patiently, lovingly and relentlessly for us to catch our breath and relax.

At these times of rest, we look down and see what we have had to overcome just to get this high. God does not let us know how far we still have to go. But we still get overwhelmed, thinking about the remaining climb. We gaze at the top, keep looking up, and taking one step at a time. At some levels, the rest is nice with balconies and water fountains. At other times, only by faith do we keep climbing. We know that climbing back down would be a mistake but we think about it sometimes. Other times, we just want to sit right where we are without continuing in the climb, just hang out for a while in one place. We pass those on our way up the stairs who are hanging out in their spot, and then sometimes we step aside to let someone pass by who is heading back down the stairs. Those of us who continue upward know that the temptation to stop is not a valid option because we hear the Lord's call to go higher. In our heart of hearts, we know that when we finally arrive, we will be so happy we kept climbing. There is no hurry, but no advantage to slowing down either. So we pray for a steady pace, understanding each level and making a conscious decision to move up and onward.

Where are you at today on the staircase to God? Are you hanging out, taking a break, or are you seriously contemplating heading back down? Keep a couple of things in mind: fix your eyes upward towards the top and listen for God's voice. Not only is the Lord calling you to come, to keep moving, but He is also waiting for you. And the best part of all of it? He is with you in every step you take. So today, lift your head up and get moving. The prize awaits at the end.

March 14

Today's Reading: Deuteronomy 22-24; Mark 14:1-26
Today's Thoughts: Sixth Sense

"Teach us to number our days aright, that we may gain a heart of wisdom." Psalm 90:12

Presently, I have a house full of teenagers and the energy level is very high. The one thing that I have noticed the most about teenagers is their love for stimuli. Everything associated with them is loud and bright and moving very fast. They can text on their cell phone while emailing on facebook while changing the song on the ipod while sorting through a homework schedule and talk to me all at this same time. They are being prepared for a much different life as the world has definitely changed.

My teenage years were very different. Not only because we called the home phone number to talk to a friend but because there were no computers at all! How can you compare the changes that the computer has brought with roll down windows and rotary dial phones? Change has a whole new meaning in today's world. But changes in the "things" are not my real concern for the next generation. I am concerned about the many things that compete for their attention. It is exaggerated in the life of a teenager but it affects us all. The world has become so loud in our ears (cell phones, ipods, ichat, music) and so bright in our eyes (facebook, internet videos, television, text messages). These things are not something a teenager will outgrow. This way of life will only increase with advancing technology. I am NOT against change or technology. I am against constant, continual noise. We need to be proactive in protecting our minds from the noise or we will miss out on the things that matter most: a personal, walking talking relationship with the Lord.

I came to know the Lord at 14 years old and heard Him from the first time I read the Bible. I learned quickly that if I was going to seek the Lord, my mind needed to learn to be quiet. I guess it is like developing a sixth sense. When you tone down your other senses, your awareness of God becomes more evident. It is like a blind person who depends more on their sense of hearing and a deaf person who depends more on their sense of sight. When everything is shaking and moving, it is more difficult to be in tune with your sixth sense of spirituality. If we do not develop ears to hear the Lord and eyes to see Him, we will miss out on the greatest relationship there is on this earth regardless of our age or the stage of life especially since spirituality is not a "sense" but it is a lifestyle.

March 15

Today's Reading: Deuteronomy 25-27; Mark 14:27-53
Today's Thoughts: Whose Voice Are You Hearing?

***So when the woman saw that the tree was good for food, that it was pleasant to the eyes, and a tree desirable to make one wise, she took of its fruit and ate. She also gave to her husband with her, and he ate.* Genesis 3:6**

It does not always seem clear to us when God speaks. However, when we become living sacrifices who continually worship, pray and seek the Lord, we become more willing and able to hear His voice. As we draw near to the Lord, He draws nearer to us and is faithful to speak in a way that brings peace, confirmation and often, He brings changes to our surrounding circumstances. We have to ask ourselves whom are we listening to? There are so many voices speaking and our minds seem to have a voice of their own. But which voice do you follow? Adam heard the voice of God when He gave the one rule of living in the Garden of Eden. He was told by God to not eat from one tree. But Adam listened to the wrong voice when faced with a choice of disobeying God's command. Eve bought the lie of the serpent's voice and mankind has suffered the consequences ever since.

Many people in our lives give us many conflicting messages. We each have a responsibility to know what the Lord is telling us, to listen to Him and to act on those commands. We have to make the Lord's Word our guide and not deviate from His truths regardless of how appealing other additional information may sound. If you have come to Christ, you have His Holy Spirit in you. He is there to lead you and guide you in the ways He has called you.

The Holy Spirit anoints you with His presence and with spiritual gifts which are supernatural. These gifts are not based on your skill or experience. So, you need to be dependent on the Lord to fulfill your calling. Your calling may be completely different than what your family thinks of you. Your calling may be completely different than what you thought it would be. Be sure to follow *His* voice. Be confident, knowing that He knows what is best. If you listen to other voices than God's, you could suffer consequences that will have a negative effect on your life. Take your thoughts captive, stay in His Word and trust Him to lead you in the way you should go.

March 16

Today's Readings: Deuteronomy 28-29; Mark 14:54-72
Today's thoughts: The Gift of Salvation

For by grace you have been saved through faith, and that not of yourselves; it is the gift of God, not of works, lest anyone should boast. Ephesians 2:8-9

Why is a gift of salvation so hard to accept? Why do so many people look for their own way to heaven? Why do people try to make God fit into their own belief system? Regardless of the many who reject these two verses, they are truth. There is only *one way* to heaven and *the only Way* is by receiving the free gift of salvation through Jesus Christ. Despite the growing and ever present number of religious groups which believe otherwise, this truth will be the only one that matters when it is all said and done. Because in essence, we as mankind have a hard time accepting the "gift." The old saying that *there is no free lunch* sets us up to trust only in those things we earn. Trust is the key word. Where do we put our trust? We either trust in the Lord or we trust in ourselves.

Many today try to blend a combination of the two. They trust the Lord to give them the gift of salvation because they work so hard to please Him. And these works and efforts can really look good. They attend church regularly, serve when needed, give as required, and try to live overall lives that maintain Christian moral values. But they give themselves too much credit. God does not want anyone to do anything for Him because He has already done everything for us. So how do we accept His gift? We need to realize that the gift has been given by grace, and to receive the gift, all we need to do is believe. We just need to believe in our hearts that Jesus died on the cross for all of our sins and that without Him we are spiritually dead and destined to spend eternity separated from Him. Believe in Jesus, not in man's good works.

Are you tired and weary of trying to please God? Even if you have professed faith and accepted Jesus as your Lord, you still might be trying so hard to please Him. Today, we need to accept His promise and rest in Him. Let's live today to the fullest in God's gift by trusting Him in all things and all areas of our lives. Start by talking to Him. Give your heart to Jesus, give Him your burdens today and He will show you mercy. Worship and praise Him and your heart will begin to sense His joy and peace. Why would any of us desire to follow any other way?

March 17

Today's Reading: Deuteronomy 30-31; Mark 15:1-25
Today's Thoughts: Crush, Kill and Destroy

Be sober, be vigilant; because your adversary the devil walks about like a roaring lion, seeking whom he may devour. **1 Peter 5:8**

Satan is alive and well. His mission has been the same since the Garden of Eden: to crush, kill and destroy (just like the android on *Lost in Space*). He is good at twisting concepts and manipulating ideas into truths that are not real. Satan's battlefield is in your mind and starts within your circumstances. He frustrates your plans, trying to crush that peace within your heart. When your attitudes don't match, you have just become a hypocrite and Satan loves to make hypocrites out of Christians. Once you are aware of this scheme (which for some of us is a lifetime), you learn to pray and resist him by drawing near to God. You do this by learning how to discipline your behavior and choosing not to allow the fruits of the Spirit to be crushed. Praying, submitting to God and meditating on scripture are the next steps. But then Satan's next plan of attack hits your mind. Now, Satan cannot read your mind but he can discern what you are thinking by your words and behaviors. He can also put his thoughts into your mind. I have found that many times I don't need the devil to put bad thoughts into my mind because I can do that all by myself. I am capable of taking my own self out of the ministry or walking away from the Lord, with very little effort or attack from him.

So "be on guard" for Satan is seeking someone, like you, to devour. If you look weak, think weak, act weak, and your words express a weakness in your faith, then you will surely become his next target. Peter warns us to be sober and vigilant. Both words have similar Greek meanings: to watch and keep awake. Obviously this is more difficult for us because we don't see into the spiritual realm to understand what's going on. If we disregard the work of the enemy, we allow him to have the advantage over us.

After we pray, the Lord opens our eyes and gives us discernment to know what's really going on. Too many Christians are apathetic and complacent. The enemy doesn't have to bother with an apathetic and complacent Christian. That is just where he wants us to be, but God has so much more for us. The Lord has won the battle; we just have to put the armor on and pray, "Lead us not into temptation." But when we are tempted, remember 1 John 4:4, "Greater is He who is in us than he who is in the world." The Lord wants to give us an abundant life on this earth so why not start living it today!

March 18

Today's Reading: Deuteronomy 32-34; Mark 15:26-47
Today's Thoughts: Surrender in Thanks

In everything give thanks; for this is the will of God in Christ Jesus for you. **1 Thessalonians 5:18**

What do you do when nothing but uncertainty and insecurity fill your mind? How do you overcome the sensation that your entire body is burning as a result of fear and panic? Whether we admit it or not, all of us will have to deal at times with circumstances that overwhelm us. We all go through problems that take us by surprise, come out of nowhere and make us feel trapped that we have to surrender to its way. It has been in these times that I have cried out to God from the depths of my heart, "Why?" My thoughts range from "What have I done wrong to deserve this?" to "Where is God?" My mind sorts through the circumstance until I am exhausted, while my body burns like a fire that melts my insides. For some of us, we share the pain we are suffering because we are searching for answers and trying to sort through our thoughts and emotions. For others, we attempt to control and mask the thoughts and emotions so no one will suspect the struggles we face.

Regardless of how we look or feel in a time of crises, we need to learn how to get through those times with the Lord. The circumstances may seem to trap us, but getting and keeping control are not the answers. The only way to find freedom when life has pinned you down and chained you in is through *surrender*.

The word "surrender" has mixed connotations. It is an easier word to accept than "submit" but it really does have the same meaning. During those times that our mouth is praying out of panic, our inward soul needs to learn to surrender and submit to the sovereign, all powerful, ever present God. In other words, your heart needs to speak words of praise and thanksgiving even though your flesh (including your mind) is pleading for God to rescue you. It is in this place of praise that an overwhelming peace begins to cover your heart, mind, soul and quenches the fire within you. To give thanks in everything is a hard action to apply. The Christian who can truly apply this principle to his or her life demonstrates a real maturity in their walk. It is in this place that the chains of oppression will be broken, setting the captive free.

Pray about it: *Oh Lord, teach me how to sincerely praise You in the good times so that a wellspring of thanks may truly overflow from my heart during those frightening, overwhelming trials. I want to walk in power and peace. I know that they stem from a heart of praise. Help me to have that heart of thanksgiving for everything You are doing in my life. In Jesus name, Amen.*

March 19

Today's Reading: Joshua 1-3; Mark 16
Today's Thoughts: Just Step Out in Faith

***And as soon as we heard these things, our hearts melted; neither did there remain any more courage in anyone because of you, for the Lord your God, He is God in heaven above and on earth beneath.* Joshua 2:11**

Joshua sent two men to the enemy city of Jericho to spy out the land. Of all of the places they could end up staying, they lodged in the home of a harlot named Rahab. Once the king of Jericho heard of their whereabouts, he sent his men to go and find them, and the first place they went to was Rahab's home. Instead of turning them in, she hid them and protected them from being caught. In return for her loyalty and faith in Israel's God, she and her family were spared when the city walls of Jericho came down.

It is interesting to note that up until this time, neither she nor the inhabitants of Jericho had actually seen any of the Israelites. But they had heard of them. They heard of their conquests in the desert and how the Lord had parted the Red Sea. At this point, the Red Sea parting had occurred some forty years earlier. The fear of the Lord caused this response from Rahab; she knew which side she needed to be on. God had prepared the way by moving on Rahab's heart even before these two men entered Jericho. The Israelites were about to take over the Promised Land, not because of their strength, but because of the Lord's promise and power.

The same is true for us today. As Christians, we are blessed to live in God's promises. The Bible is filled with His promises for us. Sometimes, the new land is hard to see and even harder to walk through, but if we just step out in faith, we will see how God has taken care of every step we need to take. As the Lord leads you in His purposes for your life, you will see how He has prepared even those around you. As God would continue to tell Joshua, "Be strong and of good courage; do not be afraid, nor be dismayed, for the Lord your God is with you wherever you go." (Joshua 1:9) Take this message with you today and go forth in all that the Lord is leading you to do. The Jordan River will part for you when the Lord sends you in.

March 20

Today's Reading: Joshua 4-6; Luke 1:1-20
Today's Thoughts: Wild Grapes

What more could have been done to My vineyard that I have not done in it? Why then, when I expected it to bring forth good grapes, did it bring forth wild grapes? Isaiah 5:4

God used this story about a vineyard to describe the condition of His people. God was saying that He did everything possible to expect the best from His people. He provided them with the best land, complete protection, and the choicest weather conditions; but they chose to live according to their own wants and desires instead of His.

This same thing happens with the people of God today. He has provided for us the best pastors and teachers to be heard through internet, televisions, radios and mass book replication. He has given us freedom of speech and the mandate to learn to read. The Lord has over-educated us spiritually, but for some reason very few are truly seeking God and meeting His expectations. We go to church on Sunday to make ourselves feel better, but are we going to really know and serve God better? Our intentions are important to God who knows the thoughts and intents of our hearts.

I say this because I see more wild grapes in the church today instead of an abundance of good ones. We have sporadic fruit production instead of mass amounts. God has done everything for us but only a few are really sharing in the harvest. The others are willing to encourage and applaud those few. But it's not about applauding but about sowing and reaping. We use our hands to applaud as well as to sow and reap; however, the angels are given the job to applaud. We are told to be the laborers in the harvest. We need to become fruit inspectors of our own vineyard. If the grapes are good, praise God! If the grapes are wild, ask the Lord of the harvest to send you out as a laborer. He is more than willing to answer that prayer and He expects great grapes in abundance from you.

March 21

Today's Reading: Joshua 7-9; Luke 1:21-38
Today's Thoughts: Success in God's Rest

"I will feed My flock, and I will make them lie down," says the Lord GOD. Ezekiel 34:15

Often times, there is nothing I would rather do after eating a good meal than to lie down. But, instead, I think of all of the things that need to get done. I keep myself going using food as the energy source to fuel my activities. I used to believe that this busy lifestyle was just a part of the American way and that to be successful, one must stay productive. After dedicating my time to God's Word instead of to keeping up with our society's ideas, I began to realize that God has a different definition of success. The Lord says that He will feed us and He will make us lie down. He wants us to rest!

Does God want us to stop working and stop being productive in our lives? No. A healthy balance of work and recreation are essential to our overall well-being, but when the work or activity takes control of our lives, we have a problem. Our society is filled with sick people at many levels. We are sick from stress, from bad diets, and from all kinds of unhealthy habits. Where does our Lord fit in to our lives? Where are His provisions in these things? Maybe we have ignored His warnings telling us to slow down and to come spend time with Him. How long can we keep going at a pace that excludes His Word and His rest?

Our God knows us well. He created us. He knows that He has to make us lie down or we will just keep going. Just know that our Lord will do whatever it takes to get us to stop and pay attention to Him. A life that goes along with the world's activities will face serious troubles until the Lord brings you back to Him. God knows what is best for you. Trust Him to feed you and give you rest. Today is the day to seek the Lord, repent and turn back to Him. Maybe you need to eat a good meal at His table, then lie down and spend that time just talking to God. Give Him your stress and your troubles and let Him be your God. He will give you rest. He will make you whole. And then you will find the true success He has for you.

March 22

Today's Reading: Joshua 10-12; Luke 1:39-56
Today's Thoughts: Missing the Lord

You will show me the path of life; in Your presence is fullness of joy; at Your right hand are pleasures forevermore. **Psalm 16:11**

Have you found yourself all alone at times? Your home may be filled with people and your schedule jammed packed with events but you feel all alone. Why is that? I have come to find myself in that place frequently. I really believe that the loneliness is a result of understanding the companionship with the Lord. Only Jesus can fill our cups, only Jesus can be our source of fellowship, and only Jesus can love us in the way that brings rest and peace to our souls. After knowing what it is to have true fellowship with the Lord, nothing (and no one) can take His place. When I find myself alone in a crowded room, I realize that I'm missing Jesus. He has not left me, for He promises to never leave me nor forsake me. Instead, the emptiness is a result of not finding time to be alone with Him.

Frequently in the Bible, Jesus spent time alone with His Father. He knew that He needed that uninterrupted time alone in prayer and fellowship. If Jesus needed time alone with God, how much more do we? The Lord promises that in His presence is fullness of joy. Having experienced that joy, we long for quality time alone with Him. Once we have tasted of the Lord, we know that nothing else tastes as sweet.

Ask the Lord to help you find time every day to be alone with Him. He hears our prayers and He desires to have fellowship with us. Jesus has made a way for us to come to Him freely. He will make a way for us to walk with Him daily if we are truly seeking Him.

March 23

Today's Reading: Joshua 13-15; Luke 1:57-80
Today's Thoughts: Physical Infirmity

You know that because of physical infirmity I preached the gospel to you at the first. And my trial which was in my flesh you did not despise or reject, but you received me as an angel of God, even as Christ Jesus.
Galatians 4:13-14

The Galatians were just one of the churches that Paul started on his missionary journeys. They were Gentiles who converted to Christianity, all because of Paul's passion to preach and teach them the gospel of Jesus Christ. Through the power of the Holy Spirit, Paul was able to preach the gospel despite his own physical infirmities. Speculation as to Paul's true afflictions (i.e. his thorn in the flesh) have been debated and discussed by many theologians, but Paul does not give specifics or details that draw attention to his weaknesses. He, instead, uses them as a means to give glory to God. In our weakness, God is strong.

How often do we allow our physical ailments to bring glory to God? It seems that the most concerning issues receive the most attention and prayer, and we are more than ready to seek the Lord's guidance in those cases. Life-threatening diseases are often wonderful opportunities to bear witness of God's awesome power, regardless of the path. But what about the day-to-day annoying ailments? A headache can make many of us cranky all day. A cold gives us just cause to be cranky for a longer period of time. What happens to our witness in those times? Do we give the daily annoyances to the Lord in prayer or do we let them rule our flesh? Paul knew how to give everything over to the Lord and he did not let those things stop him; he used them to further his mission for Jesus Christ.

Remember today that God is in everything that happens in your life. No matter what ails you, God is with you. He will use your afflictions to bring Him glory and to bless you and others. Sometimes God does the greatest work through our weaknesses and pain, but that only happens when we take our eyes off ourselves and put them on Jesus. Try putting your focus on the Lord instead of your pain, and let His Holy Spirit be your strength in weakness.

March 24

Today's Reading: Joshua 16-18; Luke 2:1-24
Today's Thoughts: The Promised Land

But there remained among the children of Israel seven tribes which had not yet received their inheritance. Then Joshua said to the children of Israel: "How long will you neglect to go and possess the land which the Lord God of your fathers has given you? **Joshua 18:2-3**

Joshua led the children of Israel into the Promised Land. He led them in battles against the existing nations in the land God promised them. The Lord continued to bless the Israelites in victories as they continued to take more territory, just as He had promised. In the beginning, their faith was strong. But as years passed, there was still the need to drive out the other inhabitants from their own allotted territories. Seven tribes hesitated to go forth and take their inheritance. The gift lay before them, so why the delay? Why did Joshua have to force them to go forward and take their gift?

Maybe they were afraid of being on their own for the first time. Maybe they were afraid of the unknown. Maybe they just did not know what to do with their own land. What about us? How many times do we neglect to receive the gift that God is giving us? We pray and pray for something and then one day, God answers our prayer. But the answer involves stepping out in some way, maybe having to do it alone. We can see the answer clearly before us, but we hesitate to go forth. Maybe we are afraid, just as the Israelites were. Maybe we are more comfortable just praying for something instead of actually receiving it. Are we always ready for the change that can come when we pray for things?

I pray that I don't miss out on what God has for me. I pray for more faith when those fears creep into my thinking. I pray for wisdom to know how to pray for those things that will involve changes in my life. Think about these verses today and how they apply to your life. Is there something you are neglecting to possess? God has a life of promise lands for each one of us. Don't let fear or hesitation keep you out of them.

March 25

Today's Reading: Joshua 19-21; Luke 2:25-52
Today's Thoughts: Not of Works, But of Faith

What good is it, my brothers, if a man claims to have faith but has no deeds? Can such faith save him? Suppose a brother or sister is without clothes and daily food. If one of you says to him, "Go, I wish you well; keep warm and well fed," but does nothing about his physical needs, what good is it? In the same way, faith by itself, if it is not accompanied by action, is dead. **James 2:14-17**

Martin Luther struggled with the book of James because he felt that the book discredited grace. After he, himself, tried to work so hard to prove his own faith, Luther came to the conclusion that salvation is by grace alone. He also realized that even our works are an act of God's grace. Verses like those above stumbled him because earlier in his life he tried to earn his salvation and to prove his faith through works, instead of grace. It is not works that save or justify or affirm your faith; it is a result of your faith that you desire to work for the things of God. When Luther ended up exhausted, burned out and lacking peace in his works, then he let faith be his guide to do what God called him to do. The Reformation is a result of his faith, not of his work. The Reformation resulted as a work of God through the faith of Martin Luther.

As the Bible says, God causes all things to work together for good. Today, there are countless churches and denominations reaching the world through the means and convictions that God has placed on each one's heart. God works through movements in His churches, but His works are still accomplished through individuals. That work is a result of the outpouring of love we have for the Lord. His burden is easy and His yoke is light. Titus 3:5 says "*not by works of righteousness which we have done, but according to His mercy He saved us,*" meaning that we do works of righteousness but those deeds are a result of God's mercy, saving us and then working in us through faith. Today, allow the Lord to develop and strengthen your faith. Rest in His saving mercy and He will accomplish His work through your faith.

March 26

Today's Reading: Joshua 22-24; Luke 3
Today's Thoughts: Of Things to Come

But the Lord is the true God; He is the living God and the everlasting King. At His wrath the earth will tremble, and the nations will not be able to endure His indignation. Thus you shall say to them: "The gods that have not made the heavens and the earth shall perish from the earth and from under these heavens." He has made the earth by His power, He has established the world by His wisdom, and has stretched out the heavens at His discretion." **Jeremiah 10:10-12**

The prophet Jeremiah had a really tough job. He had to prophesy to the people of Israel the dreadful things to come if they did not put away their false gods and serve the one true God. His message was not a feel-good, motivational pep talk. The Lord clearly told Jeremiah exactly what to say to them and that they would hate him for it. They had no interest in hearing God's word. They refused to believe in the destruction to come. Besides, there were other prophets who gave them the opposite message of hope and prosperity. They chose to believe them instead.

The Bible tells us of things to come. The prophecies that have yet to be fulfilled will one day come to pass. Are we listening with ears to hear? There are those who decide not to believe in such things and even though they believe in the Bible, they do not necessarily believe in *all* of the Bible. Whether we choose to believe it or not will not change the truth of what it says to us. It is the inerrant, infallible word of God. One day, in the twinkling of an eye, life as we know it will change forever. Maybe this will happen in our lifetime, maybe not. But it will happen to each of us, when we suddenly stand before Jesus the moment we die.

Do you know Jesus Christ as your Savior? Do you believe that He is coming back one day? Do you believe that His Word is truth and all that it says is truth? Too many Christians today live in such compromise that the truth is hard for them to define. One person's "truth" is not necessarily someone else's. Our world is one big melting pot of tolerance and gray areas; but God's truth is constant and non-changing. We as Christians are called to shed the light of God's love in this dark world of darkness. Jeremiah did his best to give the people a chance to repent and return to the Lord. Even though they were God's chosen people, God would still have to deal with their sin. As Christians, we too are still called to repent and return to the Lord when we walk away. If we do not, the storms of life alone will threaten our faith and even our lives. We are in desperate need of a Savior—every day of our lives. Praise God that He gave us His Son and that the times ahead are in His hands. Just make sure that you are in His hands as well. Give Jesus your heart, your life and your all today.

March 27

Today's Verses: Judges 1-3; Luke 4:1-30
Today's Thoughts: Verses vs. Voices

***"The devil led him to Jerusalem and had him stand on the highest point of the temple. "If you are the Son of God," he said, "throw yourself down from here. For it is written:" 'He will command his angels concerning you to guard you carefully; they will lift you up in their hands, so that you will not strike your foot against a stone.'" Jesus answered, "It says: 'Do not put the Lord your God to the test.'"* Luke 4:9-12**

Many times my mind taunts me with thoughts like "If you are really God's child, then why doesn't He do this for you?" or "If God has really given you this spiritual gift for the body of Christ, why aren't you any good at it?" Over and over, I hear thoughts that make me feel guilty or condemned and insecure in my Christian walk. I know the Word of God and yet, **my mind doesn't seem strong enough to just stay focused on the verses to overcome the voices.** In essence, Satan attempts to discredit us just like he tried with Jesus. Satan said to Jesus, "If you are the Son of God"… *prove it!* These are the same kind of words we as believers hear in our heads too. We question if we are really saved, we question God's ability to use us and we question that the promises given to us in His Word are really for us today. In other words, we find ourselves *testing* the Lord's ability to save us, protect us, use us and speak to us; thus, saying to Him, "Prove it." But Jesus didn't do that.

Jesus confidently used the Word of God to attack Satan's use of the Scriptures directly. It is amazing that the **Living Word of God** (Jesus) quotes the **Written Word of God** (the Bible) to attack the words of the enemy.

Lately, I have been struggling with a trial that I have failed in the past. The circumstances and people have changed but the trial is the same. I felt myself heading down the same path as in the past. Crying out to the Lord, I said to Him, "OK, Lord, I've been here before and have lived through the consequences of failing this test. I know that You have been faithful regardless of my faithlessness. I know that You love me and You have every right to test my heart and try my motives to see which way I will go. Lord, I choose You. I choose to not test You or question You in this. I know that Your promises will come to pass and You don't have to _prove_ anything to me." With that prayer, I knew that the temptation to fail had been lifted. Jesus is our example as He was tempted in the same types of trials (with different circumstances and people) but overcame them all. We, too, can overcome the taunting voices of the enemy to live a life pleasing to God by staying on His side and by fighting His way using His words.

March 28

Today's Reading: Judges 4-6; Luke 4:31-44
Today's Thoughts: An Answer to Prayer

Then he said, "O Lord God of my master Abraham, please give me success this day, and show kindness to my master Abraham. Behold, here I stand by the well of water, and the daughters of the men of the city are coming out to draw water. Now let it be that the young woman to whom I say, 'Please let down your pitcher that I may drink,' and she says, 'Drink, and I will also give your camels a drink'--let her be the one You have appointed for Your servant Isaac. And by this I will know that You have shown kindness to my master." Genesis 24:12-14

The prayer that Abraham's servant prayed is called a "fleece." The *New Unger's Dictionary* describes a fleece as "the wool of a sheep, whether on the living animal, shorn off, or attached to the flayed skin." The miracle of Gideon's fleece (Judges 6:37-40) consisted of the dew having soaked the fleece once, without any on the ground, whereas at another time the fleece remained dry while the ground was wet with dew. Literally, fleece means the wool of a sheep. However, today we call certain prayers a "fleece" because Gideon used a "fleece" (or the wool of sheep) to have God confirm His will in a specific manner to him: the fleece needed to be wet even though the ground was dry and then the fleece needed to remain dry even though the ground was wet.

Is it ok to pray for a fleece today? I have asked God for many fleeces because I wanted God's clear direction in making decisions. There have been prayers that I have prayed for years without any need of a sign. But there have been others that I have prayed and received verses and confirmation, but I still wanted more clarification and direction. At this point, I pray a fleece and I have clearly seen God move to show me His will. Know that when you pray a fleece, you have a higher accountability to God. I have learned not to flippantly pray for a fleece. I ask for wisdom, I pray for wise counsel from others, I search His Scriptures and pray continually before resorting to a fleece. I am this hesitant because sadly I have to say that God has been more faithful to answer my fleeces than I have been faithful to be obedient with His answer. Dealing with the consequences of not following the Lord was more difficult than not knowing His will in a matter and just trusting Him for the outcome.

God is a personal God and He wants to answer specific prayers. He will lead you. He will answer you. He loves for you to seek His specific will. And He promises to answer when we call. If He doesn't answer your fleece, then it wasn't time to ask for one. Keep trusting as you ask, seek and knock. The door will be opened and you will know the answer, if you really want to know it or not! But, once you know, be careful to be obedient!

March 29

Today's Reading: Judges 7-8; Luke 5:1-16
Your Thoughts on Today's Passage:

March 30

Today's Readings: Judges 9-10; Luke 5:17-39
Today's Thoughts: Refusing to Believe

Aware of their discussion, Jesus asked them: "Why are you talking about having no bread? Do you still not see or understand? Are your hearts hardened? Do you have eyes but fail to see, and ears but fail to hear? And don't you remember? When I broke the five loaves for the five thousand, how many basketfuls of pieces did you pick up?" "Twelve," they replied. "And when I broke the seven loaves for the four thousand, how many basketfuls of pieces did you pick up?" They answered, "Seven." He said to them, "Do you still not understand?" Mark 8:17-21

From Mark 8, we see how Jesus has compassion for 4,000 people who have been following Him for three days. He feeds them, not only spiritually through His teachings, but also physically as well. It is the second time that He has performed this kind of miracle in front of his disciples, the people following Him and those who oppose Him. The Pharisees were hard-hearted to such an extreme that they could not see or appreciate the spiritual or physical miracles. Instead, they continue to insist on a sign. The answer Jesus gave them was pretty amazing. He did not rebuke them by asking, "What do you call feeding 4,000 with only 7 loaves, with another 7 baskets left over?" No, Jesus does not try to reason with them. He allows them the right to continue in their position of doubt and skepticism. He doesn't even speak in His own defense. He allows them to make a choice – to believe or not believe. Many times today, He deals with us in the same way. He is a gentleman, He will not force us to be obedient.

Sometimes we must deal with similar types of attitudes with the people in our lives – doubting and skepticism. Maybe these people have seen the Lord perform miracles in your life but they still harass you and want you to prove yourself. They might have even partaken of the bread that Jesus miraculously provided for you. Or maybe Jesus has provided miraculously for them as well but they refuse to acknowledge His provision as a miracle. As hard as it can be for us, the responsibility to change them or reason with them is not ours, but the Lord's. It is our responsibility to live in His power and presence, speaking His truth to others while leaving the results to God. We can take comfort and rest in the fact that God does not need us to defend Him. Keep praying for them diligently. And when the struggles come, ask the Lord to give you understanding, for He will continue to provide for you both spiritually and physically.

March 31

Today's Reading: Judges 11-12; Luke 6:1-26
Today's Thoughts: Learning to Wait

And he said to the elders, "Wait here for us until we come back to you. Indeed Aaron and Hur are with you. If any man has a difficulty, let him go to them." **Exodus 24:14**

Moses had led the children of Israel out of Egypt, through the Red Sea, into the wilderness and now to the foot of Mount Sinai. As everyone settled in at the base of this mountain, God called Moses to come up to the top. As Moses prepared to leave, he gave specific instructions to the elders. "Wait here…" Wait. Hang out. Stay put. Do not go anywhere. Sounds like a simple message. But as is often the case, the people became restless and impatient. How long were they supposed to *wait*? Maybe they thought Moses would head up the mountain and then come right back down, kind of a short trip. Their measure of time did not meet God's measure of time. How often does that happen to us? Very frequently in my experience!

We live in a world that is fast-paced, with instant messaging and real-time technology. We wait for very few things. We get impatient and anxious over the smallest time delays. Waiting in line to do anything seems futile and a waste of time. We pray for self-control just so we are not completely rude or offensive to those around us, especially those who may be holding up our schedule. But just as in the days of Moses, there are consequences to the behavior that comes from such impatience. The children of Israel decided to build a golden calf, throw a wild party around it, and make complete fools of themselves before God. We may not build a golden calf while waiting in traffic but many of us (if honest) must admit that we can get pretty foolish in how we handle our frustrations. Horns beeping, people yelling, gestures and faces display hostility…what has happened to us? The short answer: we have no concept of what it means to *wait*.

As Christians, we serve a God who is never in a rush. The Lord is patient and steadfast. If we treat our prayers like a slot machine or a 9-1-1 call, we will most likely be disappointed. The Lord will wait. So often, He is waiting on us to come to Him, or to come back to Him. He does not respond according to the world's pace. He wants us to follow Him, not the world. Waiting on the Lord brings strength to our character. As His child, we should be diffusing the hostility with love. There is no better time to be a witness for Jesus than when we are being pressured to act like everybody else. Stop today and think about this message. If you find yourself in a line at a store, in traffic or anywhere that you are told to "wait," pray for God's strength to give you patience. Pray for the person standing before you in line or the Cashier before you get to her. Pray that it be an opportunity to glorify Him and be His witness.

April 1

Today's Reading: Judges 13-15; Luke 6:27-49
Today's Thoughts: The Right Desire

*Now the serpent was more cunning than any beast of the field which the Lord God had made. And he said to the woman, "Has God indeed said, 'You shall not eat of every tree of the garden'?" And the woman said to the serpent, "We may eat the fruit of the trees of the garden; but of the fruit of the tree which is in the midst of the garden, God has said, 'You shall not eat it, nor shall you touch it, lest you die.'" Then the serpent said to the woman, "You will not surely die. For God knows that in the day you eat of it your eyes will be opened, and **you will be like God, knowing good and evil.**" So when the woman saw that the tree was good for food, that it was pleasant to the eyes, and **a tree desirable to make one wise**, she took of its fruit and ate. She also gave to her husband with her, and he ate Then the eyes of both of them were opened, and they knew that they were naked; and they sewed fig leaves together and made themselves coverings. And they heard the sound of the Lord God walking in the garden in the cool of the day, and Adam and his wife hid themselves from the presence of the Lord God among the trees of the garden. Then the Lord God called to Adam and said to him, "Where are you?"* **Genesis 3:1-9**

Was Eve's desire to be "like God" wrong? No. Many Christians today want to be more like God. We pray prayers like, "Let me follow You. Let me be conformed into Your image. Make me more like You. Please increase and let me decrease." Those are biblical prayers from hearts that truly seek the Lord. So why was the punishment so severe for Eve's desire? Eve was wrong in thinking that being disobedient would achieve her desire. Eve was deceived that she could become "wise" like God by doing it her way, instead of remaining in line with His ways. Eve exchanged the peace of God for a piece of fruit and severed her relationship with the Lord. As a result, that day she died spiritually, which led to an ultimate physical death. Adam and Eve were cast out of the garden as another consequence.

We should not be too critical of Eve because we struggle with the same temptations. When we make choices that are against God, we share the same feelings of fear, guilt and shame. And as we run away and attempt to hide "from the presence of the Lord," He kindly and graciously calls to us and seeks us out. But God wants us to seek Him. He wants us to come to Him with a repentant heart and honest words. Despite where you may be with the Lord, He loves you. He doesn't want you to be afraid or to try to fix the problems you have caused yourself. He will cover you with the blood of Christ, and by His kindness He will lead you to repentance. Come to God, confess and be restored. He has so much more for you but stay in a close relationship with Him. If you choose to do it your own way, you are the one who will lose.

April 2

Today's Reading: Judges 16-18: Luke 7:1-30
Today's Thoughts: Discipline of Love

***"My son, do not regard lightly the discipline of the Lord, nor faint when you are reproved by Him." * Hebrews 12:5**

I can see the anger that comes upon my children when my husband and I need to discipline them. They just hate it. At that moment, they can immediately justify, rationalize, and minimize whatever they are guilty of doing. As a parent, I want to give them the benefit of the doubt. For I know that disciplining them may bring extra consequences to me too. If they need to stay home all weekend, then I need to stay home too. Punishment takes its toll on whoever is involved. However, God tells us that disciplining our children shows them that we love them.

God takes our discipline very seriously too. He does not want us to regard it lightly. By not yielding to His discipline, it delays blessings. God has a heart to bless us, not to hurt us. We need to be disciplined to handle the blessings. Our character needs to be developed in such a way that we can receive more from the Lord without it destroying our integrity or witness. So by regarding His discipline lightly, we will not learn as quickly or as willingly and then we will not be able to receive what God wants to give to us. That hurts God's heart. God gave His Son, Jesus, and He wants to keep giving. He cannot separate from His giving nature. But He also loves us so much as to not allow us to hurt ourselves with His gracious gifts.

If you are being stretched, bent, and even burnt right now, ask the Lord to help you learn His lesson in it. If we learn the first time, we don't have to return to this place of discipline. And remember that "All discipline for the moment seems not to be joyful, but sorrowful; yet to those who have been trained by it, afterwards it yields the peaceful fruit of righteousness." (Hebrews 12:11)

April 3

Today's Reading: Judges 19-21; Luke 7:31-50
Today's Thoughts: Hope

Be of good courage, And He shall strengthen your heart, All you who hope in the Lord. Psalm 21:34

The word "hope" is mentioned repeatedly throughout the Psalms to bring encouragement in the Lord. Most of us speak of "hope" interchangeably with other words such as dreams, wishes, desires and goals. We pray and hope, dream and hope, and you can probably think of other words that you use in connection with "hope." The Bible most commonly uses the word "trust" in association with its definition. To hope in the Lord means that we are putting our trust in Him. In searching how often "hope" is used, I found that it comes up most frequently in the book of Job and the Psalms. One book describes pain and suffering; the other, praise and worship. But hope is the common thread in both.

When we are down, we are desperate for something or someone to place our hopes in or upon. We need to know that tomorrow will be better. We need to believe that lives are changing toward the good. If we are sick, we hope to be healed. If we are in financial crises, we hope for restoration and prosperity. If we are in any kind of trouble, we hope that everything works out positively. We hope for a tomorrow that is better than today. Without hope, where would we be? How could we keep going? Hebrews 11:1 says that "faith is the substance of things *hoped* for, the evidence of things not seen." We must trust in the only One who brings real hope – Jesus Christ. Though we cannot physically see Him, we must place our faith in His love for us. Jesus is our only hope, for only He has the power to change our tomorrows.

Do you need hope for today? Pick up God's Word and ask Him to show you His hope for your situation. Regardless of where you are, He already knows everything about it. No trouble or trial is too great for our Lord. We have but one requirement: to place all of our hope in Him. There is no hope in the world, only fleeting moments. But God gives us hope in the eternal life that awaits us. The Apostle Paul wrote in 2 Corinthians 4:17 that our light affliction is but for a moment compared to the eternal glory that awaits those of us in Christ Jesus. Put your hope and trust today in Jesus. Ask Him to fill you with His Holy Spirit to comfort you with His hope of glory.

April 4

Today's Readings: Ruth 1-4: Luke 8:1-25
Today's Thoughts: A Loyal Friend

But Ruth said: "Entreat me not to leave you, Or to turn back from following after you; For wherever you go, I will go; And wherever you lodge, I will lodge; Your people shall be my people, And your God, my God. Where you die, I will die, And there will I be buried. The Lord do so to me, and more also, If anything but death parts you and me." **Ruth 1:16-17**

Naomi had two sons who married Moabite women. After Naomi's husband and sons died, she decided to leave Moab and return home to Bethlehem. She told her daughters-in-law to stay behind and to go back to their own homes. One stayed but the other one, Ruth, refused to leave Naomi. Ruth's reply to Naomi's repeated attempts to get her to stay behind is written in today's verses. Her words speak for themselves. Commitment, love and devotion are exhibited through Ruth's pledge to go with Naomi.

Ruth settled in Bethlehem and married an Israelite named Boaz. She had a son who would be the grandfather of King David. Her determination to not give up on Naomi was no doubt part of God's plan for Ruth. She had a purpose in God's kingdom as she was chosen by God to be in the lineage of Jesus Christ. Many people today use the book of Ruth as a study on love, commitment and friendship.

How do you compare to Ruth in loyalty to your friends and family? Are you as devoted to Jesus as she was to her mother-in-law? Or are you more easily swayed by others than you think? Friendship at any level requires commitment and devotion, but especially in our friendship with Jesus. We express our love for Christ as we express our love for His people. He takes our friendships very seriously. Do we? If you have never read the book of Ruth, take time today and read through the four short chapters. You will be blessed and touched by her story and by God's sovereignty in all that happens here on earth. And, remember to be a loyal friend to those closest to you.

April 5

Today's Reading: 1 Samuel 1-3; Luke 8:26-56
Today's Thoughts: *A Plan for You*

So Samuel grew, and the Lord was with him and let none of his words fall to the ground. **1 Samuel 3:19**

Before Samuel was born, his mother Hannah prayed for a miracle from God to open her closed womb. She promised the Lord that she would dedicate her baby to Him, and she kept her promise. After Hannah weaned Samuel, she took him to live with Eli the priest, in the house of God. As a child, Samuel began to hear the Lord's voice. He spoke God's words as instructed. As Samuel grew up, he did as the Lord commanded him and was well-known among the people as the Lord's prophet. It is easy to see that God had a clear plan for the life of Samuel, even before he was conceived. And Samuel is seen by some even today as one similar to Moses in his heart and character before God and the people of God. What a blessed life in the Lord!

I used to read this story and be a little envious of Samuel's relationship with God. Samuel was divinely chosen for a specific purpose at a specific time. And he heard the Lord's voice so clearly that he thought it was Eli calling for him (1 Samuel 3:3-10). Then the Lord reminded me that I have even more than Samuel had – I have the Holy Spirit living in me. I have the Word of God written in its entirety. I have the risen Lord Jesus Christ as my personal Savior. What else could I possibly need or want? The issue is not about what I have or do not have in the Lord; the issue is whether or not I live my life like I *believe* it.

God's Word tells us that He has a plan for our lives. God's Word tells us that He knew us before we were formed in our mothers' wombs. God's Word tells us everything we need to know about who our Lord is, just so we can trust and believe in Him. Before you were born, God saw the entirety of your life, from beginning to end. He is never surprised or caught off guard by your behavior or sins. His desire for each of us is that we grow in Him, that we love Him with all that we are, and that we give Him all that we have. Then, His words will be spoken through us and our lives will unfold His divine purpose, all for His glory. Ask the Lord today to help you find strength and confidence in knowing He has a plan and purpose to bless your life from beginning to end.

April 6

Today's Reading: 1 Samuel 4-6; Luke 9:1-17
Today's Thoughts: Are You Hiding by the Baggage?

Therefore they inquired further of the L*ORD****, "Has the man come here yet?" So the L****ORD**** said, "Behold, he is hiding himself by the baggage." 1 Samuel 10:22***

One day, Saul went out to search for his lost donkeys. The next day, Saul was anointed to be the first king of Israel. Saul found something for which he wasn't searching. Even though he was anointed by Samuel, Saul hid himself by the baggage when it came time to reveal his anointing as Israel's first King. The Scriptures tell us that he was empowered by the Holy Spirit and changed internally. But on the outside, not much changed. He went back home and plowed the fields until one more day. At that point, Saul was ready to accept his calling and outwardly became what God had already changed in him inwardly.

The same thing happens to us. Many times God calls us to do something that we weren't searching for either. We know that the prayers that we are praying seem impossible. We know that something is changing on the inside of our hearts, but to break free and become that person on the outside is beyond our doing. We then struggle between the God-given prayers on our hearts and the person we are in the flesh. How can the two become one?

And when? The answer is time or should I say, "in His time." Regardless of what the Lord has put on your heart, you can't receive all He has for you until you have the character to keep you there. In the passing of time, the Lord is molding you through different circumstances, conflicts and experiences as training for the real deal. Take every encounter you go through seriously. Praise Him through the difficulties and maintain self control when He pushes you to your limit. He will exalt you in due time. As the Scriptures say in Galatians 6:9, "And let us not lose heart in doing good, for in due time we shall reap if we do not grow weary."

April 7

Today's Reading: 1 Samuel 7-9; Luke 9:18-36
Today's Thoughts: Everlasting Mercy

And I give them eternal life, and they shall never perish; neither shall anyone snatch them out of My hand. My Father, who has given them to Me, is greater than all; and no one is able to snatch them out of My Father's hand. **John 10:28-29**

During my sophomore year of college, I took a class on religious literature. We studied a variety of poetry, sermons and even the Psalms. One piece of literature I remember most was written by Jonathan Edwards called "Sinners in the Hands of an Angry God." The professor told us that Edwards did not preach this sermon from memory but held his hand-written notes inches away from his glasses, straining to see the words in a dimly lit church. The intensity of the message moved the listeners so much that they crawled up the center aisle with hearts of repentance and remorse, longing to be saved. Jonathan Edwards is still known and quoted for the message he delivered that day.

I have seen a lot of play on words with Edwards' title like, "Sinners in the hands of a loving God." But the remix is not nearly as powerful or life changing as the original. We are all sinners held in God's hands. This God we serve is known as a consuming fire, all powerful and holy. If not for His mercy, we would all be consumed. If we understood who we really are, we could not get out of bed in the morning. God's mercy allows us to handle our own thoughts. We try to justify ourselves as we compare the extent of our sins with others. But the Book of Romans says that only Jesus can justify us. It is Jesus' justification that covers our sins, forgives our sins and blesses us despite our ability and extent to commit sin. Many Christians are walking around thinking that they have crucified their flesh, and that they are living a Spirit-filled life. They feel good about themselves because they do not think they are committing any of the top sins (murder, adultery, stealing). But they have no power; they just have their own rationalizations. God knows better because God sees the heart. If we really understood our sinfulness, we would be flat on our faces every day begging for forgiveness, understanding that God has every right to be angry.

We serve an awesome God. Praise God that He desires for us to live in hope and not in condemnation. His mercy surrounds us and His love blankets us. God wants to give us life in abundance. He does not desire for any to be destroyed. God desires to hold us up, not to push us down. If you are having thoughts that bring you away from God, that is condemnation. If you are having thoughts that are leading you towards the Lord—to be forgiven and restored—that is conviction.
God wants to show you His loving kindness and mercy. Turn to Jesus to be saved from your sins. Being in His hands is the safest place of all.

April 8

Today's Reading: 1 Samuel 10-12; Luke 9:37-62
Today's Thoughts: Don't Look Back

But Jesus said to him, "No one, after putting his hand to the plow and looking back, is fit for the kingdom of God." Luke 9:62

I have come across this verse four times in a week so I am getting the feeling that the Lord wants my attention. There are so many things that compete for our attention. The home, the kids, work, our health, keeping the car tuned up, relationships and of course, the things that entertain us. If we continue to just try to stay afloat without thinking or praying, we find ourselves surviving instead of thriving. It is so easy to not get rid of bad habits because we at least know what to expect and what to do. When God starts helping us to focus on something better and healthier, we are at a loss for awhile on how to live with the change. That's why it is so much easier to look back.

The Israelites really struggled with this when leaving Egypt. They struggled with looking back so much that they never made it forward. All they could think about was the comparison between the luxuries in Egypt with the barrenness of the desert. Even though God completely provided miracles for them and freed them from the cruelty of slavery, they could not stop thinking of the comforts of their bad lifestyle over the freedom of their new lifestyle. And they never learned to work it out with the Lord. They just complained about it. As much as we hate to admit it, we struggle with the same things at times.

If we really think about it, life was not better before we became a Christian. The difference is that we didn't have anyone to blame our issues on then, but now we can blame them on God. How silly but how true! Life is so much better knowing the Lord. We have the power of prayer, the indwelling Holy Spirit, the promises of hope for an everlasting life. We have fellowship with Him, and we should have a great attitude. But we need to learn how to put the Lord first in everything and have Him lead us, instead of figuring out our own needs and asking the Lord to meet them. By giving our lives to Jesus, we should never look back. There is nothing back there worth living for any way. If you are struggling with bringing the past into the future, ask the Lord to help you become more fit for the kingdom of God. Confess that sin and ask the Lord to help you open your heart to receive all He has for you. Then move forward in His Name!

April 9

Today's Reading: 1 Samuel 13-14; Luke 10:1-24
Today's Thoughts: Submit & Resist

Therefore submit to God. Resist the devil and he will flee from you.
James 4:7

On numerous occasions, we have received comments from women who have issues with the word "submit," specifically as to how wives are supposed to submit to their husbands. I remember someone in my small group who literally walked out during a discussion on the biblical teachings of wives submitting to their husbands. Since she made more money, did more work and handled practically all of the family responsibilities, how dare someone tell her to submit to anybody, especially her less-than-equal husband? Who is she really taking a stand against – her husband or the Lord? Our first act of submission is to the Lord.

Today's verse does not address the issue of wives submitting to their husbands but it does address two other areas in which we all need to obey. The word "submit" actually means to come under an authority and to obey that authority. For all of us the first Authority is the Lord. When we submit to Him, we are giving Him full authority over our lives. The devil wants to harass us and attack us from all sides. By submitting to God, we are able to resist the devil. We do not realize that by refusing to submit, we are actually empowering more of the devil's attacks against us. We hinder the power of the Lord to work through us fighting off the attacks.

If you have areas of your life in which you struggle against submission, take them to prayer and ask the Lord for wisdom to help you understand. Do not let the devil deceive you into thinking that submission equals weakness. He knows that it really equals a power that can cause defeat for him and victory for the Lord!

April 10

Today's Reading: 1 Samuel 15-16; Luke 10:25-42
Today's Thoughts: Seek His Heart

But the LORD said to Samuel, "Do not consider his appearance or his height, for I have rejected him. The LORD does not look at the things man looks at. Man looks at the outward appearance, but the LORD looks at the heart." 1 Samuel 15:7

I don't blame Samuel for thinking that Jesse's first son was God's chosen man to replace Saul as King. Saul was described as "a choice and handsome man, from his shoulders and up he was taller than any of the people." Samuel anointed Saul as King but Saul's heart turned from following God. God was grieved that he had made Saul the King of Israel. Now Samuel was to anoint someone after God's own heart. Like Samuel, we judge by appearances. It would be natural to think that the outward appearance mattered because of Saul's appearance. However, God exhorts Samuel to listen to Him, instead of his own instincts.

David was anointed that day. He was the youngest son, out tending the sheep. He too was described as "ruddy with beautiful eyes and a handsome appearance," but God was choosing David for his heart.

To do God's work, we have to be sensitive to His leading. If we continue to do what we think is right by our own natural inclinations, we might not be representing God's heart. We can't read or judge someone's heart. But if God has our heart, we can discern His ways. We need to pray for the Lord's discernment. Today, be sensitive to pray before you make any decisions. Then, wait to hear if the Lord is ready to answer. He will make known His will if you are willing to seek His heart.

April 11

Today's Reading: 1 Samuel 17-18; Luke 11:1-28
Today's Thoughts: Pray with Persistence

And He said to them, "Which of you shall have a friend, and go to him at midnight and say to him, 'Friend, lend me three loaves; for a friend of mine has come to me on his journey, and I have nothing to set before him'; and he will answer from within and say, 'Do not trouble me; the door is now shut, and my children are with me in bed; I cannot rise and give to you'? I say to you, though he will not rise and give to him because he is his friend, yet because of his persistence he will rise and give him as many as he needs. "So I say to you, ask, and it will be given to you; seek, and you will find; knock, and it will be opened to you. For everyone who asks receives, and he who seeks finds, and to him who knocks it will be opened. **Luke 11:5-10**

God tells us over and over in the Bible to pray. There are many reasons and advantages to praying: we develop a relationship with God, we know that God hears us, we can hear back from God, we can receive peace and we may receive what we want. God loves us and wants for us to tell Him all of our desires. The parable of Luke 11 gives us a different slant on prayer. Jesus is telling us "how" to pray, not "why" to pray. The key point in these verses is to be persistent in prayer and diligent to keep praying.

When my oldest child was a newborn, he would sleep all day. I would look at him and think such nice thoughts. I would have thoughts of unconditional love and dedication to commit my life to raising him. But as soon as all the lights were out in the house, he would wake me up by crying and screaming. My thoughts were not as nice then. I was tired from labor and my body did not want to get up to address his cries in a dark, quiet house leaving my soft, warm bed. If he had not screamed, I might have slept right through his cries. But because of his persistence, I got up and tried my hardest to meet his needs the fastest and easiest way possible. It was not always because I had a sacrificial love that I got up to meet *his need,* but often because of *my need* to get back to sleep. That is Jesus' point in Luke 11.

We all have those newborn cries and screams within us. The Lord tells us that He hears our prayers and will answer them according to His will and what is best for us. But sometimes we need to know the depths of our own desires, and we need to cry out to the Lord with an enduring persistence. Keep asking, keep seeking and keep knocking...the key is persistence. When we see God answer those prayers, our faith is increased to continue praying for other things as well. Through persistence, we are rewarded.

April 12

Today's Reading: 1 Samuel 19-21; Luke 11:29-54
Today's Thoughts: God's Dwelling Place

***And let them make Me a sanctuary, that I may dwell among them.
Exodus 25:8***

Does it matter to you where you live? Are you satisfied with your house or do you hope for something more, something better? For many of us, our physical dwelling place tends to keep us longing for a something a little nicer. If we live in the same house for a long time, we probably have redecorated most of it more than once. As humans, we get bored and want a fresh look at the things around us, especially our homes. As physical structures, our homes deteriorate in the elements and require repairs and fix-ups over time. Regardless of the reasons, we all must deal with where we dwell. So what about where God dwells?

Where does God dwell? The verse above tells us that God had a place made just so He could dwell with His people. But this place was no ordinary dwelling; it was a holy place, a sanctuary where His people could enter into the presence of the Lord. This sanctuary was built to the finest detail using the best fabrics, stones and precious metals. As the Israelites transported this mobile dwelling place throughout their travels in the desert, they never needed to upgrade, redecorate, or fix-up...even after 40 years! The Presence of the Lord kept all things restored, refreshed and renewed.

But were the Israelites completely satisfied with the things given to them by God? How could these people have been anything but eternally grateful and in awe of the Lord's miraculous power in their lives? But the Bible tells us that they were stiff-necked and continually complaining. What about us? Are we satisfied with what God has given us? Even more so, what about His promise we have that His Holy Spirit lives within us? John 1:14 says that the Word became flesh and tabernacled (dwelt) among us. Jesus came to earth to dwell with us. He left us His Holy Spirit who dwells *in* us today. Paul tells us in First Corinthians 6:19 that our bodies are the "temple" of the Holy Spirit, so we are now God's dwelling place, for those who have accepted Jesus as Savior.

We should be very thankful that the Lord does not get bored with His temples, or His dwelling places today. What if He became dissatisfied to the point of looking for something better? He is God and He has every right to want more, to want the best. And thankfully for us, that something better is just what He has in store for us, not here but in heaven. Jesus promised that He would go to prepare a place for us (John 14:2). The key for us: get our eyes off our earthly houses and fix them upon Jesus, our eternal home with Him.

April 13

Today's Verses: 1 Samuel 22-24; Luke 12:1-31
Today's Thoughts: Every 15 Minutes Living Dead

For God so loved the world that He gave his one and only Son, that whoever believes in Him shall not perish but have eternal life. John 3:16

Every 15 Minutes is an intense two-day program designed to caution high school juniors and seniors against the dangers of drinking and driving. The program's name was conceived from the statistic that every fifteen minutes someone in the United States dies in an alcohol related traffic accident. This program was initiated at the high school which my children attend. It did not only impact the high school students; however, I was really surprised by how it impacted me.

A private meeting was called for 25 high school students and their parents. After arriving and hearing the introductory announcements, the parents were told that our child was chosen to become one of "the living dead." Our requirements for participation were to send in baby and current pictures, fill out an obituary, write a goodbye letter to your child, and attend the funeral service. Seriously, I glared at my son trying to hold myself together and said, "I don't think I can do this." Throughout the entire program, I definitely prayed more about the emotional components than the actual physical participation.

Looking back on the 24 hour period, I can tell you that the most challenging part for me was writing my son's good bye letter.
I was restless for quite awhile before I could pick up the pen. When I prayed, God did give me the words to write and He revealed to me many amazing insights. I realized that God is not the God of the dead, but the living (Matthew 22:32). When we are in Christ, we are always alive, even if physically dead. Jesus links heaven to earth. I literally saw a mental picture of Jesus with His arms stretched out, one hand was touching heaven and the other was touching earth. He was the bridge that connected the living on earth to the living in heaven. Death had no relevance or significance because of Jesus.

The term "living dead" had a whole new meaning. There is hope for the believer as Jesus' words from **John 5:24-25** came back to mind, "I tell you the truth, whoever hears my word and believes him who sent me has eternal life and will not be condemned; he has crossed over from death to life." Jesus fills in the great chasm. He continued to say, "I tell you the truth, a time is coming and has now come when the dead will hear the voice of the Son of God and those who hear will live." How can the dead "hear?" The answer is "with spiritual ears." Life on earth is not just about the body but about the Spirit. The Spirit is what lives forever.

April 14

Today's Reading: 1 Samuel 25-26; Luke 12:32-59
Today's Thoughts: Vengeance

David said to Abigail, "Praise be to the LORD, the God of Israel, who has sent you today to meet me. May you be blessed for your good judgment and for keeping me from bloodshed this day and from avenging myself with my own hands." **1 Samuel 25:32-34**

Have you ever found yourself in a situation in which you truly felt that you were being paid evil for good? You reached out to help someone or extend certain acts of kindness to your neighbor only to discover later that they would not do the same in return. Or, worse, that same person comes against you to cause you harm in some way. As humans, our response in the flesh tends to lean towards revenge. When they were in the fields together David and his men had protected Nabal's men who were tending their sheep. Even though David was in need of provisions, he never took anything from them. When David heard that Nabal was shearing his sheep, he sent messengers to ask Nabal for some food and whatever else he could spare. Nabal's response was harsh. No way would he going to help David. David became angry and, as a man of war, David immediately sought retaliation. But God intervened through Abigail, Nabal's wife, before David could act.

Abigail intervened on behalf of her husband, without Nabal knowing it. The sovereign Lord led Abigail to David to prevent him from taking steps that would lead to bloodshed. The Lord impressed upon Abigail the foreknowledge that David would one day be king, and as king he would not want to bear such a burden of guilt. You see, God knows all things. God allows all things that happen to us. We may not understand why suffering seems to be such a permanent and prominent part of our lives, but God does. It is hard to deal with people who hurt us, especially when we have been kind to them. But God is with us through every situation, just as He was with David. Abigail was sent by God to speak words of wisdom to David. David was wise enough to listen. David gave his vengeance to the Lord and the Lord dealt with Nabal (he died a few days later).

Remember today that our Lord is sovereign. God is in control of all things. He knows every circumstance of our lives. The decision for us is how we will choose to respond to our circumstances, especially when we must deal with being hurt or betrayed by someone. Know that God will work in your life through every situation. If someone is sent to help you, ask the Lord to give you wisdom and discernment to listen to them. Maybe God has an Abigail in your life. Or, maybe you could be an Abigail to someone who seeks revenge. Regardless of where circumstances place us, our only answer is to seek the Lord's guidance and allow Him to handle it on our behalf.

April 15

Today's Reading: 1 Samuel 27-29; Luke 13:1-22
Today's Thoughts: Do Not Fear

Only be strong and very courageous, that you may observe to do according to all the law which Moses My servant commanded you; do not turn from it to the right hand or to the left, that you may prosper wherever you go... Have I not commanded you? Be strong and of good courage; do not be afraid, nor be dismayed, for the LORD your God is with you wherever you go." Joshua 1:7, 9

. Fear is a natural emotion that can be good when used for protection but debilitating when allowed to control us. Fear can hinder our judgment and potentially prevent us from moving forward to receive all that God has for us. We too are fearful of many things, from losing control to the unpredictability of our future. God was very straightforward and serious when counseling Joshua about fear.

Courage means "to show oneself strong" or "to be alert." Other words associated with courage are "agile, quick and energetic." We tend to associate those words as personality types, not godly traits. If God commands us to be strong and courageous, then He desires for all His children to have such traits. He also gives us the same promises He gave to Joshua. Three times, the Lord told Joshua to be strong and of good courage. He then commanded Joshua to obey, while promising him that He would go with him wherever he goes. And Joshua did.

God tells us the same thing and He has given us the same promises. Hebrews 13:5 says "I will never leave you nor forsake you." The Lord has also given us His Holy Spirit that lives within us – meaning that wherever we go *God goes too.* And He tells us that all authority has been given to us through Jesus. What excuse do we have? Are we as apt to be "agile, quick and energetic" as we saw Joshua to be? We have no excuse to not receive all that God has promised us as His children. We have no excuse to not fully believe His Word and study His Word and live out His will. God told Joshua to be strong and courageous in taking the inheritance as well as in living according to God's ways. It takes courage to live for God and strength to uphold His commands. Both strength and courage are needed, to have both His inheritance and obedience.

What is stopping you? If it is fear, ask Him to give you courage. If it is apathy, ask Him to motivate you to be diligent. Tell the Lord that you want all He has purposed in His heart for you. Tell Him that you do not want anything to separate you from His will, including yourself. Tell Him to help your unbelief and empower you to push forward. And remember, do not be afraid for the Lord your God is with you wherever you go.

April 16

Today's Reading: 1 Samuel 30-31; Luke 13:23-35
Today's Thoughts: God's Eternal Promise

I set My rainbow in the cloud, and it shall be for the sign of the covenant between Me and the earth. It shall be, when I bring a cloud over the earth, that the rainbow shall be seen in the cloud; and I will remember My covenant which is between Me and you and every living creature of all flesh; the waters shall never again become a flood to destroy all flesh. **Genesis 9:13-15**

I was taught as a child that the rainbow symbolized God's promise that He would never again destroy the earth with water. Every time it rained, I looked for the rainbow and I remembered God's promise. Since moving to Southern California, I have seen fewer rainbows, mainly because it seldom rains here. But as the winter brings seasonal rains, I find myself once again searching the skies for that rainbow. What a joy to see that beautiful arc of colors when the sun breaks through the clouds!

As beautiful as the rainbow is, there is still nothing as beautiful as another symbol that Christians look toward. The symbol of a cross representing *the* cross of Christ. Just as the rainbow is a sign of God's promise to not destroy the earth with water, the cross is a sign of God's eternal promise of salvation. As Jesus, Son of God, died on a cross, the cross would forevermore represent God's covenant of grace and love. The cross is a symbol reminding us that Jesus died for our sins. For those who believe in Jesus, the cross represents a covenant of everlasting life with God in heaven.

Rainbows and crosses are just two ways that God reminds us of His covenants, or promises, to us. What other ways does God remind you of His promises for your life? Be encouraged and hopeful because the Lord will never let you down, He will keep every promise. Sit down with His word and ask Him to show you His promises for you today. You might be surprised at the reminders that He will give you. And never let a rainbow or a cross go unnoticed…take a moment when you see them and give God praise for keeping His promises.

April 17

Today's Reading: 2 Samuel 1-2; Luke 14:1-24
Today's Thoughts: Wisdom Cries Out

Wisdom calls aloud outside; She raises her voice in the open squares. She cries out in the chief concourses, At the openings of the gates in the city She speaks her words: "How long, you simple ones, will you love simplicity? For scorners delight in their scorning, And fools hate knowledge. Turn at my rebuke; Surely I will pour out my spirit on you; I will make my words known to you. Because I have called and you refused, I have stretched out my hand and no one regarded, Because you disdained all my counsel, And would have none of my rebuke, I also will laugh at your calamity; I will mock when your terror comes, When your terror comes like a storm, And your destruction comes like a whirlwind, When distress and anguish come upon you. "Then they will call on me, but I will not answer; They will seek me diligently, but they will not find me. Because they hated knowledge And did not choose the fear of the Lord, They would have none of my counsel And despised my every rebuke. Therefore they shall eat the fruit of their own way, And be filled to the full with their own fancies. For the turning away of the simple will slay them, And the complacency of fools will destroy them; But whoever listens to me will dwell safely, And will be secure, without fear of evil." **Proverbs 1:20-33**

I awoke one morning to a phone call from my mom, who lives in Tennessee. She called to tell me that she was okay, just in case I had seen the news and worried if she had survived. The night before, tornadoes had ripped through my family's small town and devastated a large portion of the area. I thanked God that she was okay, at least physically. Deeper though, I wondered how much more was headed our way – in terms of storms. Has anyone noticed that there seems to be a pattern of climate changes and weather related catastrophes? Are these "wake up" calls from God?

Today's verses speak about our attitude towards God's wisdom. Are we choosing to fear the Lord, or are we turning from Him? To fear the Lord is the beginning of wisdom (Proverbs 9:10). Today we need wisdom from above to live in this world as Christians. We must keep our eyes fixed upon Jesus so that nothing else matters. The tighter we cling to the world's goods, the harder it is to lose them. Storms will come in our lives; they will continue to challenge and test our hearts. Read the verses above and pray that your heart seeks the Lord first in all that you do. Ask Him to give you wisdom. Pray for "the fear of the Lord" to bring safety and protection in the midst of the storms. Do not stop praying.

April 18

Today's Readings: 2 Samuel 3-5; Luke 14:25-35
Today's thoughts: Count the Cost

***For which of you, intending to build a tower, does not sit down first and count the cost, whether he has enough to finish it.* Luke 14:28**

How many of us sit down and count the cost as we undertake a project? I tend to forge ahead using whatever means necessary to get started, but I find that most of my efforts are front-loaded.

I started organizing my office recently and as I was going through various stacks and boxes, I realized there were some projects either in the early stages or only halfway finished. Even though these ideas were based on good intentions, somewhere along the way they lost momentum, or maybe I lost the momentum. Or maybe I never really sat down and counted the cost of what it would take to finish. In other words, I put more thought into starting the task instead of finishing it. My intentions were good but did I really consider all of the factors? Maybe not. And as a result I have yet to finish many things I had set out to do.

Jesus is warning us through this illustration to take some steps in the beginning that will ensure the right course to the completion. For us as Christians, we need to take these words seriously. Consistency and perseverance strengthen character and teach us how to be wise builders, whether in our homes, work, or church activities. The next verse says that for those who do not finish, they will be mocked and scoffed at by others. We need to realize that others are watching us and often, they are expecting our failures. The goal for us, regardless of the size and scope of the project, is to first, "sit down" and "count the cost."

Do you have an idea for a new project? Have you considered the cost? Before jumping in head first, stop and sit down. Go to the Lord in sincere prayer asking for His guidance. Try to lay out the complete plans from start to finish and find confirmation for these plans in God's word. If we truly seek His will and plans for us, He will show us what to do and how to finish them. So often, we are in much more of a hurry than God is. Be prepared to wait on Him. Finishing something we started is tremendously satisfying. Knowing that the Lord has led our steps through the entire task brings both earthly and spiritual fulfillment, because the journey means as much as the destination.

April 19

Today's Reading: 2 Samuel 6-8; Luke 15:1-10
Today's Thoughts: Standing Firm

Be on your guard; stand firm in the faith; be men of courage; be strong. **1 Corinthians 16:13**

A couple of months ago, I shared my faith with my neighbor. The conversation went well (from my point of view). The next thing I knew she hosted a luncheon and invited all the neighbors but me. After hearing about the event I felt sad and discouraged. I thought about what I had said to her and wondered if I had offended her. As I prayed, the Lord spoke to my heart and reminded me that persecution comes in various forms. I am responsible to speak as the Lord leads me; but I cannot always expect the outcome to be as positive as I would like. The Lord softened my heart to pray for her and to ask for another chance to speak to her again, not in anger for being left out of the party, but in simple love for her.

It is not always easy to be strong in the Lord and to stand firm and not give up. Are you standing firm? We have to learn to praise God and share Him and talk about Him, especially when times are calm. Today we have the Holy Spirit and we have free speech. If we cannot stand firm for Him now when times are easy, how will we handle persecution leading to death when times become tough? God will protect you in His way with His results. Nothing can separate you from the love of God. **Do not allow your pain, pressures, problems or even persecution to separate you from pleasing Him.**

April 20

Today's Reading: 2 Samuel 9-11; Luke 15:11-32
Today's Thoughts: Ask Your Father

"The older brother became angry and refused to go in. So his father went out and pleaded with him. But he answered his father, 'Look! All these years I've been slaving for you and never disobeyed your orders. Yet you never gave me even a young goat so I could celebrate with my friends. But when this son of yours who has squandered your property with prostitutes comes home, you kill the fattened calf for him!' " "My son," the father said, "you are always with me, and everything I have is yours." Luke 15:28-31

I was thinking about the difference between the prodigal son versus his brother. One of the biggest differences between the two was in their ability to ask. The prodigal son asked for his share of the inheritance. His father graciously gave it to him. Not long after that, this younger son left his father's house and "squandered the wealth in wild living" (as quoted in the NIV). The older son kept faithfully working and obeying. When the younger son came home with a heart of repentance, the older son was angry because he felt as if he had been mistreated. But really, their father would have been just as fair. One asked and the other didn't.

Without concentrating on the poor choices that the younger son made, I have been meditating on the relationship of two boys, living under the same roof, and their dad. What gave one the ability to ask? We see the same thing occur in Christianity today. Some of us have learned to ask our Father for more blessings and fruits, willing to take more risks. While others are obedient but angry that nothing extra or miraculous is done for them.

Which are you? Many of us do not pray this way: "*Lord, I know that I don't deserve one thing for I am a sinner and Jesus has done everything for me already. But I ask for your blessings and for more of the Holy Spirit to do bigger and greater things. Not because of any works of righteousness do I ask, but because of Your mercy. Use me, empower me and fill me to be poured out as a blessing on others as You bless me. I want more Lord and You told me to ask.*"

But when you learn to pray this way, miracles happen.

April 21

Today's Reading: 2 Samuel 12-13; Luke 16
Today's Thoughts: Faithful with Little, Given Much

"Whoever can be trusted with very little can also be trusted with much, and whoever is dishonest with very little will also be dishonest with much. So if you have not been trustworthy in handling worldly wealth, who will trust you with true riches? And if you have not been trustworthy with someone else's property, who will give you property of your own?" **Luke 16:10-12**

There is an ice cream man who has come to sell ice cream in my neighborhood for at least 8 years. His van is beat up with bald tires, poor paint, bumps and bruises on the exterior, well-worn and torn up seats inside, and a very loud stereo blasting the ice cream theme song. The music sings to "Do your ears hang low? Do they travel to and fro? Can you tie them in a knot? Can you tie them in a bow?" The song plays over and over, very loudly. And I can hear that song from blocks away. The ice cream man can barely speak English but has a great countenance and has smiled for 8 years, with the facial wrinkles to prove it.

My heart has broken for this man. I have told my children to go and buy ice cream to support him. It seems like a hard job for not too much in return, but his attitude is so good and he is so faithful to do his job well. Well, the other evening I went for a walk. I heard that familiar sound down the block. As I was walking by, to my surprise, the ice cream man got a brand new truck! It was the same man, same products, same advertisements on the side and the same song but a new truck. I was so happy for him.

God spoke to my heart while walking past the truck and explained something to me. When we receive the Holy Spirit in power, God uses the same person with the same personality and same body to be empowered in a whole new way. This man was faithful with little, so he has been given much (Luke 16:10). We too will receive more and more opportunities to be used by God in the power of the Holy Spirit if we are faithful with the smaller tasks He has given us. By being faithful in the little tasks, He can trust us with bigger ones. The choices we make in every day living matter to God. Are we smiling? Are we content? Are we faithful? Same person, same product, same advertisements but with a new power and passion that comes from a faithful heart. God is so good. But how we choose to live the every day job is up to us.

April 22

Today's Reading: 2 Samuel 14-15; Luke 17:1-19
Today's Thoughts: Asking in Confidence

This is the confidence we have in approaching God: that if we ask anything according to his will, he hears us. And if we know that he hears us-whatever we ask-we know that we have what we asked of him. 1 John 5:14-15

When we teach on prayer, we often use these verses as references. Regardless of the audience to whom we are speaking, almost everyone is familiar with these verses. In the gospels, Jesus tells us to "ask" for things. Matthew 7:7 says, "***Ask***, and it will be given to you; seek, and you will find; knock, and it will be opened to you." A few chapters later, Jesus says, "And whatever things you ***ask*** in prayer, believing, you will receive." (Matthew 21:22) When you study the gospels, you will notice that Jesus makes it clear that asking and believing go hand in hand. In other words: when we pray, we need to pray (or ask) in faith. John gives us a clearer picture of that faith by using the word "confidence" in how we approach God in prayer.

At a recent seminar on how to use a prayer journal, we discussed this word "confidence." How many of us come to the Lord in confidence? It opened an enlightening discussion about what confidence before the Lord really looks like and how can we have it. One of the many areas discussed was how confused we get over praying "according to his will." Many people feel that if they are not praying God's will, then they cannot approach Him in confidence. They try to figure out what God wants them to pray *before* they come to Him in prayer. That approach is confusing and can lead to misguided, fruitless prayers.

Then what does it mean to pray "according to his will?" How can we ever truly know God's will? Let's not put the cart before the horse; start with the basics of prayer. Start with asking. Go to the Lord in honesty and humility and tell Him where your heart is. Tell Him that you are not sure what His will may be for you but that you want to know. Tell Him that you are not sure what to pray but these are the things on your heart. Ask the Lord to help you pray in confidence and to approach His throne of grace with boldness (Hebrews 4:16). The Lord already knows your wants, needs and desires. He just wants you to come and fellowship with Him. Open your heart to Him honestly; do not try to package things before coming to Him. Just ask, and allow His Spirit to do the rest.

April 23

Today's Reading: 2 Samuel 16-18; Luke 17:20-37
Today's Thoughts: Nothing Can Separate Us From His Love

For I am convinced that neither death, nor life, nor angels, nor principalities, nor things present, nor things to come, nor powers, nor height, nor depth, nor any other created thing, shall be able to separate us from the love of God, which is in Christ Jesus our Lord. **Romans 8:38-39**

How do you respond to sudden changes in circumstances that did not result to your advantage? Examples such as when your car suddenly breaks down, your vacation leave does not get approved, your kids get sick the day you had something planned or you lose your cell phone. I have found that there are two main reactions we can have:

God, I am so sorry. I will seek You now. What do You want from me?

God, what did I do wrong? I thought You loved me.

Both cover each end of the spectrum. If we don't walk with the Lord daily, we blame the incident on ourselves and feel guilty. If we do walk with the Lord, we feel abandoned by Him and question His character.

Jesus told us that the rain falls on the just and the unjust alike, meaning that a change in circumstances can happen to all of us. However, the problem comes when we choose to lose our perspective of who God is in the process. God wants us to love Him and enjoy His fellowship. He never stops loving us. We can face any change in circumstance better if we include the Lord in the problem, instead of blaming Him for the problem. Next time something suddenly changes pray, "Lord, help me to stick with You through this and please cause this to work together for good." Nothing can separate us from His love so we shouldn't separate ourselves from Him.

April 24

Today's Reading: 2 Samuel 19-20; Luke 18:1-23
Today's Thoughts: Seasonal Changes

But his delight is in the law of the Lord, and in His law he meditates day and night. He will be like a tree firmly planted by streams of water, which yields its fruit in its season, And its leaf does not wither; And in whatever he does, he prospers.
  Psalm 1:2-3

My aunt came from the east coast to visit. Our days were filled with theme parks, shopping malls, and numerous activities. One evening, as we were planning for the next day, she said, "Can't we see what the weather does and then decide?" That question took me by surprise. Those of us living in Southern California tend to not factor in the weather until we actually see it raining. Why? Because there are few seasonal changes here. Even when the weather forecaster predicts rain, wind patterns change frequently and we often do not get the predicted rain or inclement weather.

Her statement, however, had greater implications. She was tired and ready to slow down and change the pace. On the east coast, weather is a natural boundary that helps adjust schedules and lifestyles. Although the sunny skies are an advantage for most daily events and plans, Southern Californians have to make a conscious effort to rest and slow down. God knows that we need cycles of rest as much as periods of productivity. He factors into our lives spiritual seasons of sunny skies, rainy days, snowed in days and even some hurricanes, tornados and thunderstorms. Why? Because in each of these events, God wants to show Himself faithful and build our faith to trust in Him.

Make a commitment to understand the ways of the Lord. If every day were perfect, there would be no dependency on the Lord or desire to be with God in heaven. Factor in time to spend with the Lord before your day starts. He will prepare you for the weather changes in your spiritual life for the day. The Lord's desire for you is to be stable and consistent, even when the seasons are not. By considering God in your day, you will find peace and joy regardless of what the weather brings.

April 25

Today's Reading: 2 Samuel 21-22; Luke 18:24-43
Today's Thoughts: Emotions

For he who sows to his flesh will of the flesh reap corruption, but he who sows to the Spirit will of the Spirit reap everlasting life. And let us not grow weary while doing good, for in due season we shall reap if we do not lose heart. Galatians 6:8-9

To bear fruit in horrific circumstances takes the ability to separate ourselves from those circumstances. We have to learn to pull ourselves out of the situation to see it from God's perspective, instead of our own. It takes discipline and self-control to live for the Lord despite the situations and events in our lives. Many times, I have had to say, "Ok, God is powerful and has the ability to change these circumstances if needed. God is wise and has the ability to give me the wisdom to deal with these circumstances if needed. God is loving and knows what I need better than I do." When I can apply these three attributes of God: powerful, wise and loving, to my circumstances, I am able to trust God in all situations and events and thus, able to bear fruit in spite of them.

The same thing is true for our emotions. The way we feel can hinder us from doing the will of God and bearing fruit. Our emotions can clearly impact our actions. Here are some examples:

- If we do not feel like cleaning the house, we will not do it even if it needs to be done.
- If we do not feel like being nice today, we will not be nice even though the consequences will linger beyond the day.
- If we are led by our emotions, we will do or not do what we feel like doing. But in order to bear fruit and stay in the will of God, we need to be able to separate ourselves from our feelings and do what God has called us to do in spite of our selves.

To pull ourselves out from the circumstances, we can remind ourselves of the attributes of God. To pull ourselves out from our emotions, we need to be in the Word of God. We have to claim verses as promises to keep us focused on doing things that please Him, even if it does not please our emotions at the time. It is in praying for self-control to maintain a steadfast life that brings Him glory. Every day and every decision counts. Choices we make today matter forever. Keep sowing those seeds of righteousness and in time, you will reap a healthy harvest.

April 26

Today's Reading: 2 Samuel 23-24; Luke 19:1-27
Today's Thoughts: Careful in Counsel

They continually say to those who despise Me, 'The Lord has said, "You shall have peace" '; And to everyone who walks according to the dictates of his own heart, they say, 'No evil shall come upon you.' "For who has stood in the counsel of the Lord, And has perceived and heard His word? Who has marked His word and heard it? **Jeremiah 23:17-18**

Jeremiah was a prophet called by the Lord to give His message to the people. Unfortunately, the message from God was not good news. The Lord told the people of their impending destruction and captivity. His words were severe and too hard for them to accept. In contrast to Jeremiah's prophecies, others proclaiming to be prophets told the people not to fear, that they would have peace. These false prophets used the Lord's name to declare their worthless predictions, and the people chose to believe the false prophets instead of Jeremiah. But these so-called prophets had no relationship or counsel with the Lord God Almighty. They would be held accountable before the Lord for their actions as they blatantly deceived the people.

This message applies to us today as well, especially when we speak on behalf of the Lord. We must be sure that our counsel or encouragement to others has been prayerfully taken before the Lord. We must stand in the counsel of the Lord and hear from Him before we offer counsel or advice to anyone else. This message is not only critical to those of us in ministry roles, but also true for any Christian who truly desires to help others. We must realize our responsibility before the Lord. When someone is hurting while going through difficult trials, our human response is to comfort and encourage them. But we also must be in prayer for them and ask the Lord to give us His wisdom in how to help. If we are not standing before the Lord, then our words may give false hopes or assurances that are not of the Lord. They could end up placing their hope on our words instead of God's Word.

Regardless of how little or often you counsel others, make sure you heed the words of these verses and take everything to the Lord first. Whether in ministry or other roles, our hearts as Christians will desire to help others, but we must always recognize our inadequacies. Only God knows what He is doing with people and we must walk with Him daily to know how He wants to use us to help them. If we are not careful, just as in Jeremiah's day, we can do far greater harm by offering hope and assurances that are not of the Lord. Take everything to Him in prayer and ask Him to give you Scriptures for any counsel you may give. Always stand on the Word of God. And when in doubt, pray for the self-control to listen instead of speaking.

April 27

Today's Reading: 1 Kings 1-2; Luke 19:28-48
Today's Thoughts: Back to Egypt?

Woe to those who go down to Egypt for help, And rely on horses, Who trust in chariots because they are many, And in horsemen because they are very strong, But who do not look to the Holy One of Israel, Nor seek the Lord! Isaiah 31:1

A reference to going *back to Egypt* is made often in the Old Testament. The Israelites spent over 400 years in Egypt. Much of their time in Egypt consisted of hard labor and enslavement. They cried out to God and He answered their cries. The book of Exodus describes their departure from this life of bondage as Moses led the way. But, as soon as they encountered trials on their journey, the people were quick to want to go *back to Egypt* and return to their old way of life. Why would anyone want to go back to a life that was filled with bondage and suffering? Because they at least knew what to expect and sometimes, that is more comforting than having to face the uncertain future.

Isaiah 31 says, "Woe to those who go down to Egypt" and who are looking for help in places outside of the Lord. The same is true for us today. As Christians, we have God's Holy Spirit who is our Helper and our Comforter. Trials and hard times are inevitable in our lives because we live in a fallen world that brings sorrows and pains. Though we will be tempted at times to give up and to go back to our familiar territories, we must press on towards the upward call of Christ Jesus. Do not put your trust in the strength of men or money or worldly power, but put your trust in the Lord. Turn your eyes to Jesus and He will lead the way.

April 28

Today's Reading: 1 Kings 3-5; Luke 20:1-26
Today's Thoughts: The Good Soil

"And the one on whom seed was sown on the good soil, this is the man who hears the word and understands it; who indeed bears fruit, and brings forth, some a hundredfold, some sixty, and some thirty." Matthew 13:23

Gardening is definitely not a hobby for those of us who want nice looking finger nails. After planting flowers for a few hours, my appearance seemed to reflect gardening as a contact sport. I had dirt everywhere. I spent most of my time digging and scrapping and hoeing and massaging the dirt. Placing the flowers in the ground was the easy part.

I was talking to my daughter while planting. I said to her, "Jesus talks about seeds falling on good soil or dirt. What do you think it takes to recognize good dirt over bad dirt?" She said, "Well, I think good dirt is darker than bad dirt. It would be darker because it should be wetter than bad dirt." Her observation was very good. The areas that we planted with moister soil did have a darker color because the sprinklers covered that area better. We also found healthier plants in the darker dirt.

I also thought about this spiritually. As we acknowledge our sinfulness, we understand the darkness of our hearts. It is then through the moistening of the Holy Spirit to wet our hearts to receive Jesus as our Savior, that we are redeemed from our sin. That kind of soil is then ready for the Lord to plant His Words in our hearts to bring forth flowers or fruits of righteousness in His time.

The first step is the preparation of the soil. Praise God that He is willing to get His hands dirty for us! We might not like seeing how dark the dirt is but it is in understanding our sinful state that Jesus can do a beautiful work in the end. Remember, seed…time… and harvest!

April 29

Today's Reading: 1 Kings 6-7; Luke 20:27-47
Your Thoughts on Today's Passage:

April 30

Today's Reading: 1 Kings 8-9; Luke 21:1-19
Today's Thoughts: God Gives the Increase

So then neither he who plants is anything, nor he who waters, but God who gives the increase. Now he who plants and he who waters are one, and each one will receive his own reward according to his own labor. **1 Corinthians 3:7-8**

As Christians, we have different gifts that produce different fruit for God's work. We can accept this in Scripture but we question why some Christians are more successful in ministry than others, such as when one's church is so much larger than another's. We tend to look at the "successful person" and try to pattern ourselves to be like that person, instead of keeping our focus on the Lord. Paul's point emphasizes the fact that ultimately God is the reason for any of our success.

Paul is making a piercing statement that should penetrate each of us right into the depths of our heart. The focus of ministry needs to be about God—period. Without God, there is no increase, no success, and no value to our work. Without God, there would be no gifts, skills, or fruits. Without God, there would be no *us*. Because of God, one person is given certain talents while someone else is given the complement to those gifts, both designed by God to fulfill His purpose only (1 Corinthians 12:11). In these verses, Paul is planting while Apollos is watering, both chosen by God to perform His service.

In Matthew 15:13 Jesus says, *"Every plant which My heavenly Father has not planted will be uprooted."* Unless God is our Farmer, nothing that we plant will ever bear fruit. Most of the time, however, we forge ahead with the planting and then ask God to bless our work with good fruit. Jesus says that the Father will not allow those plants to remain; He will remove them completely.

How much time do we waste in trying to complete our plans, meet our standards, control our progress, and make our deadlines? How do you personally plant and water specific areas of your life? Is God your Farmer? If you are feeling frustrated, fatigued, and frazzled, comparing yourself to others, then you are missing what God has for you. You must realize that He has given you the talents and skills to plant and water, but all for His purposes and to His glory. Without Him, there will be no growth and no success.

May 1

Today's Reading: 1 Kings 10-11; Luke 21:20-38
Today's Thoughts: Desiring to Grow

"As newborn babes, desire the pure milk of the word, that you may grow thereby," 1 Peter 2:2

This spring season, several birds decided to reside under the eaves of my home. I have enjoyed watching the activities of the parents in relation to their babies. I am continually in awe of God's awesome provisions, especially in nature, and especially for those that can appear so helpless. The other day, I was watching some of the baby birds take their first flights. There were about four of them hanging out on the rain gutter, side by side. One, then another, then another would take off, flying in dizzying patterns with unsteady wings. Just when I thought one might fall, he (or she) would somehow get back up to the original starting point. I could almost sense its own relief when it made it back to safety.

For us, we are not so different from these baby birds learning to fly. When we first come to Christ, we are ready to jump out and take off in our new life. And the Lord loves our hearts that are completely devoted to doing just that—going for all that He has for us. However, just as we had to grow and mature physically as a baby into an adult, we must grow and mature spiritually in our walk with Christ. As we are growing spiritually, we need the pure milk of God's Word. We grow stronger and faster when we are feeding on His Word every day. The more we study the Word, the more we are able to take those flights of faith.

Paul and Peter both warned the Christians in their day to make sure that they stay in the Word of God. There is no substitute and no quick fix to gain wisdom and knowledge in the Lord. We will continue to age physically and mature in one sense, but our spiritual maturity can be stagnant and stumped, despite our chronological age. Some Christians still need the pure milk while others are devouring real meat (I Corinthians 3:1-3). Put the Word in your heart and mind and pray that your spiritual maturity is evidenced in the ways that God uses you to take longer and more steady flights with Him and for Him every day.

May 2

Today's Reading: 1 Kings 12-13; Luke 22:1-30
Today's Thoughts: Looking Deeper

But the Lord said to Samuel, "Do not look at his appearance or at the height of his stature, because I have rejected him; for God sees not as man sees, for man looks at the outward appearance, but the Lord looks at the heart." 1 Samuel 16:7

We attended the Awards night that was held at the school my kids attend. This is a ceremony to reward the students who have had an outstanding effort or have been most improved in performance throughout the year. Awards are given in all the subjects. The ceremony ends with character awards like the Nehemiah Award for the student leaders.

To our surprise, one of the students who my family prays for on a regular basis to become a Christian received the "most Christ-like" award. I couldn't figure out how that happened. We know that this student struggles with any belief in Jesus Christ and yet, this student was awarded the most Christ-like award. It was a reminder to me that there is a clear difference between personality traits and Holy Spirit fruits. Somehow, as believers, we become confused with the definitions. We judge by outward appearance and personality. When leaving the ceremony, I wondered if Jesus would have been voted to receive His own award.

All I know is that I want to receive Jesus' reward in heaven. I want to hear Him say to me, "Well done, good and faithful servant." I pray that I bring Him glory and represent Him. I know that I might not receive any awards for it here on earth. But I want to be Christ-like on earth because the Holy Spirit has empowered me to live in a manner that pleases Him alone. I want to be God's girl…His teacher's pet…Jesus' dearly beloved. God promises to reward those who diligently seek Him. I can do that. And you can too. Ask the Lord with me to pray to be His best and to be willing to be Christ-like because you want to be like Christ. He doesn't look at outward appearances; we can't fool Him, He judges the heart.

May 3

Today's Reading: 1 Kings 14-15; Luke 22:30-46
Today's Thoughts: Get in the Ark

And behold, I Myself am bringing floodwaters on the earth, to destroy from under heaven all flesh in which is the breath of life; everything that is on the earth shall die. But I will establish My covenant with you; and you shall go into the ark--you, your sons, your wife, and your sons' wives with you. And of every living thing of all flesh you shall bring two of every sort into the ark, to keep them alive with you; they shall be male and female. Of the birds after their kind, of animals after their kind, and of every creeping thing of the earth after its kind, two of every kind will come to you to keep them alive. **Genesis 6:17-20**

The parallel of the story of Noah's ark and our lives today is closer than we may realize. The people in Noah's day were living for their own pleasures and chose not to follow God. The Bible says they were selfish and eating, drinking and giving to marriage. As Noah worked on the ark, he witnessed to the people and told them of what was coming. They did not believe him. I can only imagine the mockery and ridicule that Noah endured as he worked on the ark. Why would a guy build a boat when it had never rained on the earth? Because God said so and Noah believed God.

Over a hundred years later (about 120), the ark was finished and God called Noah's family and the animals into the ark and shut the door. They were saved from the destruction that was coming. They were safe in God's arms. Today, we live in a world that resembles the days of Noah. Man is selfish, wicked and living life without much thought to what the Bible says is coming. We are told that one day Jesus Christ will return and the world as we know it will be burned up, utterly destroyed. The only hope we have is in the One who is coming to save us…Jesus Christ. Jesus is our Ark. For all who enter into His ark will be saved. All we have to do is believe that Jesus died on the cross to pay the price for our sins and to invite Him into our hearts to save us.

There are many "Noah's" today preaching the gospel and telling people to repent because one day Jesus is coming back. How many will be saved when destruction finally comes? Only the Lord knows when that day will be, but let's pray for a sense of urgency for those who have not yet entered into the Ark -a relationship with Christ. Pray to be a witness…pray that you those closest to you are safe in God's arms.

May 4

Today's Reading: 1 Kings 16-18; Luke 22:47-71
Today's Thoughts: Following Him

And Elijah came to all the people, and said, "How long will you falter between two opinions? If the Lord is God, follow Him; but if Baal, follow him." But the people answered him not a word. 1Kings 18:21

Do you ever find yourself faltering between two opinions? Have you ever felt torn between decisions and not sure of what to do? Maybe we get torn sometimes because we are focusing more on the *what*, than on the *Who*. Whether we want to admit it, everything comes down to a choice. We must choose between who, what and how almost everyday of our lives. The Bible makes it clear that we are to choose whom we will follow. There really is no middle road.

The problem lies in the fact that we do not like to think of ourselves as choosing to follow a false god, like Baal. But when we turn from following the Lord, we are turning to follow something else. We get caught up in life's activities and look for God only when we need Him. Did we just choose those activities over the Lord? Probably so. How do we make sure that we choose God first in everything?

We must make the Lord our priority everyday. We must start our day with Him, praying about our schedules, asking for His guidance, reading His word for instruction, and worshiping Him with grateful hearts. We must learn to practice these things every day.

If we put these actions into practice, then we will find ourselves following God, without faltering. God knows our hearts and He wants us to want Him more than anything else. He wants us to stop choosing the *what* in our lives and start choosing the *Who*, Jesus Christ. Start your day with the Lord and ask Him to guide you. Beware of choosing to serve the false gods of this world. The Lord will help you if you just ask Him.

May 5

Today's Reading: 1 Kings 19-20; Luke 23:1-25
Today's Thoughts: Do You Need a Restored Soul?

He restores my soul; He leads me in the paths of righteousness For His name's sake. **Psalm 23:3**

Many of us have read Psalm 23 so often that we can recite the verses, but the blessing comes in really meditating over them. If we are not careful as Christians, we will read such verses and miss what the Spirit is saying to us. Today, verse three jumps out the most to me. "He restores my soul" is the key phrase upon which I am meditating. What does "He restores my soul" really mean to me today?

Only the Lord can truly restore our soul. Our soul is that part of us that houses our emotions, our will, and the internal makeup of who we are. Our soul tends to wander off down selfish paths seeking to fulfill self-driven desires. We all struggle with this. The farther we wander, the more distant we get from God. Much of our wandering is subtle and hidden, even from our own senses; and this fact makes these wanderings often the most dangerous to us. But God knows all about us and He always wants us back. He knows that by restoring us back to Him we will be willing to be led down His path of righteousness. His paths of righteousness, not our own paths, bring us peace.

Ask the Lord today to reveal any areas of your heart that need restoration. Do you feel dull or numb in certain areas of your life? Tell the Lord about those places in your heart where you sense a lifelessness and ask Him to restore them back to Him. Restoration is not always just about life's obvious issues; it can be very much about life's less recognized ones. Take your time to really read Psalm 23 and listen to the Lord's impressions on your heart. The Word of God is the most powerful Word you will read or hear today.

May 6

Today's Reading: 1 Kings 21-22; Luke 23:26-56
Today's Thoughts: Trust and Obey

And all the prophets prophesied so, saying, "Go up to Ramoth Gilead and prosper, for the Lord will deliver it into the king's hand." Then the messenger who had gone to call Micaiah spoke to him, saying, "Now listen, the words of the prophets with one accord encourage the king. Please, let your word be like the word of one of them, and speak encouragement." And Micaiah said, "As the Lord lives, whatever the Lord says to me, that I will speak."
1 Kings 22: 12-14

One day, the king of Israel said to the king of Judah, Jehoshaphat, "Will you go with me to fight at Ramoth Gilead" (verse 4). Jehoshaphat agreed to go and fight with him to get this land back. However, Jehoshaphat suggested that they hear from the Lord first and get His Word on what they should do. The king of Israel summoned his prophets and they all (about 400) proclaimed victory for the nation and king of Israel. Jehoshaphat was not so convinced and inquired of another prophet, Micaiah, who was known for speaking the truth. The king of Israel did not want to hear the truth; he wanted to hear that his plan would work. In the end, despite the king's efforts to the contrary, the king of Israel died in the battle. His heart was too hardened to hear and obey the words of God, even when those words would have saved his life.

How often do we ask for the truth but really do not want to hear it? How often are we told the truth yet refuse to act on it? These issues usually only matter when the truth is bad news. Good news is great to hear and much easier to respond to. Selective truth is not our option, especially in our walk with the Lord. The Lord is always ready and willing to lead and guide us. But are we willing to do what He tells us to do? When His answers do not match what we expect, we must make the decision right then to obey His Word. Otherwise, we allow the enemy time to deceive our thoughts. Pride, selfishness and a hard heart destroy our obedience to God. The bottom line: we ask, God answers, we obey. None of this, "but God" stuff will work.

If our hearts are truly set on following the Lord, then our relationship with Him will be one of surrender, faith and obedience. God is so good to listen to us and to grant us answers to our prayers; however, God is still God. Be reverent in how you seek His will for your life and be honest with Him about your struggles in how to obey. Do not try to manipulate His answers or escape His requests for obedience. Be willing to surrender your pride, be honest about your fears and seek the Lord's help in the next step of whatever He is telling you to do.

May 7

Today's Reading: 2 Kings 1-3; Luke 24:1-35
Today's Thoughts: Seek an Answer

It was Mary Magdalene, Joanna, Mary the mother of James, and the other women with them, who told these things to the apostles. And their words seemed to them like idle tales, and they did not believe them. **Luke 24: 10-11**

I was listening to a very prominent and respected pastor speak on the subject of the woman's role in the church. He emphatically stated that women are not to teach or ever be considered as an authority figure (pastor, deaconess, etc.) in any church setting. He used several verses as references to make his point that women have no real position in the church except to support their husbands, and of course, to take care of their children. Even though his words seemed quite chauvinistic, I knew I needed to hear from the Lord on what He wanted me to think about this sermon.

I prayed about this issue for several hours. The next morning, the Lord led me to Luke 24. As I read the first eleven verses, I clearly saw the role of the women as messengers bringing the Good News of Jesus' resurrection to others. In this case, the women came to the tomb and received from the angels the message of the Lord. They returned to tell the men. The men chose to not believe them. What was the Lord trying to tell me? I kept praying and seeking His will on this. As I was sharing this story with my ministry partner, she told me that Anne Graham Lotz was given these same verses when she began to speak and teach the Bible in different forums. The Lord confirmed to her that He would give her His message to give to the people. I knew at that moment why the Lord had directed me to these verses, and that He was confirming to me the same message.

My point in this devotional is not to focus on the woman's role in the church, but to focus on how the Lord will speak to our hearts concerning the topics on them. The Holy Spirit led me to answers in God's Word and then confirmed them through my friend. Though the topic itself is one of controversy, the main point is that we are Spirit-led in applying His ways to our lives. Regardless of how we have been trained or taught, our final authority must always be the Lord, His Holy Spirit, and His Word.

May 8

Today's Reading: 2 Kings 4-6; Luke 24:36-53
Today's Thoughts: Old and New Testament

He said to them, "This is what I told you while I was still with you: Everything must be fulfilled that is written about me in the Law of Moses, the Prophets and the Psalms." Then he opened their minds so they could understand the Scriptures. **Luke 24:44-45**

Jesus is speaking to His disciples about the things that were written about Him before He came to earth. These writings are prophecies we find in the Old Testament. Many Christians tend to focus on the New Testament more than the Old Testament. They think that the Old Testament is not necessary or as applicable to our lives today. Of course, as Christians, we love to read the gospels and the teachings of Jesus as well as the prophecies of end time events. And the New Testament gives us clear instruction for today's concerns and issues. But we will miss so much if we only stay in the New Testament. The New Testament gives us instruction while the Old Testament gives us illustrations for daily application.

The Old Testament lays the foundation for the New Testament. It is in the Old Testament that prophecies were recorded of Who and what was to come in the future. The Old Testament included the Law of Moses, the Prophets and Psalms—all of which contain specific stories, descriptions and events that point to Jesus. These were written so that the people would recognize Jesus when He appeared. The Old Testament also chronicles the history of God's people as His plan of redemption for all mankind is laid out book by book. By studying the past, we can better appreciate God's plans and purposes for us today as we gain a clearer vision of the work of God in their lives.

It has been well said that Jesus is concealed in the Old Testament but revealed in the New Testament. Everything written about Jesus in both Testaments will be fulfilled. That is why Jesus instructed His disciples in the Scriptures while He was with them during His earthly life and after His resurrection. They needed to understand the significance of the past. Jesus did not come to abolish the Old Testament but to fulfill it. We too need to embrace the whole Bible. He opens our minds to understand all Scriptures so that we will know Him better as we learn to apply His Word to our lives every day. The Old and New Testaments are one story by one Author—Jesus, the Living Word. Ask the Holy Spirit to open your mind and to give you a desire to read and apply the Bible to your life today. Try a reading plan that incorporates a little of both, the Old and New Testament every day. You will be amazed at what you learn. God's Word is truly awesome.

May 9

Today's Reading: 2 Kings 7-9; John 1:1-28
Today's Thoughts: Comfort in Spiritual Sight

So he answered, "Do not fear, for those who are with us are more than those who are with them." And Elisha prayed, and said, "Lord, I pray, open his eyes that he may see." Then the Lord opened the eyes of the young man, and he saw. And behold, the mountain was full of horses and chariots of fire all around Elisha. **2 Kings 6:16-17**

Have you ever found yourself in a situation where you felt outnumbered? At times, many of us feel overwhelmed, either by people or by circumstances. Elisha and his servant found themselves outnumbered as the Syrian army encircled their city. When Elisha's servant saw the number of horses and chariots ready for battle, he became immediately afraid. But Elisha saw their situation through spiritual eyes, not physical ones. Elisha saw God's army and what an army it was! The keys to this story, however, lie in how Elisha prayed and how God answered. Elisha asked the Lord to open the eyes of his servant so that he could see into the spiritual realm. God then opened his eyes. How relieved that young man must have been!

Paul says in Ephesians 6:12 that "we do not wrestle against flesh and blood, but against principalities, against powers, against the rulers of the darkness of this age, against spiritual hosts of wickedness in the heavenly places." These spiritual forces are at work continually, even though we cannot physically see them. However, there is also a spiritual army fighting against them, God's army. In those times when we feel outnumbered and overwhelmed, we are to take comfort in knowing that our Lord and Savior is always protecting us.

How are you feeling today? As Christians, we will fight in battles, some harder than others, but battles and warfare are part of this life. The issue for us is how to handle them. Do we succumb to fear and paranoia? Or do we ask the Lord to open our eyes so that we may see? I think we should opt for the second choice. Our first line of defense should include prayer. Tell the Lord your fears and feelings, and ask Him to show you His plan for your protection. Sometimes the reason we go through battles is for our spiritual growth. It is through having our eyes opened that we truly begin to see the Lord working. Start asking Him today— you might surprised at all He shows you.

May 10

Today's Reading: 2 Kings 10-12; John 1:29-51
Today's Thoughts: Just Believe

Jesus answered and said to him, "Because I said to you, 'I saw you under the fig tree,' do you believe? You will see greater things than these." John 1:50

What does it take for you to believe? For some, we want the small things like a peaceful day. For others, we want our bodies healed. And for others, we want God to appear to us and give us direction on what to do next. All of these requests fall on the lines of appeasing our flesh. Our flesh doesn't want to struggle or guess; we want clarity, peace and a life that is pain free. But that is why many do not believe.

At times, God does perform these kinds of miracles, but God doesn't need to prove He is God as much as we need to prove we believe He is God. Despite how we feel, what we see or how we think, "Without faith, it is impossible to please God" (Hebrews 11:6). Faith goes beyond the senses to a deep understanding and knowing that He is God in the midst of a hassled day and in the hurts from life's circumstances. Faith is believing without seeing.

The Lord asks us today, "Do you believe?" If the answer is yes, He will spiritually open your eyes to see in faith what He has for you. He might not show you through a burning bush or through an earthquake, but it is that still small voice that will testify within your spirit His will and ways for you. Sometimes He says, "Wait." Other times He will say, "Go." But every time, the Lord Jesus Christ will say, "Just believe." He knows what is best. Trust Him today with that issue you want to see God work through so badly. Give it to Him. Lay it down at His feet. Just believe and you **will** see greater things than these.

May 11

Today's Readings: 2 Kings 13-14; John 2
Today's Thoughts: Miracles in the Everyday

Behold I do a new thing; now it shall spring forth. Do you not perceive it? Isaiah 43:19

There are many times that I just want to quit. There is no real reason but I just don't feel like continuing on the same path. I want to see new things, do new things, feel new things and think on new things. I look for the miracles to come and, at times, I seem to end up more miserable than when I started. Sounds like moments of a mid-life crisis, doesn't it? Maybe some of you deal with similar feelings at times. Wouldn't it be great to have a relationship that changes from being filled with tension to all of a sudden being filled with those loving feelings? Wouldn't it be great to suddenly wake up every morning early, bright-eyed to spend an hour with the Lord and to hear Him answer you? Wouldn't it be great to be able to eat whatever you want without worrying about gaining weight or needing to exercise? That would cause a whole new way of thinking and feeling and doing and even living. Those miracles can happen–and do. I have seen the Lord do those things, which is why I look for those kinds of miracles every day. But I have come to realize that God uses the boundaries in our lives to teach us victories in entirely new ways. I hate to admit it, but I have come to learn that the victories in life come mostly in the everyday things.

I was reading the book of Zechariah along with the book of Revelation and cross-referencing similar passages. I knew that the Lord was speaking through them to me. I got this great feeling that He was going to be doing something great in my life that day. As time went by, I was looking for the miracle. And God is so faithful; I realized that He did do a miracle that day. There was something I have dreaded to do every week, but all of a sudden, I was excited to go and participate. It was a miracle. I didn't have any of those feelings of dread or insecurity, just excitement. Why did God choose to reveal His blessing to me in this way? Because He values our everyday lives and wants to have us enjoy His blessings in them. That new thing may seem so small but it yielded so much peace. God is in the business of granting peace.

Look for the Lord in the little things. He wants your revelation of His works to become bigger as you look for Him in the every day. And believe me, living on the same path but now with peace is living in the miraculous.

May 12

Today's Reading: 2 Kings 15-16; John 3:1-18
Today's Thoughts: Rich Fulfillment

Oh, bless our God, you peoples! And make the voice of His praise to be heard, Who keeps our soul among the living, And does not allow our feet to be moved. For You, O God, have tested us; You have refined us as silver is refined. You brought us into the net; You laid affliction on our backs. You have caused men to ride over our heads; We went through fire and through water; But You brought us out to rich fulfillment. **Psalm 66:8-12**

Sometimes we have the misconception that being a Christian should be all about riches, fulfillment and abundance. As Christians, we know the Lord personally and have been given everything we need for life and godliness. We are the children of the King. What else could we possibly need that He has not already given to us? But why do we face such hard trials in life?

The Psalmist explains the real life version of receiving fulfillment. "For You, O God, have tested us," God tests the hearts and intents of His people. He allows trying circumstances to squeeze us: As if we are "in a net"; we carry affliction "on our backs", often while submitting to others whose motives and intentions are wrong. God brings us through terribly trying situations to test and grow our faith in Him.

These trials are confusing and difficult. But we can we see God's love and care through them because when we have come through the fires of this life, and walked through the tests and trials, the crown of eternal life awaits us. Fulfillment and abundance have to do with the victory in overcoming the most terrible trials. That satisfaction comes when we know that we are still with the Lord regardless of the circumstances. There is a place of rich fulfillment and abundance on the other side of the tests. All the riches of the world cannot be compared with the love, joy and peace we get from God. Read through Psalm 66 and really praise the Lord for the trials of life. Praise and worship brings the greatest fulfillment and abundance of them all.

May 13

Today's Reading: 2 Kings 17-18; John 3:19-36
Today's Thoughts: God Makes Things Right

And we know that all things work together for good to those who love God, to those who are the called according to His purpose.
Romans 8:28

Have you ever wished you could take back something you just said? Or, have you ever thought "If only I had known then what I know now, I would have done things differently?" More than once in my life, I have wished I could go back and do something over to make it right. Even with our best intentions and motives, there are times when our words come out the wrong way and we find ourselves feeling badly. We usually know it immediately. I have learned from my own experiences that the Holy Spirit will convict and impress upon me that I should have handled a situation differently. One of the Greek definitions of the word "conviction" means to bring to light, to cause shame, or to show one's fault. One of the Holy Spirit's jobs is to convict us when we say or do things outside of God's will. This includes speaking when we should have been listening or speaking from our own wisdom instead of the Lord's.

When I find myself in a situation where I feel badly about what I have just said or done, I start praying. I begin by confessing my sin and asking for forgiveness. The Lord promises that He will forgive us our sins if we confess them (1 John 1:9). Repentance is definitely the next step because we need to acknowledge that we do not want to continue sinning against Him or others in this way. Next, we have a responsibility not only to pray for ourselves but also to pray for the person(s) involved because our sins do affect others. And that is one of the reasons I love Romans 8:28. I pray this verse to the Lord when I know I have made a mess of something. I start asking the Lord to work it together for good despite what I have done or said.

I can honestly say that the Lord has always been faithful to these prayers. You see, we are going to say and do things that we wish we could take back. We are going to hurt others, even when our intentions are good. We will fail and fall short; we need a Savior. We are sinners saved by grace. So today, put the responsibility back on our Savior and Lord. Ask the Lord to work "all things" together for good and apply this verse to those times when you know you have just made a mistake. Ask forgiveness, and then start praying for God to make it right. Only He can make things right. God's love, mercy and grace cover us every moment of our lives and He gives us His Word to help guide us. Memorize this verse today and ask the Lord to bring it to your mind in times of trouble.

May 14

Today's Reading: 2 Kings 19-21; John 4:1-30
Today's Thoughts: True Worship

The woman said to Him, "Sir, I perceive that You are a prophet. Our fathers worshiped on this mountain, and you Jews say that in Jerusalem is the place where one ought to worship." Jesus said to her, "Woman, believe Me, the hour is coming when you will neither on this mountain, nor in Jerusalem, worship the Father. You worship what you do not know; we know what we worship, for salvation is of the Jews. But the hour is coming, and now is, when the true worshipers will worship the Father in spirit and truth; for the Father is seeking such to worship Him. God is Spirit, and those who worship Him must worship in spirit and truth." John 4:19-24

The Samaritan woman was really asking Jesus, "Where is the appropriate place to worship God?" Her people, the Samaritans, worshiped on Mount Gerizim while the Jewish people worshiped at the temple in Jerusalem. Jesus was very clear that the Jews had the appropriate place of worship at that time because all of the laws of God and prophets of God came through the Jewish race. However, Jesus was quick to explain to her that the "where" would soon change. When Jesus died on the cross for our sins, His death restored us to be able to have a personal relationship with God. We do not need to go to a certain mountain or a specific church to worship Him. Because of Christ, God goes with us—everywhere. So wherever you are, you worship.

The next part Jesus explains is "how?" How are we to worship God? His answer is "in spirit and in truth." Because God is Spirit, no matter how good our efforts or good deeds are, sinful man falls short in being able to please a Holy God. We must have the Spirit of God to commune with a God who is Spirit. Therefore, accepting Christ to receive His Spirit is the first step to true worship. The next step is to be trained in the ways of the Spirit. This training process is called sanctification, or being set apart for the works and service of God. Jesus said in John 17:17, "Sanctify them by the truth. Your word is truth." We learn the ways to remain in the Spirit by reading, studying, meditating and applying His truth of the Bible.

We need to be born of the Spirit of God, to receive the Spirit of God, in order to understand the ways of the Spirit of God, written by the Spirit of God in the Bible. The Bible is the truth that sanctifies us (or sets us apart and makes us holy) for His service. The only way to make it through the sanctifying process is by relying on the Spirit of God. Our service is one of worship as we become living sacrifices. A true worshiper is one who worships in spirit and truth, every day and every way. A true worshiper worships daily, from the depths of their heart in their prayer closet to the choices they make every minute of the day.

May 15

Today's Reading: 2 Kings 22-23; John 4:31-54
Today'sThoughts: Father and Son

Jesus answered them, "I told you, and you do not believe. The works that I do in My Father's name, they bear witness of Me." John 10:25

A friend of ours called the other day. He travels extensively and works with incredibly gifted people. He has been blessed beyond measure with gifts and talents that help others get the message of the Gospel out through media. His phone call, however, was not about any of those things. This time he shared with me the greatest blessing that surpassed everything that he does at work. While home from traveling, his little boy asked him many questions about God, Jesus and eternal life. Everything else competing for Randal's attention did not compare to the joy of this kind of conversation between him, as a father with his son.

While listening to him share this story, I couldn't help but think about another relationship of a Father and Son. When Jesus was on this earth, He would speak openly, affectionately and lovingly about His Father. When the Father spoke through the heavens about His Son, He spoke words that confirmed Jesus' authority on earth and the Father's ultimate support. They clearly had a relationship founded on love. As believers, Jesus gives us the Holy Spirit who points us to Jesus, as Jesus draws us to the Father. Their relationship completely blesses each other.

But it doesn't stop there. The relationship of the Triune God also blesses us by allowing us to have a relationship with them. That is exactly what happened with our friend and his son. This little 6 year-old boy asked his dad to pray with him to come to know Jesus. Their relationship as father and son is based love as it led the way to a personal relationship with the Father and Son who first loved us.

We pray that in your relationships, you point others to the only relationship that really matters. Ask the Lord for wisdom in your words and in the timing. He will open the doors for you to speak about Him as clearly as He opened up the doors of heaven for you.

May 16

Today's Reading: 2 Kings 24-25; John 5:1-24
Today's Thoughts: Your Sabbath

Jesus said to him, "Rise, take up your bed and walk." And immediately the man was made well, took up his bed, and walked. And that day was the Sabbath. The Jews therefore said to him who was cured, "It is the Sabbath; it is not lawful for you to carry your bed." John 5: 8-9

God established the last day of the week—sundown Friday to sundown Saturday—as a day of rest. God said in Exodus 31:13-17 that in six days the Lord made the heavens and the earth. On the seventh day, he rested and was refreshed. The Israelites were to observe the Sabbath as a covenant with God. They were to put aside their normal business of sowing, reaping, treading the wine press, carrying burdens and unnecessary travel. They were to call the Sabbath a delight, a holy day of the Lord (Isaiah 58:13). The Sabbath was created so that the people could rest from the busyness of life and find refreshment and restoration in the worship of God. He did not restrict the deeds of necessity (like eating), service to God (the priests' roles) or acts of mercy (kindness to others including helping animals). God did not say He stopped working in helping, healing, saving or doing good deeds. Even the Jews agreed that God was "working" because the universe remained intact. A common saying among the Jews was that God did not stop creating or judging on the Sabbath. And so the Jewish religious leaders felt the liberty to become good at judging on the Sabbath also. The Jews wrote volumes of books to define the limitations of work, yet created more burdens in the process. However, the real burden was that they had taken the Sabbath, God's gift to man, and had transformed it from a day of rest to a day of rules, regulations and restrictions.

Jesus came to set us free and to restore the Sabbath to what it was meant to be. Jesus defined the Sabbath differently. Jesus said in Mark 2:27 that "The Sabbath was made for man, not man for the Sabbath." The Sabbath was given so man would rest. The Sabbath was not given for man to worship the day of rest, but to rest in the worship of the Lord, by stopping his everyday work. Jesus got the religious leaders' attention by continuing to heal and touch lives on the Sabbath. He healed an invalid of 38 years on the Sabbath. The Jewish leaders scorned the newly healed man for following Jesus' instructions by carrying the mat he had laid on for 38 years. This man's burden was the sin that bound him, not the mat he carried on the Sabbath.

Because of Jesus' work, we entered into rest spiritually (Hebrews 4). We no longer need to try to keep laws and rituals seeking God's approval. We are saved by grace. Jesus has done it all. Take your Sabbath and spend time with the Lord and let Him give you rest.

May 17

Today's Reading: 1 Chronicles 1-3; John 5:25-47
Today's Thoughts: God Never Forgets You

"Remember these things, O Jacob, for you are my servant, O Israel. I have made you, you are my servant; O Israel, I will not forget you. I have swept away your offenses like a cloud, your sins like the morning mist. Return to me, for I have redeemed you."
Isaiah 44:21-22

Jacob (whose name was changed to Israel) was chosen by God, but not because he was sinless or above reproach from man's perspective. His descendants were just like us- real people with real problems making real mistakes. But God used this group of people to bring Jesus into the world and to give us a relationship back to God. Even though we have been dealing with the same sin issues since creation, God does not separate Himself from us because of our sin. Jesus made a way for us to be united with God forever. We do not have the ability to become perfect while on earth but we can be perfected (or conformed) into God's image because of Christ.

It is because of the cross that all things are made new. We need to cling to the cross because it is there that we have been set free. We are forgiven, reconciled back to God and called out to be used of Him. I need to do nothing else but to accept that my sins have been cleansed and be willing to accept how the debt has been paid. It is of the Lord that we have been called. We are His witnesses established for His glory and sanctified for His purposes.

Nothing we do can separate us from Him but the same is not true of us. We tend to hang on to our sins, which lead us away from the Lord. Jesus has cleansed us from all sin—past, present and future. Our thoughts should not be on our sins as much as on our Savior. We need to be thanking Him for how good God is that He made a way for us to be with Him forever.

    Relax.

    Take in a deep breath.

    Think about how much grace has abounded to you.

    Talk to the Lord about the sin issues in your life that keep pulling you away from Him.

    Thank Him for never pulling away from you.

May 18

Today's Reading: 1 Chronicles 4-6; John 6:1-21
Today's Thoughts: The Spiritual Fight Against Fear

Now the Lord said to Joshua: "Do not be afraid, nor be dismayed; take all the people of war with you, and arise, go up to Ai. See, I have given into your hand the king of Ai, his people, his city, and his land." Joshua 8:1

The Lord told Joshua to go up to Ai. Actually, He was telling Joshua to go back to Ai. Since their earlier defeat at Ai, the men of Israel have been through a tragic experience because of the disobedience of one of their men, Achan. The punishment for Achan's sin no doubt left a fearful impression upon all these men, who must now be prepared for war. So the Lord tells Joshua to "not be afraid, nor dismayed." In other words, He is saying, "Get back up and let's take care of the business at hand. Let's complete the job we have started." The Lord promises to be with Joshua and his men, even going so far as to tell Joshua that He has already given the king, the people and the land into their hands.

In the second part of this verse, the Lord tells Joshua to "see." This is not a physical seeing because the battle has not yet even begun. This form of seeing requires spiritual eyes of faith to believe that what the Lord is saying is true, even though it cannot be physically seen at this time. For Joshua, this is an important step of faith. He has just been dealt a severe blow through what happened with Achan and now he must get back up. He must continue to fight. He must take each step with the Lord.

God is faithful and just. He will complete the work He has started. Do not let fear, anger or discouragement keep you from following the Lord. Get back up, trust Him and take that next step. The enemy will do whatever he can to keep you from taking that next step, but the Lord will help you to "see" that He has delivered you and will lead you forward.

May 19

Today's Reading: 1 Chronicles 7-9; John 6:22-44
Today's Thoughts: Doing the Works of God

Then they said to Him, "What shall we do, that we may work the works of God?" Jesus answered and said to them, "This is the work of God, that you believe in Him whom He sent." John 6:28-29

The other day I had the pleasure of speaking with a *new* Christian. It was a pleasure because of her enthusiasm and zeal for the Lord. Her eyes were bright and her smile was wide as she talked about Jesus. I found myself smiling back at her as she began sharing her testimony. She is reading, studying and working hard at knowing the Lord more deeply. Why? Because she loves Him and wants to serve Him with her life. But as she sincerely poured out her heart, I realized how hard she was working at trying to do all the right things. I thought about how easy it is for us to get focused on the work: trying to please God, trying to do deeds for Him and trying to know Him more.

I love what Jesus says in that the work of God is that we *believe* in Him. Doing God's work boils down to believing in His Son. It is not about our efforts or good deeds. We do not impress God with our works. God wants our hearts set on Jesus. He wants our lives surrendered to His will. He tells us to walk by faith, not by sight (2 Corinthians 5:7) and that without faith it is impossible to please Him (Hebrews 11:6). But faith means that we give up control, we let go and we allow His Holy Spirit to lead us. Everything from that point on becomes about our belief in Him, not about our works for Him.

I often tell people, new Christians and older ones, to just relax. When I sense that their walk with the Lord is becoming one of frustration and confusion, I encourage them to stop working so hard at it. We need to learn how to let God be God. We need to put the responsibility of our lives back on Him. We are trained by the world to think and act for ourselves, but the Lord says to cast our cares upon Him. What is the key for us today? We must believe in Jesus. We must believe so confidently in His Word that we make no decisions without it. Ask yourself: Am I trying to work the works of God? If you are frustrated or disillusioned, you might be trying too hard instead of resting in Jesus. Ask the Lord to help you let go and to help you believe in faith that Jesus will do all He wants to do in your life. Let Him "work the works of God" in your life today as you believe in Him.

May 20

Today's Reading: 1 Chronicles 10-12; John 6:45-71
Today's Thoughts: He Wears Your Picture Badge

So Aaron shall bear the names of the sons of Israel on the breastplate of judgment over his heart, when he goes into the holy place, as a memorial before the Lord continually.   Exodus 28:29

Have you ever watched the news broadcast of a missing person? As I saw a report on a missing teenager, the family being interviewed all wore picture badges displaying the person's face. They not only wanted this face in public view for possible recognition but also they wanted this reminder close to their own hearts. Just having this type of memorial helped this grieving family pursue their hopes of finding their child.

As discussed in Exodus Chapter 28, Aaron wears a garment that has the names of each of the twelve tribes of Israel etched in precious stones and placed over his heart. This robe is worn by Aaron as he enters the most holy place in the presence of the Lord. Aaron represents the people to the Lord and is a continual reminder of who they are to the Lord. Today, we have Jesus Christ who is our High Priest and who bears our names before the Lord. We, too, are considered as precious stones before the Lord as our names are placed over the heart of our Savior Jesus.

How many of us truly realize how precious we are to the Lord? When life's ups and downs seem to dominate our time and thoughts, we may not realize that we have a High Priest who is continually presenting us and our needs to our heavenly Father. Sometimes, we, as believers, stray from our Father's presence, maybe fall out of fellowship or a regular time of prayer and Bible reading. During those times, I can picture Jesus going before the Father with a picture badge over His heart reminding the Lord of His child who has gone astray. More than anything else, the Lord desires a personal relationship with us and never wants us to be lost or to be away from Him.

Where are you today? Are you in close fellowship with Jesus? Regardless of where you are, know that if you are a child of God, born again in Jesus Christ, then your High Priest is continually presenting you to the Father. You can have the hope and peace today that Jesus is carrying your picture over His heart as a constant memorial. But, even more than that, He is living in your heart and wants to be with you always. Can you say the same? Are you carrying His picture in spirit and truth over your heart today? Are you spending time in worship and fellowship with Him? None are missing in the Lord today…for "I once was lost but now I am found, was blind but now I see" (*Amazing Grace*).

May 21

Today's Reading: 1 Chronicles 13-15; John 7:1-27
Today's Thoughts: The Power of Praying Scripture

For this reason I bow my knees to the Father of our Lord Jesus Christ, from whom the whole family in heaven and earth is named, that He would grant you, according to the riches of His glory, to be strengthened with might through His Spirit in the inner man, that Christ may dwell in your hearts through faith; that you, being rooted and grounded in love, may be able to comprehend with all the saints what is the width and length and depth and height--to know the love of Christ which passes knowledge; that you may be filled with all the fullness of God. **Ephesians 3:14-19**

It is an amazing blessing to learn to take Scriptures and make them our personal prayers. The third chapter of Ephesians gives us a great example of doing this because it is a real prayer that Paul prayed for the recipients of this letter. Let's make it our prayer today.

Dear Lord, I bow my knees to the Father of our Lord Jesus Christ, from whom the whole family in heaven and earth is named, that You would grant me, according to the riches of Your glory, to be strengthened with might through Your Spirit in the inner man, that Christ may dwell in my heart through faith; that I, being rooted and grounded in love, may be able to comprehend with all the saints what is the width and length and depth and height—to know the love of Christ which passes knowledge; that I may be filled with all the fullness of God. I ask this prayer of Him who is able to do exceedingly abundantly above all that I ask or think, according to the power that works in me, to You be the glory in the church of Christ Jesus to all generations, forever and ever. Amen.

Take time to study, meditate and write your own prayers from Scriptures that speak to your heart. God's Word is powerful and life-changing, especially when we make them personal and pray them back to Him.

May 22

Today's Reading: 1 Chronicles 16-18; John 7:28-53
Today's Thoughts: Too Hard for the Lord?

And the Lord said to Abraham, "Why did Sarah laugh, saying, 'Shall I surely bear a child, since I am old?' Is anything too hard for the Lord? At the appointed time I will return to you, according to the time of life, and Sarah shall have a son." Genesis 18:13-14

Abraham was 100 years old and Sarah was 90 years old when Isaac was born. Both Abraham and Sarah were well past child bearing age. Most likely, they could not imagine having a baby at such an old age, but God had a different plan. The Lord knew the exact time in which Sarah would have Isaac and it would clearly be a miraculous blessing to her life. Throughout the years, as the Lord kept telling Abraham of this promise, Abraham kept believing and waiting on the promise to be fulfilled. Twenty-four years after God first promised an heir to Abraham, three men showed up at Abraham's door. The Lord was one of the three and He again restated His promise that they will bring forth a son, a baby boy. Sarah reacted by laughing. The Lord responded with a question: "Is anything too hard for the Lord?"

Abraham was a man of faith. God said it and Abraham believed what God said. Simple, pure, powerful faith. Sarah believed Abraham and trusted his words but when the Lord showed up, she lacked in her faith to believe. She even laughed at what God said. For us, sometimes it is easier for us to believe a person or to believe in a person than to trust in God. If God gives us a promise we can have a hard time trying to work it out ourselves. The hardest part is waiting on the Lord to fulfill it. We wait, we wonder and sometimes we worry but God is always faithful to keep His promises. Nothing is too hard for the Lord. Nothing. If He says it, then He will do it.

May 23

Today's Reading: 1 Chronicles 19-21; John 8:1-27
Today's Thoughts: Lift Up Your Eyes

Song of Ascents. I will lift up my eyes to the hills-- From whence comes my help? My help comes from the Lord, Who made heaven and earth. He will not allow your foot to be moved; He who keeps you will not slumber. Behold, He who keeps Israel Shall neither slumber nor sleep. The Lord is your keeper; The Lord is your shade at your right hand. The sun shall not strike you by day, Nor the moon by night The Lord shall preserve you from all evil; He shall preserve your soul. The Lord shall preserve your going out and your coming in From this time forth, and even forevermore.
Psalm 121:1-8

Today's psalm is one of my favorites. I read it and meditate on the words when I need encouragement from the Lord. Are you in need of encouragement today? Do you feel as if trouble surrounds you? Maybe you feel as if the weight of the world is on your shoulders. Let's be encouraged together as we look at what these verses are saying. First of all, we need to look up and "lift up" our eyes to the hills. We need to look up at the Lord and believe that He is coming to help us. Acknowledge that our Lord is the One who made heaven and earth with amazing power and majesty. He is strong enough to help us. I love the verse that says He "shall neither slumber nor sleep." Our Lord is watching over us constantly. He is our keeper.

This message is not just for today but also for eternity. He preserves our soul and protects us from all evil. He preserves our every move and keeps watch over us everywhere that we go. Because of His Son, Jesus Christ, all who believe in Him are preserved for eternity in heaven. Some days seem longer than others here on earth, but be encouraged and know that the Lord of heaven and earth has His eyes upon you every moment of every day. Take comfort in the words of this psalm and meditate on it throughout times of trouble. Consider these precious words as promises from God, and trust that He always keeps His promises.

May 24

Today's Reading: 1 Chronicles 22-24; John 8:28-59
Today's Thoughts: To Please Him

And He who sent Me is with Me. The Father has not left Me alone, for I always do those things that please Him." John 8:29

Jesus is a man who came from His Father to do those things that always please Him. And because of Jesus' obedience to His Father, His Father became our Father. It has bothered me to think that it pleased the Father to send Jesus to the cross. When I think of love, I do not want to equate it with suffering. But my perspective remains earthbound. God did not equate the cross with suffering as much as the cross was equated to glorification. The cross glorified the Son in restoring and bringing everything back under the authority of Christ. The cross brought salvation to us. As a result, we too suffer on earth for the greater call of glorification for eternity.

Because of Jesus' death, we can say with confidence that the Father has not left us alone. He has given us His Holy Spirit to intercede for us, counsel us, teach us and comfort us. We are never alone. But Jesus' words tug on my heart the most when He said, "I always do those things that please Him." Oh, how I pray that I can live a life that pleases Him. To hear Him so clearly, and know Him so dearly that nothing distracts or takes away from the call to please Him is the cry of my heart. Suffering becomes less and less the focus as my desire to please Him becomes more and more.

Oh Lord Jesus, You are my Savior and Lord but you are also my example. I want to be like You to please You. Let the things of earth become strangely dim in the light of Your glory and grace as I daily seek Your face. Amen.

May 25

Today's Reading: 1 Chronicles 25-27; John 9:1-23
Today's Thoughts: Inquire of Him First

Some men came and told Jehoshaphat, "A vast army is coming against you from Edom, from the other side of the Sea. It is already in Hazazon Tamar" (that is, En Gedi). Alarmed, Jehoshaphat resolved to inquire of the Lord, and he proclaimed a fast for all Judah. The people of Judah came together to seek help from the Lord; indeed, they came from every town in Judah to seek him.
2 Chronicles 20:2-4

Think about one difficulty you are facing right now. It might have to do with your family, your finances, conflicts at work or even an accident you recently had. All these things can be considered as armies coming against you. I know from personal experience that any circumstance that has overpowered me is an army coming against me. I might come up with my own schemes and plans, but nothing I do is going to fix it. In desperation and anxiety, I cry out to the Lord.

Crying out to the Lord as a last resort is a lot different than "resolving to inquire of the Lord" from the start (as Jehoshaphat did). God honored Jehoshaphat's prayer and then he honored the Lord by singing praises to God before the battle even began.

Circumstances are difficult in life. Many times, we wonder what is really going on and why is the Lord allowing this to happen? All the Lord wants is for you to include Him. Circumstances can be hard but His answers to your prayers are not. Go to God today first. Tell Him, "Lord, I don't know what to do but I am looking to You." He not only hears, sees and answers, but He assures the victory.

May 26

Today's Reading: 1 Chronicles 28-29; John 9:24-41
Today's Thoughts: One Thing Will Remain

***Heaven and earth will pass away, but My words will by no means pass away.* Matthew 24:35**

Time passes and life goes on, but there will come a day when the world as we know it will cease and pass away. Heaven and earth will be gone. In Noah's day, even as he built the ark and told the people of what was to come, they continued on as usual. They ate, drank, married and did all the things in life just as they always had done. Then the rains came, and within 40 days every living thing was under water, except for Noah and his family. Noah's family was safe in the ark. They listened to God's words and were saved.

The Bible is filled with prophecies concerning the last days. Many biblical scholars who study prophecy fully believe that we are in the last days and that at any moment our lives will be forever changed. Jesus could return at any moment. Are you ready? Or do you think that your life will just keep going? God did promise us that one thing would continue. He told us that His Word will never pass away. His Word through His Son, Jesus, will save us. All we have to do is believe in Him to live forever with Him. Our bodies will die but our eternal life will go on and on. We need to live for that day when we hear the Lord say, *'Well done, good and faithful servant; you have been faithful over a few things, I will make you ruler over many things. Enter into the joy of your lord'* (Matthew 25:23).

May 27

Today's Reading: 2 Chronicles 1-3; John 10:1-23
Today's Thoughts: God is Keeping Watch

"---shall be keeping watch over the doors" **2 Chronicles 23:4**

As I was leaving my house to go out of town for a few days, I said a prayer and asked the Lord to watch over my home while I was away. I was leaving behind my husband and my dog. Though the two of them are fairly self-sufficient, they are not always the most attentive to certain needs in the home. I prayed for the Lord to see that the daily activities would continue without me being there to take care of them. One of my greatest concerns involved Buddy, my Bassett Hound. He loves to wander off when given the opportunity and although I have "doggy-proofed" the escape routes, my husband leaves them open at times (like garage doors). So, with that in mind, I prayed for the Lord to watch over Buddy and keep him safe, regardless of what may happen.

I came home to learn that Buddy did indeed get out. My husband did leave the garage door open—all day! A neighbor told him that Buddy was literally sitting in the street, as cars slowed and honked for him to move. As the Lord watched over Buddy, however, He made sure that the neighbor escorted Buddy back home and put him in our yard. Needless to say, I was at first shocked and upset that this had happened, but all I really could do was ***praise the Lord***. God is good. I know that the Lord even had me pray that prayer for protection specifically for Buddy because He would get the glory when I found out what had happened.

Take time today and pray that the Lord will watch over your home, your family, your business, everything that you have. We never know what a day will hold. We need to practice praying for certain things everyday and protection is a big thing to pray for. We are promised that "the Lord is faithful, who will establish you and guard you from the evil one" (2 Thessalonians 3:3). And that "His name is our shelter, a strong tower for the righteous to run to" (Psalm 18:10). Say a prayer as you leave your house today and sense the peace of God who will be watching over your doors.

May 28

Today's Reading: 2 Chronicles 4-6; John 10:24-42
Today's Thoughts: No Favorites With God

***For there is no partiality with God.* Romans 2:11**

As school is coming to an end, my second grade daughter is very aware that she hasn't received the student of the month award all year. She is troubled by this. I hear her saying to me over and over, "I just want to be the teacher's pet. My teacher has two pets and I am not one of them." When I asked her to explain, she said, "Mom, you know that you are the teacher's pet when the teacher says, 'this student has worked very hard this year' and gives them an award. They haven't worked any harder than a lot of us." She continued, "The teacher makes up that excuse because the teacher just likes them more than everyone else."

It breaks my heart to hear her say these things but she is probably right. There are some people that we get along with better than others. There are certain personalities that fit better. Isn't that why we get married? We are convinced we can live with "this person" the rest of our lives.

Well, I take great comfort in knowing that God does not play favorites. The Lord loves each of us exactly the same and not one of us has a more special part of His heart than another. He chooses the gifts He gives to us for our good, not His. The Lord has no partiality. If we ask for more of the Holy Spirit or for more wisdom or to be used more, He works with us in answering those prayers. The same is true for those who are saved. In Romans 10:12-13, Scripture says, "For there is no distinction between Jew and Greek; for the same Lord is Lord of all, abounding in riches for all who call upon Him; for 'Whoever will call upon the name of the Lord will be saved.'"

I know from personal experience that I am not saved because of my sinlessness or good behavior. I am thankful that the Lord continues to rescue me and answer my prayers despite my behavior. I am grateful that His grace continues to cover me as He uses me more and more just because I ask Him to. God wants our hearts. Our behavior won't always follow right away. But God doesn't play favorites, and we can rest knowing that He loves each of us so abundantly that one day we will each receive the "Teacher's pet" award.

May 29

Today's Reading: 2 Chronicles 7-9; John 11:1-29
Today's Thoughts: Seeking God

If My people who are called by My name will humble themselves, and pray and seek My face, and turn from their wicked ways, then I will hear from heaven, and will forgive their sin and heal their land.
2 Chronicles 7:14

Do you really want God to hear your prayers? Would you like to know how to get God to listen to you? Read today's verse and notice what God Himself is telling His people to do to get His attention. The children of Israel had turned from God. They practiced idolatry of all kinds, worshipping other gods, disobeying God's laws and living their lives according their own desires. Regardless of the depths of their sin and wickedness, the Lord showed them mercy. God wanted to give them a chance for redemption but the people needed to come back to Him.

The Lord clearly tells the Israelites what to do: humble themselves, pray, seek, and turn. In return, the Lord says that He will hear, forgive, and heal. Right here, in this one verse, we receive a message from the Lord that is just as relevant today in our lives as it was back then. First of all, we are told to *humble* ourselves. If you notice the order, humility comes before prayer. God wants us to come to Him in submission, in lowliness and brokenness, recognizing our desperate need for His saving grace. Then, we *pray*. We pour out our hearts to the Lord, asking, seeking and knocking for His response. In prayer, we must seek not only answers, but we must *seek His face*. To seek God's face demonstrates our love for Him, our desire to know Him, and our relentless pursuit of His attention. We need more than just words; we need intimacy and fellowship with Jesus. Finally, our prayers must include confession and repentance of our sins. Again, not just in words asking forgiveness, but in a true desire to change. To change means that we are willing to *turn* from our "wicked ways" and start following the Lord's ways by obeying His Word.

Regardless of where you are today in your relationship with the Lord, be assured of one thing: He wants more of you. If you feel distant from God, He wants you to know He is not distant from you. If you feel close to God, He wants to be even closer to you. If you humble yourself before Him, pray and seek His face, and are willing to turn from your sins, then He promises to hear, forgive and heal you. Just ask Him. Humble yourself before the Lord today and open your heart to Him. Be sincere, honest and vulnerable before Him. If you come to the Lord in this manner, you will definitely have His attention, and your heart will be ready to receive His answers.

May 30

Today's Reading: 2 Chronicles 10-12; John 11:30-57
Your Thoughts on Today's Passage:

May 31

Today's Reading: 2 Chronicles 13-14; John 12:1-26
Today's Thoughts: The Thrill of Victory

For whatever is born of God overcomes the world. And this is the victory that has overcome the world--our faith. 1 John 5:4

Did you know that this word "victory" is used only six times in the New Testament? The Greek word for victory is "nike" and it means "a conquest" or "means of success." For many of us, the word *Nike* refers to the mega-sporting enterprise that sells shoes and clothing. Victory has been skewed by society standards to now mean something of extraordinary achievements. But Jesus would come and tell us to just have faith, for through faith we will be victorious.

How many of us as Christians today have victory in our lives? How awesome to experience the thrill of victory! But far too often we face the agony of defeat. Jesus did not come to earth, give His life, and leave us His Holy Spirit, just so we could enter heaven. Yes, He came to give us eternal life if we believe in Him, but He wants us to experience heaven on earth—today. Heaven on earth in this day and age? Is such a thought really possible?

Today's verse tells us that "whatever is born of God overcomes the world." If you have accepted Jesus as your Savior, then you have been born of God and are His child. To even accept Jesus indicates a position of faith on your part, for "by grace you are saved through faith" (Ephesians 2:8). But real victory does not stop here, it begins here. Faith must be grown, matured and developed by the tests and trials of life. One reason prayer is so important is that our faith is increased every time we see God answer our prayers. And we must plant His word on our hearts and in our minds, for "faith comes by hearing and hearing by the word of God" (Romans 10:17). Christians who have weak prayer lives and no time in God's Word will face the agony of defeat more than the thrill of victory. But Christians who pray, who read the Bible, and who seek the Lord with all their heart will have victory simply because of their faith. Just remember, however, that victory in Jesus is not always defined the same as victory in the world.

Does your life reflect more victories or more defeats? Where do you spend your time and what are you seeking after? Start today by asking the Lord to increase your faith. Start praying for God's help. Confess and repent from any worldly desires that steal your time and attention away from the Lord. Begin reading His Word and praying for more understanding of how to apply it in your life. At some point, we must decide which way we want to live; and then start acting on the faith given to us through Jesus. Then, and only then, will we have victory.

June 1

Today's Reading: 2 Chronicles 15-16; John 12:27-50
Today's Thoughts: Permanent Ink

He who overcomes shall be clothed in white garments, and I will not blot out his name from the Book of Life; but I will confess his name before My Father and before His angels. **Revelation 3:5**

When my son was in the first grade, he came home from school very concerned. He said to me, "Mom, I got in trouble at the lunch tables today. My teacher wrote my name down and put it in her desk." Sensing his anxiety, I told him not to worry. I knew that the teacher would call me if she had any real concerns. He then said, "But mom, you don't understand, she wrote my name in permanent ink."

So often in life, having our name written in ink signifies a remembrance of something we have done wrong, ranging from teachers' notes to public court records. But there is one place that we all should desire to have our names written in permanent ink. That place is called the Book of Life and the ink is nothing less than the blood of Jesus. Our names are not written down to remind Him of our sins or wrongdoings but to secure our place in His heavenly home for eternity. The only way to gain entry into this Book is by accepting Jesus Christ as Savior and by believing that without His redeeming blood, our sins require a penalty of death. It is the blood of Jesus that keeps our names in His Book of Life permanently. He promises that, by His blood, He will never leave us nor forsake us, and He is permanently with us.

Are you assured today that your name is written in the Book of Life? The only way in is to accept Jesus as your Savior and to mean it with all your heart. Confess your sins and ask Jesus to forgive you. Thank Him for the work on the cross and know in your heart that without His redeeming blood, all are lost. This is one Book that you want to be sure your name is written in permanence.

June 2

Today's Reading: 2 Chronicles 17-18; John 13:1-20
Today's Thoughts: Endurance to the End

"For you have need of endurance, so that after you have done the will of God, you may receive the promise:" **Hebrews 10:36**

I turned off my computer and saw a couple of fingerprint marks right in the middle of my screen. I pulled my shirt sleeve over my hand and began to rub the screen. The marks were still there so I used a paper towel and the marks were still there. I grabbed a dish towel and rubbed some more. They were still there but were beginning to fade. I hate to admit it but my wrist began to tire and I had to stop for a moment. I worked diligently for several more minutes to get the fingerprints off my screen. Normally I would not associate this type of task with the word "endurance" but in this case, it took a while to clean my screen. It seems like such a small thing in the big picture. But, in the midst of these kinds of tasks, I sometimes hear the Lord quietly say to me, "Be faithful in (and keep working on) the little things."

We grow in our walk with the Lord by persevering in the day-to-day tasks that require our patience and endurance. The hardest tests of endurance are often disguised as the monotonous tasks in life. I must admit that I seriously considered just how badly I wanted my computer screen clean. Would I settle for just enough to get by, or would I keep going until the spots were gone completely? Sometimes I think that we are faced with the same question when God begins working on our "spots". Will we persevere with Him? Do we have the endurance to keep going through the hard times?

The writer of Hebrews mentions more than once our need for endurance. We "have need of endurance" to ultimately receive the promises of God. I wonder how often we get to the edge of the promise land and stop walking. There was a reason the Lord kept telling Joshua to go in and take the land that had been given to them. It took endurance and perseverance to keep walking, to keep taking one day at a time, and to stay focused on the goal. From fingerprint smudges to promise lands, the test for each one of us is whether or not we will keep working until the task is finished. The Lord is so good to give us the goal, and He will lead us every step of the way. However, we must be willing to not only step out but also to keep on stepping forward.

Has the Lord given you a glimpse of the promise land He has for you? Do not stop moving towards it. Pray for endurance and perseverance. The training often comes in the little tasks of the day. Take every opportunity to see a job through to its completion and learn how to lean on the Lord for support and guidance. You never know when the day will come that you take that first step into your land of promise.

June 3

Today's Reading: 2 Chronicles 19-20; John 13:21-38
Today's Thoughts: In His Time

**Yet the chief butler did not remember Joseph, but forgot him.
Genesis 40:23**

When Joseph gave the dream interpretations for both the butler and the baker, he requested that the butler tell Pharaoh about him, hoping that the king would release him from prison. When the butler was restored to his position again serving the king, he forgot to mention Joseph to him. No doubt Joseph was disappointed at being forgotten. Can you imagine how it would feel to be in prison for a crime you did not commit and then to be overlooked by the one person who could put in a good word for you? Joseph endured a great deal of anguish through his trials but he remained steadfast as the Lord continued to grant him favor.

How often do we overlook the good deeds of those around us? Have you ever promised to return a favor but for some reason never really did it? When we are in trouble or need help, we tend to pledge whatever someone wants in return for helping us. This bargaining is not always a requirement or necessity for getting help, but it often comes in times of desperation. Have you ever prayed so hard for God to answer a prayer that you vowed something back to Him in return? God is gracious in how He answers such prayers, as most of us can never keep our end of the bargain.

Joseph made a sincere plea to the butler who was about to be set free. "But remember me when it is well with you, and please show kindness to me; make mention of me to Pharaoh, and get me out of this house" (Genesis 40:14). God had a different plan for Joseph as to how he would be set free from prison, a plan that would bless his life and his family's lives in amazing ways. God has a plan for you as well. Pray against discouragement in your circumstances and ask the Lord to give you peace as you wait on His perfect timing. God is sovereign and always knows what is best for us.

June 4

Today's Reading: 2 Chronicles 21-22; John 14
Today's Thoughts: A Place for Us

"Let not your heart be troubled; you believe in God, believe also in Me. In My Father's house are many mansions; if it were not so, I would have told you. I go to prepare a place for you. And if I go and prepare a place for you, I will come again and receive you to Myself; that where I am, there you may be also." **John 14:1-3**

In John 14, Jesus has become a friend to His disciples, as well as their Teacher and Lord. They have been with each other for three years but Jesus is now saying goodbye. Jesus gave them this message to calm them and to focus them *upward instead of inward and outward*. Although He is heading to the cross, His focus is on comforting them. Verse One could be translated, "Do not let your hearts be troubled. You trust and believe in God even though you cannot see Him, so trust also in Me when you can no longer see Me." Jesus then continues to explain to the disciples that His task at hand while He is away is to prepare a place for them. The Lord wants them to know that He is coming back for them, but in the meantime, He has some very important things to do for them and for us too.

There is more to life than what meets our eyes. We need to fix our eyes and our actions toward heavenly things, for heaven is a real place. Heaven is mentioned often in Scripture: Matthew 6:9 says that heaven is where God dwells (and Jesus now sits at the right hand of the Father), First Peter 1:4 calls heaven an inheritance, Second Peter 1:11 describes heaven as a kingdom, Isaiah 66:1 says heaven is where God reigns on His throne and John 14:2 calls heaven, home. It is home for God's children. A "place prepared" in Greek means rooms or abiding places. Some of us think it means a palace for Billy Graham and a shack for the person who accepts Christ on his deathbed. But that is wrong. Jesus Christ is preparing a place for all His saints and every such place will be beautiful.

Life is not a dress rehearsal; our troubles, pains, tears and even our actions will all count when we get to heaven. Jesus told us about heaven so that we would know there is more than just the pain we experience here on earth. He wants us to believe and trust Him with our troubles. For James 1:12 says, "Blessed is the man who perseveres under trial, because when he has stood the test, he will receive the crown of life that God has promised to those who love him." And where will we receive that crown? A *place prepared in heaven called home*!

June 5

Today's Reading: 2 Chronicles 23-24; John 15
Today's Thoughts: Are You in His Flock?

***"Most assuredly, I say to you, he who does not enter the sheepfold by the door, but climbs up some other way, the same is a thief and a robber. But he who enters by the door is the shepherd of the sheep. To him the doorkeeper opens, and the sheep hear his voice; and he calls his own sheep by name and leads them out."*
John 10:1-3**

Jesus began speaking to His listeners with a metaphor that was familiar to them. The sheepfold was usually an enclosure made of rocks with an opening for the door. The shepherd (or watchman) would guard the flock at night by lying across the opening. It was common that several flocks could be sheltered together in the same fold. A group of shepherds would share the responsibility at night allowing each other to sleep in their own beds. The system was like a night shift rotation. In the morning, the shepherds would come through the door or gate and assemble their own flocks, just by their voices. Thieves and robbers could not walk through the gate so they would climb in over the walls, attempting to steal the sheep. But the sheep would not follow them because the sheep would not recognize the stranger's voice. Jesus describes the thieves and robbers as coming to steal, kill and destroy.

During Jesus' time, there were professional religious men (the Pharisees and Sadducees) who were only interested in the people for their own personal advantage. They were more concerned with perfecting themselves according to their own standards and trying to make everyone else feel guilty for not matching up. That is not Jesus. As a good shepherd, He develops a relationship with His sheep so that He knows them by name, calls them personally and leads them. The sheep's responsibility is to welcome, listen and follow the shepherd. The relationship is easy and the burden is light. There is also the sense of security and peace in being protected by the Shepherd.

We need to realize that the responsibility of our life and death is in the hands of Jesus. We do not need to carry our own burdens as an attempt to help Jesus. The Good Shepherd knows that it is His responsibility to take care of the sheep, just as the mechanic knows that the car can't fix itself. We are protected eternally with Jesus for He alone is the Author and Perfecter of our faith. He began a good work in us and He will be faithful to complete it. The responsibility is on Him. There is peace with Jesus as our Shepherd, our Savior. Our responsibility is to hear His voice and follow Him.

June 6

Today's Reading: 2 Chronicles 25-27; John 16
Today's Thoughts: Thirsting for God

As the deer pants for the water brooks, so pants my soul for You, O God. My soul thirsts for God, for the living God. When shall I come and appear before God? **Psalm 42:1-2**

Have you ever found yourself with a thirst so intense that you could not focus on anything else? As Americans, we are blessed to have fresh clean water in abundance and going thirsty is not one of our major issues. Contaminated and diseased water is a major problem in many countries today, and without constant work by those who keep it filtered and clean, sickness and death can result. God created us with a thirst for water being one of our strongest desires, for without it, we cannot live. However, the word "thirst" is not limited to just a need for water. To thirst means that we have a vehement desire and craving for something so intense that we become consumed by thoughts of it. What is it today that you are thirsting for?

The Psalmist wrote of his soul thirsting for God, "for the living God." What a beautiful depiction of love and adoration! He uses the deer as an illustration to describe how he longs for his God. As this animal is parched to the point of panting for the water, it sounds as if the deer is about to faint from thirst. The water gives the deer life again as he drinks it in to quench his thirsting body. Instinctively, the animal knows that he will die without water. The Psalmist expresses his desire for the living God the same way, knowing that he will die without the Lord. He loves the Lord so much and desires to be with Him so strongly that he just wants to "come and appear" before Him. Have you ever had that kind of desire for God? Has your soul ever longed for the Lord to the point of thirsting after Him, and panting after Him?

We are constantly telling people to seek the Lord. Through prayer and the study of the Bible, you put yourselves at His feet. Through a daily commitment to spend time with Him, you build a relationship of love and trust with Him. Through worship and praise to Him, you open your hearts to an intimacy of love and adoration. Once you have tasted of His living water, you will thirst for Jesus so intensely that all you will want is to be with Him, to see Him face-to-face. Many Christians do not experience this level of intimacy with the Lord.

If your soul is longing for relief, pray today for the Lord to give you a thirst for Him. And once you have that thirst, act on the things that will bring you closer to Him: prayer, the Word, worship and time with the Lord everyday.

June 7

Today's Readings: 2 Chronicles 28-29; John 17
Today's Thoughts: Chosen by the Lord

"My sons, do not be negligent now, for the LORD has chosen you to stand before Him, to serve Him, and that you should minister to Him and burn incense." **2 Chronicles 29:11**

After King Solomon's death, the nation of Israel was divided into two separate kingdoms. Israel had ten tribes and Judah had two. Each kingdom had its own king. The Lord had promised King David that he would always have a successor in Judah and God honored that promise as David's line of descendants continued to reign on the throne in Jerusalem. Unfortunately, not all of these kings walked in the ways of the Lord as David had done. One of the most ruthless and wicked of these kings was Ahaz. King Ahaz turned from the Lord to such an extent that he literally cut the articles from the temple in two, set up idolatrous altars in every corner of Jerusalem and even offered his own children as sacrifices. His wickedness incensed the Lord and brought His wrath upon the people for their abominable sins. But upon Ahaz's death, his son, Hezekiah, became king and Hezekiah chose to follow the Lord. Today's verse reflects Hezekiah's words to the people after they had restored the temple and were preparing to offer their sacrifices and worship to the one true God.

King Hezekiah warns the people to not be negligent for it is time to take a stand for the Lord. He instructed them to ask for His forgiveness, and to worship Him in praise and thanksgiving. The people did all that Hezekiah directed them to do and God's wrath was averted. The verse says that the Lord chose them to stand before Him, serve Him and minister to Him. In other words, they now had to act upon their loyalty to God, not just give Him empty words. But God is so awesome that He made a way for them to succeed in their service because He chose them and He led them in all they did. Jesus would later say that we did not choose Him but He chose us (John 15:16).

What can we learn from this story and how can we apply it to our lives? First of all, we need to see how God's Word speaks to us every day, regardless of the chapter or verse or book of the Bible. Even though as Christians, we may read the Bible, pray and go to church, we are capable of easily turning from the Lord. One step leads to more steps until we are farther and farther away from the Lord's teachings. We then find ourselves engrossed in sins of idolatry and wickedness, just like the Israelites. But the Lord is always with us. He is waiting for us to come back to Him and to serve Him, just as He did with the Israelites. Offer your heart to Him in prayers and worship. Repent from your sins, ask forgiveness and know in your heart that God has chosen you for His own. Do not be negligent. Seize the day.

June 8

Today's Reading: 2 Chronicles 30-31; John 18:1-18
Today's Thoughts: The Master Builder

For we know that if our earthly house, this tent, is destroyed, we have a building from God, a house not made with hands, eternal in the heavens. **2 Corinthians 5:1**

I had an opportunity to watch as the first side (panel) of a new building was put in place. The sides of the building were made of concrete and steel. I learned that the foundation of the building was laid first, then the sides were actually placed on top of the foundation. Each panel was engineered and laid out on the ground first, then lifted off the surface foundation and raised up on its side. To say the least, the size of the crane to do this kind of work was awesome. I was told that this particular crane could move up to 300 tons in one lift. The panels were massive in size and weight, but the crane seemed to move them about effortlessly. Within a few hours, the first outside walls of the building were affixed in their places.

As I looked upon this awesome structure, the Lord reminded me of Who made the materials that went into that building. The Lord of heaven and earth created all things and oversees all things that take place here on earth. I reflected on some basic truths about God and me in relation to Him. Buildings made with man's hands can and will be destroyed, regardless of their size and strength. Our physical bodies are to be considered buildings that house our souls and spirits; and they, too, can and will die at some point, regardless of how healthy we try to be. The only building that will remain is the one God builds—"*a house not made with hands, eternal in the heavens.*" This is the building we should be the most concerned about.

How much time and energy do we put into our earthly buildings? Whether they are the buildings that house our souls or the buildings that house our bodies, we all live within some sort of frame. And regardless of how strong and powerful they seem to be, there is One who handles them all as though they were but dust. Our focus today should be on the Master builder, the One who made each of us in His image. God is the One who created all things and will be the One who determines what happens to all things, even us. For those of us who believe in His Son, Jesus Christ, we have an eternal building prepared for us in heaven. Let's spend some time today working on our spiritual life, not just the physical. Spend some time today with the Master builder.

June 9

Today's Readings: 2 Chronicles 32-33; John 18:19-40
Today's Thoughts: The Arm of Flesh

"Be strong and courageous. Do not be afraid or discouraged because of the king of Assyria and the vast army with him, for there is a greater power with us than with him. With him is only the arm of flesh, but with us is the LORD our God to help us and to fight our battles." 2 Chronicles 32:7-8

Hezekiah was a king who helped focus God's people back to the pure and true worship of God. He restored the temple, tore down the high places built to other gods, reinstated God's feasts and celebrations and was known as a king who led people to celebrate the one and only true God. But now, another nation came up to war against him. Hezekiah knew that this other king had the strength and ability to overcome the Israelites in battle. This king also caused a lot of conflicts for King Hezekiah.

Hezekiah could have just surrendered to the fear caused by the tactics and overconfidence of the king of Assyria but he didn't. Instead, Hezekiah started building up his military defenses, working hard on repairing broken parts of the walls around the city and making more weapons. Then he spoke the words of these verses that encouraged the people. These words, however, were not just bursts of intimidation as with the king of Assyria, but came from a heart of passion and purpose from the Lord. By verse 20 of the story, Hezekiah and the prophet Isaiah were crying out to the Lord, knowing that He has the power to fight the battle for them. And He did. Verse 22 says, "So the LORD saved Hezekiah and the people of Jerusalem from the hand of Sennacherib king of Assyria and from the hand of all others. He took care of them on every side."

We worship a God who takes care of us on every side. It was good that Hezekiah prepared for battle, but it was better that Hezekiah prepared His heart to turn to the Lord. The arm of flesh is limited. We need to be prepared to fight as we use the weapons God has given us. The strongest weapon is a tender heart that knows the Scriptures and how to turn to the Lord on our knees. No battle is too great for the Lord.

If you are in a situation today in which you need to see the Lord on your side, get on your knees and start asking for help. God is waiting to show His power no matter what circumstance you are facing.

June 10

Today's Reading: 2 Chronicles 34-36; John 19:1-22
Today's Thoughts: Letting Go Is Hard

I am the LORD your God. You shall have no other gods before Me.
Exodus 20:2-3

Corrie Ten Boom once said that she learned to hold on to things loosely because it hurt too much when God pried them away. There are many things in our life that we get attached to: our work becomes our identity, our homes become our personality, our children become our goals. Sometimes it is hard to know who we were before these things came in and took over our lives. Where do I end and these things begin? When God challenges us on our priorities, it is difficult to truly understand what He is doing.

Because I work in the ministry, it gets confusing at times to separate my walk with the Lord from my work in the Lord. Recently, I have fallen in love with the plans of a ministry that truly helps people and honors God. I received His promises and His revelation of how to get the ministry started. Then, out of the blue, I had the sense that He wanted me to give the ministry to someone else to do. He used me to get the ministry going, but now He wanted me to let it go and not feel that I must do the work for Him. It was hard to let go because that ministry had become a key part of my life. How do you let go of something that God placed in your heart to do, so that you can honor God by not doing it?

We have to remember that nothing is ours. We are just stewards entrusted with the Master's gifts. Our calling is to be obedient. Our hearts are to have nothing above Him, including His ministries. We are to align ourselves up with Jesus, to rest at His feet and to have a relationship with Him. Jesus wants you, not what you can do.

Lord, thank You that You don't want my attention divided. I am sorry that I fall in love with the work to a point of missing Your will. Help me to walk in Your paths and stay in step with Your call. You have given me everything I have and I give them back to You. Be glorified in me. Amen.

June 11

Today's Reading: Ezra 1-2; John 19:23-42
Today's Thoughts: A Humble Heart and Willing Spirit

***Then Peter came to Jesus and asked, "Lord, how many times shall I forgive my brother when he sins against me? Up to seven times?" Jesus answered, "I tell you, not seven times, but seventy-seven times.* Matthew 18:21-22**

So many factors are involved in someone sinning against another. Many times, the thoughts and motives of the one who sinned are not as clear as the reaction of the person who was sinned against. Conflict is hard. Conflict takes a toll on both parties. Defensiveness goes up and trust goes down. Both parties though, the one who needs to forgive and the person who needs to ask for forgiveness, have their share of difficulties. The one who would forgive, deals with skepticism. They hope for real change this time, and deal with thoughts such as, "I hope this is the last time you need to ask forgiveness. I hope this time it works for good. By choosing to forgive you, I may just get hurt again." But the person who continually asks for forgiveness is also in a difficult situation. When we sincerely confess our sin to another, we have to admit to ourselves that we have hurt someone else as a result of our behavior or words. To ask forgiveness repeatedly is to admit that we do not have the ability or power to change that trait in us. So to ask sincerely means that we need to keep seeking ways to change. After a while, it is natural in the flesh to justify and rationalize the sinful behavior instead of continually trying to change. That is why Jesus takes both sides. If someone is repeatedly willing to ask forgiveness, sincerely looking for help, then we need to be willing repeatedly to restore that person back.

I am thankful for Jesus' teaching because I know that He lives by His own teaching. We sin against Him more than anyone else. If we are repentant, He is willing to forgive us - over and over and over again. We have to pray that our hearts remain soft enough to keep asking for forgiveness. His mercies are new every morning, probably because we use up all His mercy the day before.

If you need to ask forgiveness from someone, ask the Lord to give you a humble heart and a spirit willing to change. If you are being asked to forgive, ask the Lord to help you look to Him to help restore the person back in your heart without bitterness. The Lord will help you. He is on both sides. Let us pray that we keep His focus and His heart during the conflicts, and not our own.

June 12

Today's Reading: Ezra 3-5; John 20
Today's Thoughts: The Power of Prayer

Then the prophet Haggai and Zechariah the son of Iddo, prophets, prophesied to the Jews who were in Judah and Jerusalem, in the name of the God of Israel, who was over them. So Zerubbabel the son of Shealtiel and Jeshua the son of Jozadak rose up and began to build the house of God which is in Jerusalem; and the prophets of God were with them, helping them. **Ezra 5:1-2**

The Lord stirred up the spirit of Cyrus, king of Persia, to send some of the Jews out of Babylon and back to Jerusalem to rebuild the temple. "Then the heads of the fathers' houses of Judah and Benjamin, and the priests and the Levites, with all whose spirits God had moved, arose to go up and build the house of the Lord which is in Jerusalem" (Ezra 1:5). But once they arrived, adversaries rose up against them and many threats and distractions were used to thwart their work. How often do we encounter such distractions when trying to do God's work? Our adversary prowls about seeking whom he may devour (1 Peter 5:8) and doing whatever he can to stop our progress in the Lord.

What the people needed, God provided. Zerubbabel and Jeshua needed support and strength. They got it from the prophets whom God raised up to help them. Haggai and Zechariah prophesied in the name of the God of Israel with His power on their lips, a power that brought the others back to work. For us, that power is in the prayers of others. We have the Holy Spirit and God's Word to strengthen and sustain us. But, we also need prayer intercessors in those times when we feel weak and worn out, especially when the adversary seems to be coming after us.

Take time today and pray for those you know who are working for the things of God. Leadership and ministerial Christians are often on the front lines of battle where distractions are constantly interfering in their lives. Pray for them. If you are one of those on that battle line, do not hesitate to ask for prayer when you feel weak and weary. We have Jesus. We have His Word and we have His Holy Spirit, but He has also given us His body of believers to lift each other up in times of trouble. Prayer is a tool we need to use everyday—not only for ourselves, but also especially for those on the spiritual frontlines.

June 13

Today's Reading: Ezra 6-8; John 21
Today's Thoughts: More Than Imaginable

***And there are also many other things that Jesus did, which if they were written one by one, I suppose that even the world itself could not contain the books that would be written.* John 21:25**

Verse 25 is the last verse in the Gospel of John, and what a great way to end this wonderful book. John is basically saying that it would be impossible to record every single one of the "things" that Jesus did while here on earth. John only knew Jesus for three years! My mind struggles to comprehend such amazing works but my heart rejoices in knowing that Jesus is that awesome. Jesus spent three years in full-time ministry, and from studying the Gospels, it is clear that He was all about His Father's business. Jesus spent His time doing the things He was sent to do. Jesus is our only true example of living a life sold out to God. Everything He did had one main purpose: *to glorify His Father.* One cannot glorify God and live a selfish life at the same time.

Often people will ask us questions that involve God's will for their lives; they want to know how to be used by God, they want to make sure that they are doing His will and pleasing Him. How can they be sure they are doing all God wants? We hear these questions and concerns frequently from people who truly desire to fulfill the calling God has on their lives. Many times, however, people are looking for a more awe-inspiring, spiritual answer than we can give. The answers are all in the Bible. God's Word is our living handbook for how to please God and live for Him. Once we learn of His ways, then we begin to learn how to apply His ways to our lives. Just studying the Gospel of John gives us more revelations of who Jesus is and what He did than we can even fully grasp. The world could not contain the books it would take to write everything down. Think about the magnitude of that statement.

For us today, there are a couple of things we need to take from this verse. First and foremost, we must know that Jesus was and is God. He was not just a good man who had a powerful ministry. No man could do what He did. Only God could do miracles beyond what the world could even record. Secondly, for us to do the things God has called us to while here on earth, we must learn from our Teacher. Jesus demonstrated for us how to live a life completely sold out to God. And He has given us His Holy Spirit to teach, lead and guide us today to live a life pleasing to God and in the center of His will. Start reading the Gospel of John today and commit to reading it from beginning to end. You will begin to get a greater understanding of what verse 25 is really saying. Pray that you fall in love with the Lord Jesus and that you glorify His name with your life.

June 14

Today's Reading: Ezra 9-10; Acts 1
Today's Thoughts: Times and Seasons

And he said unto them, It is not for you to know times or seasons, which the Father hath set within His own authority. Acts 1:7

A close friend of mine was recruited to go to Iraq. The Navy told him that he would be in Iraq from six to eight months. He communicated with us via email and it was evident from his emails that it did not matter how long the Navy said he would be there, he was taking each day as it came—praying that another day would come for him. There were days that he was not sure if he would make it.

We think that knowing the dates, times and seasons would help us to focus. Jesus instructed us that it is not for us to know those things. If the disciples knew that Jesus would not return until after 2000 years, I am sure their focus and convictions would have been different.

For my friend, his focus could not be fixed on staying alive to return to his family. Instead, he said that he focused on the tasks that he was equipped to do for the Navy and for the purposes of Americans. The times and seasons did not matter as much as his focus for the day. That is what Jesus is asking of us today. Live a life that is fixed on Jesus each day. He has equipped us through His Spirit and asked us to do a work for the purposes of God. Those purposes include prayer, fellowship, reading the Word and loving others. It does not matter when He will return if we stay busy with the things He has asked us to do until He returns.

My friend was in Iraq for exactly seven months. The Navy told him one afternoon that he would be leaving for home that night. No advanced notice, no time to call home—just time to go. Jesus' return for us will be the same. It is not about the times or seasons but about each day. Pray that you can keep your eyes fixed on the Author and Finisher of your faith, knowing that at any day, we will soon be going home to Him too.

June 15

Today's Reading: Nehemiah 1-3; Acts 2:1-21
Today's Thoughts: The Power of Twos

As iron sharpens iron, so one man sharpens another. Proverbs 27:17

A few summers ago, we wrote a Bible study on the Book of Ezra. I learned a lot like the Books of Ezra and Nehemiah used to be Ezra 1 and Ezra 2. These two men worked together for God's common purposes. Ezra is not just about these two men though. There are multiple twos: Zerubbabel and Jeshua (Ch 3 & 4) and Haggai and Zechariah (Ch 5 & 6). There are 3 sets of twos. This pairing is carried over to the New Testament as Jesus sent the disciples out by twos. How come we do not seem to promote the "twos" today? Husband and wife couples very promoted. There is a very popular church who believes that the husband's calling is carried over to the wife (like a dual anointing). But after studying the book of Ezra, Moses & Aaron, Elijah & Elisha, Ruth & Naomi, we cannot biblically conclude that God has established husband and wife anointing only. Even in the New Testament, the example of couples in ministry like Priscilla and Aquila being good equalized out the couple of Ananias and Sapphira who represent the bad.

So, how come we do not see other groups coming together for the purposes of God? Pulling from my own experience, I really think because it is just plain hard. We already have spouses, children, obligations, extended families, neighbors and churches that our relationship box is maxed out. As I remember back 9 years ago, I said to a friend of mine, "I need a prayer partner that does not get too close. I am already maxed out with my family and my friends. I just need someone who is objective to pray with me and for me without any additional commitment or requirement on my part." When I prayed about that person, the Lord brought Bobbye to mind. She is totally different than me and we don't cross any of the same paths. Little did I know that God had a major plan for two women who were willing to pray together and be committed to Him more than to ourselves. Is it hard? Yes but with time, it gets easier.

Let's get back to praying together, witnessing together, working together so the Lord may be glorified. We are put on this earth to be united for His purposes today like Ezra and Nehemiah were in the past.

June 16

Today's Reading: Nehemiah 4-6; Acts 2:22-47
Today's Thoughts: Miracles

"Most assuredly, I say to you, he who believes in Me, the works that I do he will do also; and greater works than these he will do, because I go to My Father." John 14:12

Does God work miracles today, or did all miracles cease with the last of the apostles? This is a question that has occupied theologians for many years. I counter with my own question: "Is God dead?" A miracle is a supernatural happening. If God is still alive and still working, then there *will* be supernatural happenings. Therefore, the days of miracles cannot be over.

Salvation is a miracle. The rich young ruler, who came to Jesus seeking the way of salvation, finally went away sorrowful. Jesus turned to His disciples and said, "It is hard for a rich man to enter the kingdom of heaven. It is easier for a camel to go through the eye of a needle than for a rich man to enter the kingdom of God." The puzzled disciples replied, "Who then can be saved?" Jesus answered, "With men this is impossible, but with God all things are possible" (Matthew 19:23-26).

If a miracle is achieving something that is humanly impossible, then salvation is a miracle because it is humanly impossible for man or woman to save himself or herself. So to say that the days of miracles are over would be to deny that people can be saved today. Let's thank God that the days of miracles are not over; He is still in the business of working miracles.

Do you believe that God is the same today, yesterday and forever? If so, then pray for the impossible and believe that Jesus is still working miracles today. Put your faith in action and ask the Lord to use you in amazing ways.

June 17

Today's Reading: Nehemiah 7-9; Acts 3
Today's Thoughts: Learn From Him

I will instruct you and teach you in the way you should go; I will counsel you and watch over you. Do not be like the horse or the mule, which have no understanding but must be controlled by bit and bridle or they will not come to you. **Psalm 32:8-9**

We have all we need when we truly understand what God has provided for us through His Holy Spirit. Many times we run to our families, friends and counselors for instruction or guidance. Many times we search horizontally for guidance when the Lord in His mercy and grace has told us to come to Him. By learning from Him, we can avoid a lot of unnecessary hurts, sins and problems.

There was a long period of time when my mother and I were inseparable. We spent our days together and truly enjoyed each other's fellowship. One day, the Lord spoke so clearly to me that He desired to change the relationship. We have similar gifts and callings, and His desire was for us to minister to others, not just to each other. Instead of trusting the Lord, I told Him that I couldn't do it and He would have to do it. Two years later, the Lord did intervene and our relationship changed. Both of us were very hurt and both of us needed to cling to the Lord each day to learn a new way of life. In the midst of that change, the Lord gave me this verse to hold on to. I knew that I would need counsel and instruction. I did not want to run to a new friend to fill the void and I did not want to sin against the Lord during this time of very uncomfortable change because I had known this was His will for years.

The Lord did do a new work in both of us. She and I are in separate ministries, teaching the gospel of Christ. God had His way. We both struggled to let each other go, but in the depths of our hearts we have seen good fruit, eternal fruit, by submitting to God. If you are struggling with a relationship of which you know God is speaking to you about, trust Him. Come to Him for instruction and counsel. He will give you the desires of your heart if you truly turn to Him. Change is scary but the Lord promises to watch over you, lead you and help you as you chose to come to Him. He will fill that void. In His council, you will receive peace and in time, He will give you understanding.

June 18

Today's Reading: Nehemiah 10-11; Acts 4:1-22
Today's Thoughts: Who is God to you?

***If I have told you earthly things and you do not believe, how will you believe if I tell you heavenly things?* John 3:12**

Do you ever think you can figure God out? I mean, there are certain prayers that He should answer just the way you asked it. After all, I am his child and He only desires what's best for me and I know what's best. So I give God options of how He should answer my prayers. And through the years, I have noticed common themes of how I pray. My prayers always include peace, joy, comfort, safety, ease, love and happiness—or making all of the green lights. But my prayers never include things that bring pain, suffering, confusion, hurt or tears. Don't we all tend to pray for the good things because that is what is best for us?

There is a modern day parable about a man during a flood. He prayed, "God, protect me. I have faithfully served You for all these years and You have always kept me safe in this house." What he was really praying was, "Save me my way." Well, the rains came down and the water rose and rose. Radio warnings were blaring to evacuate homes. His neighbors all left. He didn't, he just kept praying. A county sheriff came by and said, "Hey, leave!" but he said, "No, I know the Lord will protect me and I'm staying in this house." The water got higher and this time, the Coast Guard came out. But the man still refused to go. As the water rose, the man sat on his roof, determined not to leave his home. A helicopter came by and said, "Sir please come with me, I'm here to save you." The man said, "No, only God will save me and I'm staying." Well, the man drowned and now, he is standing in heaven. The man said to God, "What happened? I thought you were going to save me." God responded, "Well, I sent the police, a boat and helicopter. What else did you want?"

This man's concept of what God should do for him and what God would do for him was very different. He didn't like God's options. Sometimes what we want God's will to be for us, and what it really is, brings us conflict. We want to serve God our way, please God our way and live for God our way. God says He protects us, so we expect protection to come our way too. Today, submit to the Lord's leading in your life. He knows what is best. Ask the Lord to make you sensitive to His will and that you can be willing to do things His way. The end result just might reveal heavenly things to you.

June 19

Today's Readings: Nehemiah 12-13; Acts 4:23-37
Today's Thoughts: A Great Power

***And with great power the apostles gave witness to the resurrection of the Lord Jesus. And great grace was upon them all.* Acts 4:33**

When the disciples were left alone as Jesus ascended into heaven, a new beginning was just around the corner. Jesus told them to go to Jerusalem and wait for the Promise. That Promise came on the day of Pentecost, a day that would change their lives forever. God's Holy Spirit came upon them with sounds of a mighty wind that came from heaven, a demonstration of supernatural power. Tongues of fire were then dispersed on ("sat on") each person (Acts 2:1-4). This event made it very clear that something amazing and indescribable was taking place amongst these people. In the days that followed, this small group of believers came together in one accord, knowing that they had been given gifts from the Lord, and that they had been called to witness to the gospel of the Lord Jesus Christ.

The first gift given involved "great power" as the apostles zealously told of their knowledge and personal experience of who Jesus Christ was and what He had done. The second gift, however, is one we sometimes take for granted but it is extremely crucial to the first one: "great grace." They were not only granted power but also they were given great grace. Without grace, they could not go forth and do what they were called to do. The same is true for us today. We may have gifts, callings, knowledge, experience and all of the right answers, but without grace, we will not get very far in our service or our usefulness to God. Why? Because we will be taken out by our own thoughts and behaviors as human beings. We become too aware of our weaknesses and we try to overcome them in our own strength. We cannot control condemning thoughts of the enemy who tells us we are not worthy. We lose faith, hope and the eternal perspective because our focus shifts to ourselves. Grace, however, steps in and says that the Lord has granted us unmerited favor in His sight. He chose us. He loved us first. He called us to do His work. He gifted us for His purposes. Not of us, but of Him, by His grace.

Grace must accompany power in your life today. Grace must be connected to your faith. Take time today and look up verses on "grace." Use your Bible concordance and spend some time reading about grace. Make a study of it and take your time, even over several days. Let the Lord minister to you through His Word. If you are suffering or struggling, you need to approach His throne of grace with boldness and receive His mercy today (Hebrews 4:16). And even though we are not worthy, the Lord still uses us by giving us the gifts and grace to minister for Him and by Him and in Him. You will be blessed!

June 20

Today's Reading: Esther 1-2; Acts 5:1-21
Today's Thoughts: True Love

Your glorying is not good. Do you not know that a little leaven leavens the whole lump? 1 Corinthians 5:6

A few years ago as I attended a church retreat, I sat at lunch with ladies I did not know. The woman in charge opened with prayer and then introduced an icebreaker to share around the tables. The question was: "What do you think true love is?" Six out of the eight of us at my table said that love is unconditional acceptance. Their definition included the ability to accept people right where they are, no matter what they look like or what they do. The seventh woman said that she thought love was telling people the truth about sin and where sin leads them. I heard the others snicker at her answer. However, according to Paul's statement in First Corinthians 5:6, only one in seven got the right answer.

The Corinthian church was accepting sinful behavior from a church couple and glorying in their ability to do so. Paul is saying that this is not good; this is not love and they should not be proud of themselves for accepting this couple's behavior. Often we get confused of what love really is when judging others' sins. We do not want to judge others because we ourselves want to be accepted for our imperfections. So we say that acceptance of sin is love. Jesus says in John 8:32, "And you shall know the truth, and the truth shall make you free." The truth is that sin destroys. The truth is that the wages of sin is death. If we truly love others, we do not want to see them destroyed by choices that the Bible calls sin. Paul then takes it a step further by saying that the consequences of accepting this sin of sexual immorality will permeate the whole church and will destroy the whole fellowship of believers. Why? Because this acceptance allows people to sin instead of helping them to overcome it. It demonstrates the wrong standards of love, sin and the ways of God.

Our choices and behaviors really do impact others, just as a little bit of leaven makes the whole lump rise. We have to know the Word of God so we can love others according to God's standards. True love speaks out with kindness and compassion, not allowing others to destroy themselves.

June 21

Today's Reading: Esther 3-5; Acts 5:22-42
Today's Thoughts: Hearing the Word of God

So then faith comes by hearing, and hearing by the word of God.
Romans 10:17

The Bible is clear in its teachings on faith. Faith is essential to our salvation. If we do not have faith to believe in Jesus Christ, then we cannot receive Him as Savior. It takes faith to believe in God. Believing in God through our faith is the victory that has overcome the world (1 John 5:4). Jesus says in Matthew 17:20 that with faith we can move mountains and that nothing will be impossible for us. The power of God's Holy Spirit lives within each believer; but do we live as if we believe it? Not without faith.

How do we get more faith? First of all, faith is given to each of us as a gift. Everyone has faith. But we can have more faith in the things of the world than in the things of God. We have faith that the sun will rise and set each day, even though we have no way of controlling its movements; we just trust that it will continue to work the way it always has. But we need to grow in trusting the One who makes the sun rise and set, the One who put it there in the first place. Therefore, our first step in gaining more faith is by getting to know more about our God, our Lord, our Creator. And the way to obtain knowledge of God is by getting into His word.

I love today's verse because it tells us to *hear* the word of God. Did you know that when you start reading His Word, you will start *hearing* His voice? When you start hearing His voice through His Word, then you will see His intervention in all areas of your life. You will pray in accordance with His will because you will know where He is leading you. You will see God answer your prayers. And the end result: more faith! Take a moment to ask the Lord to help you in your faith by helping you hear His Word. Ask for more of Him, more of His Word in your life—more faith.

June 22

Today's Reading: Esther 6-8; Acts 6
Today's Thoughts: Not to be Passed Over

***Jude, a bondservant of Jesus Christ, and brother of James, To those who are called, sanctified by God the Father, and preserved in Jesus Christ.* Jude 1**

The book of Jude is small, containing only 25 verses, and is positioned between the three letters of John (First, Second and Third epistles) and Revelation, the last book of the Bible. Jude writes his letter to the church exhorting them to beware of the false teachers who are corrupting the people. Jude covers much interesting territory in his short letter; he writes of the supernatural realm and the Archangel Michael, and then brings us back into the Old Testament with Enoch, the seventh from Adam. It is a great book, packed with a powerful message to all of us believers today.

Sometimes in Scripture, I think we pass over some of the most powerful words in verses that we take for granted. For example, today's verse is the opening introduction to Jude. It identifies the author as a bondservant of Jesus Christ and a brother of James, and then it addresses the reader. This introductory point can often be overlooked, but I want to focus on how "those" (the reader) are described. "To those who are *called, sanctified* by God the Father, and *preserved* in Jesus Christ." Those who are called are the first ones addressed. Called means appointed or invited. We have been invited to receive Jesus Christ as Savior. We were appointed for this invitation even before the world was formed (Romans 8:29-30).

Next are those who are sanctified. Sanctified means holy, purified and set apart by God. We are made holy when we receive Jesus as Savior and we are set apart from those who do not believe. Finally, those who are preserved are preserved in Christ. The root of this word in the Greek actually means to watch or keep an eye on. We are being watched and kept an eye on by Jesus Himself. As Christians, we are called for the purposes of God, we are sanctified and set apart by Him for His purposes, and we are preserved by His Son who watches over us and keeps us in His care.

What awesome comfort this verse should bring to all of us today! There is so much to glean from these words. Write it down and really pray over it as to what God is saying to you personally. Maybe you did not know that this verse was written just for you. Take time to study this short book and ask the Lord to teach you His word. And remember, even the introductions are packed with power. Do not skip over any part.

June 23

Today's Reading: Esther 9-10; Acts 7:1-21
Today's Thoughts: Looking for Love

And we have come to know and have believed the love which God has for us. God is love, and the one who abides in love abides in God, and God abides in him. **1 John 4:16**

There is a chorus of a song that says, "Looking for love in all the wrong places, looking for love in too many faces." Funny how true that is. We each have an innate desire to be loved and accepted just the way we are. We long for others to be kind, caring and selfless. We look for this kind of love from our families, friends and even neighbors, believing that we are giving this kind of love so others should give it back. However, the harder we look and the more we seek, the more discouraged and heartbroken we become. Why? Because we are looking for love in all the wrong places and in too many faces.

God is love. Love comes from God. If we are going to talk about love, look for love and try to love others, we need to learn from the Creator of love. Jesus said in John 15:13, "Greater love has no one than this, that one lay down his life for his friends." Jesus laid down His life for us. He was willing to take on the form of a man and then lower Himself to the role of a servant to express His perfect love. The right place to look for love is at the cross and the only "face" that truly expresses love is Jesus'.

Jesus loves you. There is nothing you can do to receive more of His love and there is no sin so great that you can lessen His love for you. His love is constant. It is funny that we look for love, but when we find it, it is a hard thing to accept. Ask the Lord to open your heart today to be able to receive His love for you. Rest in His love and seek His face. Fill that void in your heart by knowing the true essence of Love. Then, instead of your looking for love in the faces of others, others will find it when they see Christ reflected in yours.

June 24

Today's Reading: Job 1-2; Acts 7:22-43
Today's Thoughts: Friends' Advice

Now when Job's three friends heard of all this adversity that had come upon him, each one came from his own place—Eliphaz the Temanite, Bildad the Shuhite, and Zophar the Naamathite. For they had made an appointment together to come and mourn with him, and to comfort him. And when they raised their eyes from afar, and did not recognize him, they lifted their voices and wept; and each one tore his robe and sprinkled dust on his head toward heaven. So they sat down with him on the ground seven days and seven nights, and no one spoke a word to him, for they saw that his grief was very great. **Job 2:11-13**

Job's friends had obviously been getting reports as to what was happening to him, and at some point decided to come and see for themselves what was going on. He was so severely disfigured that his three friends did not recognize him. These friends came with the purpose to "mourn with him, and to comfort him." But if you continue reading the rest of the book of Job, you will see more criticism than compassion is given by these friends. They each used their own wisdom in giving Job advice and counsel; but their counsel is not of the Lord, it is of themselves. They do not know or understand God's ways. They ended up being anything but comforting to Job, as Job had to defend his own actions and beliefs to them. In the end, God restored Job double what he originally had, and Job was granted an amazing insight into just how glorious and holy God is. As for these three friends, it was Job himself who prayed on their behalf so that God would pardon their sins against him. They may have thought they were helping Job, but in reality they were sinning against him, and God was watching.

God is always watching us too. Sometimes God sets up circumstances just to see how we will respond. Will we turn from the Lord or will we worship Him? People around us will always have an opinion. Some of these people will be true friends who try to help. God is watching them as well. As Christians, we will find ourselves at times on both sides. When life's trials and afflictions hit us, we may feel like Job, and all we can really do is to continue to trust the Lord, to cry out to Him and to worship Him through it all. If we are watching a friend go through a difficult trial, then we must pray for the wisdom on how to help them. It is not our place to give them our personal counsel or guidance. Our role as their friend is to intercede in prayer for them and to help them in the ways that God leads us. We need to provide compassion and love, just as Jesus does to us. We need to remember that God is watching how we will respond in both situations. We must trust the Lord, seek His counsel, and wait upon Him. For in the end, God is the only One who can truly restore us.

June 25

Today's Reading: Job 3-4; Acts 7:44-60
Today's Thoughts: Love Covers a Multitude of Sin

***Then he knelt down and cried out with a loud voice, "Lord, do not charge them with this sin." And when he had said this, he fell asleep.* Acts 7:60**

Stephen was a man described as being "full of faith and power, [doing] great wonders and signs among the people" (Acts 6:8). He had been selected by Peter and the other disciples as one of seven to whom they would entrust certain ministerial duties, while preaching about Jesus. It was not long before Stephen found himself in the center of controversy, standing before the powerful Sanhedrin Council personally to account for his zealous activities for Christ. Today's verse records the last words that Stephen would speak as he is being stoned to death by the people. Stephen's prayer is similar to Jesus' prayer in Luke 23:34 as Jesus said from the cross, "Forgive them Father for they know not what they do."

Stephen had a love for the people that only comes from the Father. He was not only filled with faith and power but also with love because he had the Holy Spirit living within him. As the people gathered to stone Stephen, they lay their garments at the feet of a man named Saul. Saul became the Apostle Paul, the one who brought the gospel of Jesus Christ to the Gentiles. As Saul stood in agreement to kill Stephen, Stephen asks the Lord to forgive them all (including Saul). Did Stephen have any idea that God would greatly use Saul (who would later become the Apostle Paul)? Not at this point, but his love would cover a multitude of sins and God knew how it would all work out.

As Christians, we know that we are supposed to love and to forgive those who sin against us, but truly doing that in our hearts does not come easy for any one of us. The Holy Spirit is the One who fills us with God's love and only God's love can forgive those who hurt or betray us. Instead of waiting until something happens to us personally, we need to be proactive and to ask for the love of Christ to fill us to overflowing. Love is not something to take for granted. Love is something we must pursue. Pursue love today and consciously make it part of your prayers. Ask the Lord to give you a love for Jesus and for His people that surpasses understanding. For Jesus loves us so much that He is still asking the Father to forgive us our sins.

June 26

Today's Reading: Job 5-7; Acts 8:1-25
Today's Thoughts: Integrity and Character

"No, my lord, hear me: I give you the field and the cave that is in it; I give it to you in the presence of the sons of my people. I give it to you. Bury your dead!" Then Abraham bowed himself down before the people of the land; and he spoke to Ephron in the hearing of the people of the land, saying, "If you will give it, please hear me. I will give you money for the field; take it from me and I will bury my dead there." And Ephron answered Abraham, saying to him, "My lord, listen to me; the land is worth four hundred shekels of silver. What is that between you and me? So bury your dead." Genesis 23:11-15

When Sarah died in Hebron in the land of Canaan, Abraham went to the sons of Heth (the people who lived in the land) and requested a burial place. Abraham acknowledged that he was a foreigner in their land and humbled himself by bowing before these men to ask his petition. They, in turn, honored Abraham by calling him a prince and instructing him to take the choicest piece of land for his burial. They wanted him to take it for free but Abraham insisted on paying the fair price for the land. After much debate back and forth, an amount of 400 shekels of silver was given as the land value and Abraham gladly paid it.

Abraham was obviously a man of integrity. He was respected and honored among the people living in Canaan, even though he was considered a foreigner in their land. At this time, Abraham's descendants were yet to come. He was living in the faith of the promise from God that one day his descendants would live freely in this land of Canaan, the Promise Land. But instead of flaunting his heritage, he humbled himself and honored the current residents by insisting on paying his fair share for a burial place. In return, his integrity and character were perceived even greater in the eyes of these people. God blessed Abraham for his faith and obedience.

Our integrity is often tested through our finances. How we handle our money is a true sign of our character and it seems that Christians are under the spotlight more than others. By paying for the land, Abraham was not under obligation to the landowners. When in debt, we are obligated to the lenders. When in debt, it is hard to make wise choices because our need for money can overshadow the right decision. Pray about your finances and ask the Lord to guide you in all decisions regarding money, even in issues of debt. Integrity and character are far more valuable.

June 27

Today's Reading: Job 8-10; Acts 8:26-40
Today's Thoughts: God's Time

***But, beloved, do not forget this one thing, that with the Lord one day is as a thousand years, and a thousand years as one day.* 2 Peter 3:8**

Time is an amazing concept. We can only understand time by how we experience it and how it applies to our lives. Our lives revolve around a 24 hour day, 7 day week, 4 week month, and 12 month year, give or take a few days. We measure time in increments by celebrating markers such as birthdays, holidays, and anniversaries. We discuss time in terms of how fast or how slow it passes. We know that time is moving but we cannot see it, so we watch the sun rise and set, and we keep track of time with our watches and clocks. Our entire life is one constant span of time, uniquely purposed by God for each one of us.

Peter writes to tell us one more truth about time: God's idea of time is not the same as ours. We understand time as a constant, linear progression but God sees time without boundaries, limits or schedules. Can you imagine one day being as a thousand years, or vice versa? The Lord's definition of time does not fit into our neat box of calendars, clocks and sunsets..

As Christians, we must learn to live in God's timeline. We must surrender to His timing in all things. And we must be careful not to think that we have control over time, even our own. 2 Peter 3:10 says it best: *But the day of the Lord will come as a thief in the night, in which the heavens will pass away with a great noise, and the elements will melt with fervent heat; both the earth and the works that are in it will be burned up.* There is a day coming when this earth will burn and those of us in Christ will be safe with Him. For those not in Christ, they will spend eternity forever separated from Him in a terrible place called Hell. God is waiting even now just for one more person to come to Him, before it is too late. One day could take a thousand years to complete, or a thousand years could be as one day, and then time is up. Is there time for one more person to come to Jesus? Only God knows; and that is why our time here on earth is precious.

Maybe you know someone who needs to come to Jesus. Maybe the Lord has been telling you to share His message with them, the Gospel of Jesus Christ. Maybe time is running out and this may be their last chance. Do not let time pass by carelessly. One day the Lord will say to Jesus that time is up for people on earth. Are you ready for that time?

June 28

Today's Reading: Job 11-13; Acts 9:1-21
Today's Thoughts: The Work Never Ends

The man went away, and told the Jews that it was Jesus who had made him well. For this reason, the Jews were persecuting Jesus, because He was doing these things on the Sabbath. But He answered them, "My Father is working until now, and I Myself am working." John 5:15-17

Jesus does not ever stop working *with* us or *on* us. At the time of Christ, the religious leaders were angry with Jesus for making a sick man well on the Sabbath. They exclaimed that Jesus had no right to do such a thing on the Sabbath. To us, this statement does not make sense. God is good and He can choose to be good whenever and on whoever He wants.

Jesus showed us a quality of God's character that the Jewish religious men did not represent due to their self-righteousness. Jesus is "God with us." Through His Holy Spirit, Jesus is now "God in us." So God is continually working *on us*, and *in us*, despite the day or circumstance at hand. God's desire is that we be whole and healthy. Jesus came to break the bonds of sin and habits that keep us captive; behaviors that do not bring glory to God, but keep us focused on ourselves. Jesus is continually working on and in us and He is never in a rush. God is too gracious and merciful to bombard us with attempts to change all the areas in which we fall short all at once.

Trust Him for your shortcomings; sometimes we are our own worse critics. Recognize that God is not always worried about, or surprised by, or in a rush to fix some of the weaknesses which concern us so much. We need to be honest, be open and be willing to be obedient. The Lord will be faithful to complete the work He began, because Jesus is always working.

June 29

Today's Readings: Job 14-16; Acts 9:22-43
Today's Thoughts: Grief of Mind

When Esau was forty years old, he took as wives Judith the daughter of Beeri the Hittite, and Basemath the daughter of Elon the Hittite. And they were a grief of mind to Isaac and Rebekah. **Genesis 26:34-35**

Do you have relationships in your life that are "a grief of mind"? Think about those people who really challenge your thoughts. It is as if you just cannot get along with them no matter how hard you try and you cannot accept who they are or what they do no matter how much you pray. However, for some reason, you cannot escape the relationship either. These people quench our peace and rob us of joy. Why can we not just live life without personality conflicts?

The answer has to do with the two greatest commandments. We must love the Lord with all our hearts, minds, souls and strength and we must also love others as ourselves. Jesus says in Luke 6:32-36 that, if you love those who love you, what credit is that to you? For even sinners love those who love them. And if you do good to those who do good to you, what credit is that to you? For even sinners do the same. And if you lend to those from whom you hope to receive back, what credit is that to you? For even sinners lend to sinners to receive as much back. But love your enemies, do good, and lend, hoping for nothing in return; and your reward will be great, and you will be sons of the Most High. He is kind to the unthankful and evil. Therefore be merciful, just as your Father also is merciful.

God wants us to learn mercy, kindness, thankfulness and love. We cannot love God with everything we have and then not love others. We can have people in life that are a grief of mind but God desires that we learn to love with His heart, touch with His hands and see through His eyes. We can only do that through a dependency on the Lord through His Holy Spirit. God wants us to be more like Him and He can change us if we work with Him through all this.

To overcome these personal conflicts, try stepping back from the issues and pray that you can have a discernment to change the dynamics of the relationship. That person may not ever change but you can. Be proactive in prayer if you know that you will be interacting with that person and ask the Lord to check your spirit before you act out in the flesh. Slowly but surely, you will begin to have victory and God will receive the glory.

June 30

Today's Reading: Job 17-19; Acts 10:1-23
Your Thoughts on Today's Passage:

July 1

Today's Reading: Job 20-21; Acts 10:24-48
Today's Thoughts: Face to Face

And it came to pass, when Moses entered the tabernacle, that the pillar of cloud descended and stood at the door of the tabernacle, and the Lord talked with Moses. All the people saw the pillar of cloud standing at the tabernacle door, and all the people rose and worshiped, each man in his tent door. So the Lord spoke to Moses face to face, as a man speaks to his friend. **Exodus 33:9-11**

When I read the verse that says, "the Lord spoke to Moses face to face," I have a longing in my heart to see the Lord like that. What must it have been like to be in such an intimate presence of the Lord? I try to picture in my mind what it looked like when the cloud descended. Regardless of how hard I try, I am sure that my thoughts could never come close to the awesome-*ness* of such a meeting.

Moses was just a man, no more special than any other person. But God called him for a special purpose and God equipped him for the job. Moses had a tough job leading the children of Israel through the wilderness for forty years. I do not envy his calling, but at times, I have envied his intimacy with the Lord. These feelings of longing and envy; however, cannot remain too long within me because I have God's Spirit living within me. When I desire to be with the Lord in that intimate place of His presence, I can go to Him and be with Him.

I no longer envy Moses because I have my own personal intimacy with the Lord, and so can you. For all who come to Jesus and accept Him as their Savior are sealed with the Promise of His Holy Spirit.

Have you ever longed to be in the presence of the Lord God Almighty? Do you desire to experience Him intimately? Please know that you can find Him. The Bible tells us that if seek the Lord with all of our heart we will find Him (Jeremiah 29:13). The keys are in truly seeking Him, waiting on Him, and obeying His words to us. Find a quiet place and commit to just spending that time with the Lord. Start with worshiping Him and coming before Him with an open heart. Keep seeking and keep knocking until you find that sweet place of fellowship. Remember that He is always here for us…we are the ones who are distracted and turn away from Him. Practice His presence every day.

July 2

Today's Readings: Job 22-24; Acts 11
Today's Thoughts: The Promise of God

Peace I leave with you, My peace I give to you; not as the world gives do I give to you. Let not your heart be troubled, neither let it be afraid. **John 14:27**

Today, everyone is looking for peace. We want our desires fulfilled and we think that will bring us happiness. We hear promises of fulfillment on radios, see promises on billboards, become a part of them through television and after hearing these promises enough times, we sing about them and make daily choices because of them. What is the primary message that these slogans promote? The answer, "We've got what you need." Promises, promises! Promises of a fulfilled life, an improved life, a happier life and a more peaceful life suggest that this life is attainable through the things of this world. Commercials promise that you can "have it your way," "we do it all for you," "you asked for it, you got it," and "you deserve it." But can these promises really be true? And will these products bring us the peace we are all seeking?

Because of the inability to have these promises truly fulfilled from the world, it then seems hard to believe that all the promises from the Lord will come true. Unlike the promises of the world that are conditioned upon our spending money and time to benefit from their product, Jesus has given us the promises of God, which are free. The Lord does require something from us: our trust, belief and obedience. And in return He promises peace to His people. Our attitudes, goals and desires should not depend on the promises of the world, but on the promises of the Word. It's that peace that makes us different and gives us something that others want. That peace will affect the way we live in our actions, behaviors and attitudes. Acts 10:36 says, "You know the message God sent to the people of Israel, telling the good news of peace through Jesus Christ, who is Lord of all." Jesus promises us peace, not from this world, but from Him. His promises are not empty slogans but a way of life: for He is the way, the truth and the life.

Do you need to experience His peace today? If so, begin by asking Jesus to fill you with His loving peace. His promises will never fail you.

July 3

Today's Reading: Job 25-27; Acts 12
Today's Thoughts: Step into New Land

Now the Lord had said to Abram: "Get out of your country, From your family And from your father's house, To a land that I will show you. I will make you a great nation; I will bless you And make your name great; And you shall be a blessing I will bless those who bless you, And I will curse him who curses you; And in you all the families of the earth shall be blessed." So Abram departed as the Lord had spoken to him, and Lot went with him. And Abram was seventy-five years old when he departed from Haran. Then Abram took Sarai his wife and Lot his brother's son, and all their possessions that they had gathered, and the people whom they had acquired in Haran, and they departed to go to the land of Canaan. So they came to the land of Canaan. Genesis 12:1-5

The Lord called Abram out. He told him to leave his country, not just his home, but his country. God told Abram to leave his family and head towards a new land. Can we begin to imagine how Abram must have felt? I wonder if his family thought he was crazy. Let's see now—you are going to leave your home, family, country and head to a land that you have not seen yet? Abram did leave his home and country behind and went as the Lord directed him. Abram was a man of great faith. He believed in the promises of God and it was accounted to him as righteousness.

In the New Testament, Jesus told his disciples a similar message when he said, "Follow Me." To follow Jesus meant leaving everything else behind, including homes and families. Some did and some did not. The same is true today. How many of us are truly willing to forsake all for the gospel of Jesus Christ? Do we really have to go to such extremes in this day and age? In our hearts, we must answer those questions. Despite our behaviors and outward appearances, God knows our hearts. If we are willing to surrender all to Jesus, the Lord will do the rest.

Think upon these verses today. Maybe God has a new land that awaits you. Without a doubt, God has blessings planned for your life, planned from before you were born. Are you willing to step out of your comfort zone? If so, this could be the step that changes your life in amazing ways.

July 4

Today's Reading: Job 28-29; Acts 13:1-25
Today's Thoughts: Listen as a Disciple

The Lord God has given me the tongue of disciples, that I may know how to sustain the weary one with a word. He awakens me morning by morning, He awakens my ear to listen as a disciple.
Isaiah 50:4

Do you ever find yourself knowing deep within the depths of your heart that God wants to bring about change, and even though you have been the one praying for it, you aren't ready when He is ready to answer your prayer? Well, that happened to me today.

In the mornings, I spend time reading the Bible, praying on my knees and worshiping. Then I allow time for silence so that the Holy Spirit can minister to me in the quietness. Sometimes that's all there is...quietness, which in itself is such a blessing before the day begins. But this morning's quietness was different. This morning the Lord wanted to impress something on my heart that I didn't want to hear. My quiet time was cut short by my choice this morning. But the Lord didn't quit, and right before I left the house, I got a strange impression that I was going to have to face something that I was refusing to hear from the Lord. And that is exactly what happened. Within two hours of my devotional time and one hour from the impression, I heard things that I wished I had heard first in the presence of the Lord. And then I found myself repenting to Him and shocked that it was actually going to come about. Foreknowledge would have eased my heart regarding this matter.

God's intention is always for our best. We all have to remember that God does nothing without revealing His plans to His servants, the prophets (Amos 3:7). He desires to include us in His plans and He even places those desires within our hearts first. God wants to give to us whatever we ask in Jesus' name to bring Him glory, as He fulfills the desires He has placed within our hearts. If we are in His Word and quiet in worship before Him, we can also be included when He unfolds His plans. Today, take time to just be still before Him and listen. Quiet your heart and allow His heart to minister to you. And I am sure that your day will go better than mine did today.

July 5

Today's Reading: Job 30-31; Acts 13:26-52
Today's Thoughts: Peace in Him

These things I have spoken to you, that in Me you may have peace. In the world you will have tribulation; but be of good cheer, I have overcome the world." John 16:33

Most of us would agree that tribulations are very much a part of life. They come at different times for different reasons, but they no doubt come. Do you ever wonder just where the peace is that Jesus was speaking about? How about in your own personal life? Do you have peace in your daily routines? Jesus tells us that we will have tribulation in this world, a definite promise not just a possibility. But our only hope for peace is in Jesus. We must claim this verse as a direct promise from Him. He has overcome the world; therefore, we can overcome those things that try to take our peace by living for Him.

Yet the words sound easier to do than the reality of actually living our lives this way. For me, my life seemed to be much better when I was in control of my circumstances. If I could somehow control the factors that influenced my day, then I could produce the peace I desperately needed. For a long period of time, my day consisted of doing those things that either mattered most to me or to others who expected a level of performance from me. I believed that being in control led to greater stability and less turmoil, a win-win for everyone around me.

There were clues along the way that maybe this way of life was not so great for everyone else around me. One major clue came one night several years ago when my husband asked me if I was okay. Of course, I was okay. Our lives were good. Our marriage was fine, rarely a conflict or cross word. Our jobs were more than sufficient to meet our needs. We were young, owned our own home, had money, and enjoyed many of the world's amenities. Why was he asking me what was wrong? He began to express his concerns to me about—me. I could not listen to him without becoming defensive. After some discussion, I confessed to him that I often felt like an android, a machine on the inside but normal looking on the outside. Still, I was not ready or willing to accept that maybe I needed to change, that I needed help.

Over time, the Lord has led me to make many, many changes in my life. I was deceived into thinking that I could make my own peace by controlling my circumstances and situations. The changes have come by surrendering my control to Jesus. There is no peace in this world without the Lord. I have known Jesus as my Savior most of my life, but only when I made Him Lord of my life did I begin to know His peace. He left us His peace. May we all let go of our need to be in control and trust Jesus to help us get through all things....both good and bad.

July 6

Today's Reading: Job 32-33; Acts 14
Today's Thoughts: Follow Me

After these things He went out and saw a tax collector named Levi, sitting at the tax office. And He said to him, "Follow Me." So he left all, rose up, and followed Him. Then Levi gave Him a great feast in his own house. And there were a great number of tax collectors and others who sat down with them. **Luke 5:27-29**

A tax collector was not an honorable position during Jesus' time. Tax collectors were known for their shrewd and fraudulent ways. This passage says that Jesus went out and saw a tax collector named Levi. Jesus obviously got his attention as He commissioned Levi to follow Him. It is amazing that the passage says that Levi left all, rose up and followed Him. There is a sense that Levi had immediate action with little thought. He seemed to abandon all for the call of Christ. Think of what it took to go from a tax collector with a negative reputation, to a disciple of Jesus Christ. Questions come to mind such as, "How did Levi do that? Did he just quit his job, sell his house or what? Where did He follow Jesus to?" Levi's next steps are right there in the next verse. Verse 29 says that Levi followed Jesus to "his own house." Levi brought Jesus to his own home, so, Levi followed Jesus back home.

The first step to being Jesus' disciple is by becoming His follower at home. Not only does Jesus meet us right where we are, but also Jesus becomes part of where we are. When we come to know Him and follow Him, Jesus leads us back to our own homes. Home is our training ground. It is in our homes that we first speak out about Jesus, so the change in our lives needs to become evident at home first.

Jesus saw Levi and Jesus sees you. Jesus called Levi and Jesus is calling you. Are you willing to answer His call? Are you following Jesus in your home? Levi gave Jesus a great feast. There was preparation and planning for this party as he made a public profession that he was following Jesus' ways, starting immediately. For you today, it may be that you stop watching certain shows or stop saying certain words. It also will mean that you start sharing your faith with your family. Levi is an example of someone with the kind of commitment it takes to follow Jesus. Levi became one of the twelve apostles whose name was changed to Matthew. He was the man who wrote the Book of Matthew, leading many others to follow Christ just like he did.

July 7

Today's Reading: Job 34-35; Acts 15:1-21
Today's Thoughts: The Real Food

But food does not commend us to God; for neither if we eat are we the better, nor if we do not eat are we the worse. But beware lest somehow this liberty of yours become a stumbling block to those who are weak. **1 Corinthians 8:8-9**

How often do you think about food? What types of cravings do you get for your favorite foods? Midnight snacks and mad dashes through our favorite drive-thrus are part of our lives and culture. Hunger is a drive necessary to sustain life while appetite is a desire to satisfy our cravings for food. Food is a necessity to our physical bodies and to enjoy good food is one of our greatest pleasures. However, even with all its necessity, this earthly food has no value to us spiritually.

If Paul is saying that what we eat, or do not eat, has no effect on our relationship with God, then why can we not eat whatever, drink whatever and be merry? One reason is because the choices that we make impact others. We must realize that our actions are being watched by others and that we have a responsibility as Christians to set godly examples for others. God does care about the behaviors and attitudes we exhibit that affect those around us.

There are numerous examples of this behavior in our churches today. Movies and television choices can influence our families and friends. What kind of example do we set for those in and out of the church? Would seeing certain movies stumble others? What about drinking alcohol? Many Christians have no problem with limited alcohol consumption. But, would it not influence the Christian who feels that this type of drinking, without exception, is wrong? Some might argue that the Christian church in general stumbles non-believers, because much of our behavior is not just *in* the world, it is *of* the world.

Food is not the real issue, but it is certainly representative of how our behaviors as Christians can greatly affect those around us, even if there is technically "nothing wrong" with certain activities. Jesus gave His life for us so that we can have the real food—the bread of life. If we follow Him, our choices in this world will bring glory and honor to His name while also setting the right example for others. We should pray not to get hung up on things of no value spiritually but to get caught up in the Spirit of God so that we may be a blessing to others, not a stumbling block.

July 8

Today's Reading: Job 36-37; Acts 15:22-41
Today's Thoughts: Things to come

Thus says the Lord, The Holy One of Israel, and his Maker: "Ask Me of things to come concerning My sons; and concerning the work of My hands, you command Me. Isaiah 45:11

It has taken most of my life, my Christian life, to finally realize that if I want to know something from the Lord, then I just need to ask Him. Sounds kind of like a no-brainer, but for some reason, I spent most of my time with the Lord dealing with my personal requests. I have prayed pretty much my whole life. I have prayed very specifically at times and I have prayed in the bigger picture at times, but I have always prayed. What I did not necessarily pray about, though, were of "things to come." I think many of us tend to pray about the "here and now" issues, or the things that we want "to come," not necessarily asking the Lord to reveal what He is doing "concerning the work of My hands."

Why is this so important to understand? When I see verses that say to "ask" the Lord for something, then I know that must be His will. Today's verse even says, "you command Me." Instead of just focusing on the issues in our immediate path, let's ask the Lord about what He is doing down the road. I have been immensely blessed and have had huge increases in my own level of faith because I have learned to ask the Lord "of things to come" and have seen Him answer through circumstances or situations that have come to pass. My role is to ask, then to pray for wisdom and discernment in how to go forth in what He shows me.

We need to know that the Lord is more than willing to reveal things to us, if we are seeking Him with all of our hearts. As we grow in our walk with Christ, He will teach us His ways and mold our character into more of His image. When we spend time in His Word, we begin to see how verses like Isaiah 45:11 take on a completely new meaning to us personally. When the Holy Spirit impresses a verse upon your heart, take it to prayer and seek the Lord in what He is saying to you. Maybe God wants to reveal something to you—maybe even "of things to come."

July 9

Today's Reading: Job 38-40; Acts 16:1-21
Today's Thoughts: Teach Them Diligently

"And these words, which I am commanding you today, shall be on your heart; and you shall teach them diligently to your sons and shall talk of them when you sit in your house and when you walk by the way and when you lie down and when you rise up."
Deuteronomy 6:6-7

My children got into the car after a day at their Christian school and told me that a friend of theirs was going to hell. They said, "He lies, mom, and liars go to hell." I asked, "Why do you think that?" Then the two kids broke out in a song that has a line that says, "Liars go to hell...liars go to hell...Revelation 21:8, 21:8...burn burn burn." I thought to myself, "How can my kids believe this when we know we each fall short everyday?" Well, the subject changed and I never addressed it.

Around 4:30 the next morning, I woke up with this thought racing in my mind: "You need to teach your children about grace. They are not being taught how to live in God's grace at their Christian school. The gospel is about God's grace to sinners. Your kids don't understand how to extend grace to others or themselves."

After much prayer, I approached the subject of grace and what Revelation 21:8 says. My son said to me, "Are you telling me that I sin everyday (lying included) and God doesn't hold that against me? Are you telling me that I can do whatever I want, and because of the blood of Christ, it is covered and God sees me as righteous anyway?" Immediately I prayed, "Oh Lord, I really want my 13 year old son to completely get this but this Truth can lead to poor choices...help." And the wisdom of the Lord came upon me as I opened my mouth and said to him, "Yes. That is exactly right. However, if you really love the Lord and know Him, you won't want to sin and hurt Him. Your heart should respond back to His love and as a result, you will live a life that is pleasing to Him." Both of my kids nodded in agreement and said, "Cool."

I thanked the Lord when my daughter then said to me, "Mom, school teaches us about God but you teach us how to know Him better and love Him more." God gets the glory because I ignored it initially.

Today, we need to realize the responsibility we have to give our children a strong biblical foundation. We have to role model to them the love of Christ from our hearts and we have to teach them how to know Him better. Be sensitive to the conversations you have with your children. God will lead you in what to say...just be diligent!

July 10

Today's Reading: Job 41-42; Acts 16:22-40
Today's Thoughts: Signs of Today or Eternity?

He [Jesus] answered and said to them, "When it is evening you say, 'It will be fair weather, for the sky is red'; and in the morning, 'It will be foul weather today, for the sky is red and threatening.' Hypocrites! You know how to discern the face of the sky, but you cannot discern the signs of the times. **Matthew 16:2-3**

Jesus understood how much the weather plays a part of our daily lives. He spoke about discerning the face of the sky, trying to predict the day's events. Just think about how much we know today from the weather channel, newspapers and computers. We have local and remote forecasts available in an instant, unlike in Jesus' day. And just as we have advanced in understanding the weather, we also have advanced in understanding the signs of end times. Our lives can change as a result of a natural event, and our eternity is shaped by being tuned in with the signs of end times.

Do you live your life for today or for eternity? Everything we live for here will burn some day, but our soul has a choice to burn in the fire or live in the presence of the light of Jesus. In a moment, the weather can change completely. It is the same with our lives. Today, pick up the Bible and read something in Revelation or find a book that can update you with the signs of end times. Until I studied the book of Revelation, the thought of reading it frightened me. That fear is gone because the love of God is even so evident in the book of Revelation. He wants all to come to Him and He has made a way for all to understand. Ask the Lord for discernment to understand. We can never be sure of the future outcome of the weather, but we can know the outcome of our eternity.

July 11

Today's Reading: Psalm 1-3; Acts 17:1-15
Today's Thoughts: Delighting in Him

Blessed is the man who walks not in the counsel of the ungodly, nor stands in the path of sinners, nor sits in the seat of the scornful; but his delight is in the law of the Lord, and in His law he meditates day and night. He shall be like a tree planted by the rivers of water, that brings forth its fruit in its season, whose leaf also shall not wither; and whatever he does shall prosper. **Psalm 1:1-3**

I love the promises of this Psalm. There are promises of blessing to the one who delights in the law of the Lord, a promise of fruit, strength and prosperity in whatever we do. And what does He want us to do? He wants us to love His Word; the law of the Lord is His word. John 1:14 says that the Word became flesh and dwelt among us, which means that Jesus is the living Word. If we delight in the Word, then we are delighting in Him. So, our delight should be to desire the Lord more than anything else, so much that we meditate and think about Him day and night.

There are so many other things to think about in our lives today. From televisions to radios to computers, everything around us makes some kind of noise or sight that demands our attention. Regardless of what we are doing, we are always thinking about something. The activities and the stimuli seldom stop in our busy lives. And then we wonder why we are so tired (our leaves wither) and why we are not efficient or effective (prosperous) in our works. The answer lies in God's Word. We will find our strength when we make the Lord the priority of our thoughts, by tuning out the other sights and sounds.

God promises us His blessings if we focus on Him, keep His word close to our hearts, and make Him the center of our thoughts. Start today by reading through the Psalms starting with chapter one and reading a Psalm a day until ending with chapter one hundred fifty (150). You will see the blessings of God as you delight in His Word every day.

July 12

Today's Reading: Psalm 4-6; Acts 17:16-34
Today's Thoughts: Both Near and Far

Therefore the Lord Himself will give you a sign: Behold, a virgin will be with child and bear a son, and she will call His name Immanuel. **Isaiah 7:14**

We read this prophecy written thousands of years ago and understand that the fulfillment was Jesus, as it is quoted again in Matthew 1:22, testifying to the virgin birth of Jesus as prophesied by the prophet Isaiah.

However, there was also a closer fulfillment of that prophecy as we read in Isaiah 8:3, *"So I approached the prophetess, and she conceived and gave birth to a son. Then the Lord said to me, 'Name him Maher-shalal-hash-baz'."* The prophecy for us is about the virgin birth of Jesus, but it was also immediately about this prophetess having a son as a sign to the people during Isaiah's time. This verse is an example of near and far prophecy. "Near" to Isaiah but also predicted "far" for us.

Prophecy still works like that in our lives today. There are prophecies that have near and far fulfillments. As we read the Scriptures, we know that those verses were placed in the Bible for a set time and place. But, those same verses pop out to us as verses that we can claim for our own lives. How does that work?

God's Word is alive and His Word has the ability to reach through all places, settings, times and events. His Word is as applicable and real to us and for us today as it was when written. I have seen near and far fulfillments in verses for myself. Many times, I receive a verse (or promise) from the Lord for the circumstance I am currently facing, and then years later in my life, I see the Lord further fulfill that verse in a new way.

God wants to speak to you and He does that by prophesying through His Word. Claim those verses but then relax. If God said it to you, He will fulfill it….again and again…both near and far.

July 13

Today's Reading: Psalm 7-9; Acts 18
Today's Thoughts: Speak Without Fear

Now the Lord spoke to Paul in the night by a vision, "Do not be afraid, but speak, and do not keep silent; for I am with you, and no one will attack you to hurt you; for I have many people in this city."
Acts **18:9-10**

I am encouraged when reading statements like these. Paul was known as a man with such boldness and never-ending endurance. He spoke before kings and started churches. He was persecuted and beaten but continued to sing praises. We read that his body had the brand marks of Jesus Christ and yet, this verse allows us to look into his heart. Paul was scared.

I would guess that not many people would know that Paul needed this kind of encouragement. Paul obviously knew that his fears stemmed from his speaking, which led to him being attacked and hurt. God knew what Paul needed to keep going. The Lord assured Paul to continue, to not quit, but to speak, and then calmed his fears by saying that no one would attack him to hurt him. God knows just what we need to hear. God knows just how much each of us can take. And God knows how to assure us, comfort us, encourage us and motivate us to continue.

Are there fears you are dealing with today? Are you struggling with thoughts of quitting or stopping? God knows your heart and He can minister to you at the deepest level. Allow His Holy Spirit to minister to you today. Open His Word and listen as He speaks to your heart. Then, go and speak as He leads you.

July 14

Today's Reading: Psalm 10-12; Acts 19:1-20
Today's Thoughts: Our Teacher

But the anointing which you have received from Him abides in you, and you do not need that anyone teach you; but as the same anointing teaches you concerning all things, and is true, and is not a lie, and just as it has taught you, you will abide in Him. **1 John 2:27**

Jesus tells His disciples in John 16 that when He goes away He will send a Helper, His Holy Spirit. "When He, the Spirit of truth, has come, He will guide you into all truth…" (John 16:13). When we accept Jesus into our hearts, we are immediately indwelled by His Holy Spirit. We *believe*, then, we *receive*. One of the biggest attacks Christians face today comes from an enemy telling us that we have no real power, no victory, and no hope. In John 8:44, Jesus describes our enemy: "He was a murderer from the beginning, and does not stand in the truth, because there is no truth in him. When he speaks a lie, he speaks from his own resources, for he is a ***liar*** and the father of it."

Our Lord and Savior Jesus Christ is perfect in every way. He cannot lie. He cannot deceive. He is holy and righteous beyond our understanding. His promises are true. His faithfulness reaches to the heavens. When He says that we have His Holy Spirit, then we must believe in faith that we have an abiding relationship filled with His love and power. He promises that no eye has seen, nor ear has heard, nor has entered into the heart of man what God has prepared for those who love Him (1 Corinthians 2:9).

Do you believe Him today? Do you want to know more about the things of the Lord? As a child of God, saved and sealed for the day of redemption, you have the anointing of the Holy Spirit who will teach you all things. Let today be the day that you start learning about these things. How to start? Read His Word and study it. Pray and ask for the things you want to learn about. Ask questions that you want answered. Get up every day with an attitude of faith and do not allow the enemy to tempt you with his lies. Maybe it is time for you to go back to school…and what a great Teacher you have! And the best part…He will only give you open book (the Bible) tests.

July 15

Today's Reading: Psalm 13-15; Acts 19:21-41
Today's Thoughts: Seek God's Wisdom

***The LORD looks down from heaven on the sons of men to see if there are any who understand, any who seek God.* Psalm 14:2**

Do we understand the ways of God? Understanding God is different than believing in God. To understand means that we can come into alignment by acknowledging the decisions God makes on our behalf. We do not have to agree with the decision but because we understand that God's ways are higher than ours are, we can yield to Him. At that point, we show our trust in His ways as we continue to seek and submit to His will.

As a parent, I have been challenged to explain "why" I am making a certain decision. If my decision pleases my children, no discussion is necessary. However, if I disagree, I am sought out to discuss and debate why my choices did not match their hearts' desire. As a parent, I have the right to make final decisions as much as I have the right to explain or not explain the factors that went into making those decisions. My children then have the right to agree or disagree with my decisions but what a joy it is when they understand and accept them as is.

God wants a relationship with us. As our Heavenly Father, He knows what is best for us. He is more than willing to reveal Himself to those who seek Him. God does not want us to just believe in Him but to understand Him as well. God has not hidden Himself from us. He has given us His Son, Jesus, His Word, and His Spirit.

The Spirit of God intercedes for us, convicts our hearts and leads us into all truth. We have the opportunity to seek God and to find Him. But we also have the ability to understand Him as we continue to choose to read the Bible, pray and submit to the circumstances that He has allowed us to encounter in our lives.

If you are struggling with understanding why God is allowing certain circumstances to continue in your life, seek Him for wisdom. James 1 tells us that anyone who is lacking in wisdom should ask God for it because God will give liberally to all that ask. But when God answers, believe that answer and do not doubt. Doubting only leads us to becoming double minded. Double mindedness is a worse state than questioning God's decisions in the first place. And remember, God is looking down from Heaven to see if we understand and seek after Him.

July 16

Today's Reading: Psalm 16-17; Acts 20:1-6
Today's Thoughts: Promises of God

Are you so foolish? Having begun in the Spirit, are you now being made perfect by the flesh? Galatians 3:3

I was talking to my wise friend about the promises of God. Bobbye said to me, "I have learned to look for the promises but not to live for them." If God gave us a promise, it is His responsibility to fulfill it. Whenever we are given a message from the supernatural, it takes the leading of the Spirit to have it be fulfilled in the natural. Too many times, we take it upon ourselves to complete in the flesh the promises that were given in the Spirit. We look for the promises in the word as well as look for their fulfillment on the earth but in the mean time, we live for the Lord.

Write a prayer about your expectations concerning your personal promises of God and ask the Lord to reveal His will for you:

July 17

Today's Reading: Psalm 18-19; Acts 20:17-38
Today's Thoughts: My Mouth and Meditation

***Let the words of my mouth, and the meditation of my heart, be acceptable in thy sight, O Lord, my strength, and my redeemer.
Psalm 19:14***

It is interesting to note that the words of our mouths and meditation of our hearts are watched by the Lord. The Psalmist asks the Lord to let them be acceptable in His sight, not in His ears. So what is God looking for in our mouth and meditation?

Recently this verse kept coming to my mind to pray. But just because I prayed it didn't mean that I noticed anything different or changed in any way. Then, all of a sudden, to my complete surprise I was confronted about something I said. Immediately, the conviction became so severe that my heart started burning within me. I knew that God heard my prayer and what He was seeing in my mouth and meditation were not acceptable. He took my prayer seriously and now was addressing deep-seated issues.

When God convicts us, it is amazing how naked we feel. We naturally want to conceal our motives, justify our words and cover up our actions. Those are all "sight" feelings, like God is seeing our sin. But our only hope at that time is to repent. The more we fight to maintain our sinful state, the more time we lose in receiving His good counsel and wisdom to change. God sees the fruit of our ways. Those ways are manifested through the words of our mouth and meditations of our heart. The fruit is what God sees more than what He hears. But the Psalmist completes His prayer by addressing two very important characteristics of God: my Strength and my Redeemer. God is able to strengthen you to change and redeem your old ways into a new person with ripe healthy abundant fruit. Today, will you ask the Lord with me to allow your words and meditations to be acceptable in His sight?

...for out of the abundance of the heart the mouth speaks.
Matthew 12:34

July 18

Today's Reading: Psalm 20-22; Acts 21:1-17
Today's Thoughts: So That You May Believe

When Jesus heard that, He said, "This sickness is not unto death, but for the glory of God, that the Son of God may be glorified through it." **John 11:4**

The disciples accepted what Jesus said at face value. They probably thought something like, "Oh good, Lazarus is sleeping. He will wake up so we do not need to return to Judea where they are threatening to kill Jesus." But when Jesus said, "Ok, it's time to return," the disciples became very uncomfortable. They were now quick to challenge the change in plans. Jesus explained to them plainly, "Lazarus is dead, and for your sake I am glad I was not there, so that you may believe. But let us go to him" (John 11:14-15). Jesus had a different perspective and purpose and they knew they could not change His mind. Thomas then said what every other disciple was thinking, "Let us also go, that we may die with him" (Luke 11:16). In other words, Lazarus is dead and by Jesus returning to Judea, He may end up dead, so they might as well all go with Him so they can all be dead together. The disciples were mainly focusing on themselves, not on what Jesus was doing. They had little faith in what Jesus could do. Jesus had waited to go back to Judea until Lazarus was dead three days so that their faith in Him would be strengthened. Jesus chose to wait to return to Bethany, not because His life was threatened, or because Lazarus would get better on His own or because He had more important matters to attend too. He waited so that you may believe. **God's delays are not God's denials**. We have to trust in how He chooses to answer our prayers. His perspective is always to do what is best for grounding and strengthening our faith.

Pray about it: Lord, help me to accept your ways. Sometimes I look at impossible circumstances and I limit Your ability to intervene. Help me to believe as I trust in You for the details and delays. Amen.

July 19

Today's Reading: Psalm 23-25; Acts 21:18-40
Today's Thoughts: The Valley

The Lord is my shepherd, I shall not be in want. He makes me lie down in green pastures, He leads me beside quiet waters, He restores my soul. He guides me in paths of righteousness for His name's sake. Even though I walk through the valley of the shadow of death, I will fear no evil, for You are with me; Your rod and your staff, they comfort me. You prepare a table before me in the presence of my enemies. You anoint my head with oil; my cup overflows. Surely goodness and love will follow me all the days of my life, and I will dwell in the house of the Lord forever. **Psalm 23**

Psalm 23 is an awesome promise from the Lord, so personal to each of us. God is our Good Shepherd and He knows all about us. He knows that by restoring us back to Him, we will be willing to be led down His path of righteousness. At times, those paths will lead us through valleys of the shadow of death. Most of us hate those paths. Even though the Lord is holding us by the hand, **oh,** how we resist! It is hard to acknowledge that the God of love is truly leading us through the dark, scary valley. We may even want to blame such terrible circumstances on everything and everyone else instead of submitting to God ourselves. In time, however, we will see that it was only by going through the valley of the shadow of death that the Lord could help us break hurtful habits and behaviors, so that we can receive His peace and rejoice in His love.

Ask the Lord today to reveal any areas of your heart that need restoration. Do you feel dull or numb in certain areas of your life?

Tell the Lord about those places in your heart where you sense frustration and barrenness.

Ask Him to restore them back to Him. Restoration is not always just about life's obvious issues; it can be very much about life's less recognized ones.

Take your time to really read through Psalm 23 (again) and listen to the Lord's impressions on your heart. The Word of God is the most powerful Word you will read or hear today.

July 20

Today's Reading: Psalm 26-28; Acts 22
Today's Thoughts: What Is There to Fear?

The Lord is my light and my salvation; Whom shall I fear? The Lord is the strength of my life; Of whom shall I be afraid? **Psalm 27:1**

My dog Buddy is a consistent source of insight and amusement in my life. For you animal lovers, you know what I am talking about. God's creatures are simply unique and amazing in wonderful ways. My dog is all of those things to me and I am amazed at how the Lord teaches me certain messages through Buddy. For example, my husband and I had taken Buddy out to some property where deer and coyote roam freely. Buddy was quick to take off, running through weeds so high that all I could see is the white tip of his tail. He crawled through a barbed-wire fence, jumped over a creek, and dragged himself through thickets and brush that left him covered with briars. He was in heaven, loving every minute of his excursion, and without a care of any danger seemed to enter his mind. Later that evening when we got home, I heard Buddy whimpering outside near the back door. I thought he was hurt or that something was wrong with him. What was wrong? There was a twig lying on the concrete and he was afraid to step over it.

One minute my dog was walking through the jungle with no fear; the next, he was whimpering over a twig. Why? Because his fear was based on something unknown and unfamiliar to him, though we know that a twig is far more harmless than plowing through rattlesnake weeds out in the country. How often do we allow fear of something harmless to stop us in our tracks and keep us from going forward? How often does the Lord look at us and think we look ridiculous (as I did of Buddy)? We so often let the fear of the unknown keep us from having all that God has for us.

Do you have fears in your life that stop you in your tracks? Are you consistently afraid of the unknown? Sometimes we are far more comfortable in dangerous places just because we are familiar with them. Let's be willing to be uncomfortable sometimes and let the Lord truly lead us. The Lord tells us to trust Him and to step out in faith. The Lord promises to protect us, regardless of where we go with Him. Just as I was watching Buddy closely, just in case that twig jumped up at him—the Lord is always watching out for us too.

July 21

Today's Reading: Psalm 29-30; Acts 23:1-15
Today's Thoughts: God's Wall of Fire

"'For I,' says the LORD, 'will be a wall of fire all around her, and I will be the glory in her midst.'" Zechariah 2:5

Living in southern California is an adventure of sorts, from the surf to the desert to the mountains and so much activity in between. One of our chief blessings is the consistently nice weather. Not too hot, not too cold, usually the temperature is just the right balance for most people. However, we do have our challenges from the weather—drought, dry winds and firestorms. I personally have witnessed two devastating firestorms, the last one dangerously close to my home. I literally could see the fires blazing as the tips of the flames seemed to stretch higher with each gust of wind. It is a terrifying and awesome sight, so mesmerizing and powerful. These events remind us of how powerless we are to control the forces of nature.

Today's verse describes "a wall of fire", amazing, powerful and yet protective, not destructive. God's wall of fire encloses around us to keep us in His presence. God's wall of fire burns with the power of His glory in our midst. What a sight (or thought) to behold! Fire is used in the Bible to describe God's presence, a consuming fire of holiness. For us, we are to revere God's presence for His holiness and perfection, but to also take comfort in knowing that He is that fire of protection around us.

When I walk the beach or gaze upon a mountainside, I am reminded of God's indescribable creative power. He is truly an awesome God. His hand of protection is always upon us. His love is evident in everything around us. Take time today and reflect on the beauty of God's glory. Ask the Lord to show you things you have never seen before, maybe you will get a glimpse of His wall of fire and His glory in your midst. Spend time intimately with the Lord and you will find that you pay less attention to things that truly do not matter to your life. May He fill you with His love, joy and peace.

July 22

Today's Reading: Psalm 31-32; Acts 23:16-35
Today's Thoughts: Before You Set Out

When the cloud remained over the tabernacle a long time, the Israelites obeyed the LORD's order and did not set out. Sometimes the cloud was over the tabernacle only a few days; at the LORD's command they would encamp, and then at his command they would set out. Sometimes the cloud stayed only from evening till morning, and when it lifted in the morning they set out. Whether by day or by night, whenever the cloud lifted, they set out. Whether the cloud stayed over the tabernacle for two days or a month or a year, the Israelites would remain in camp and not set out; but when lifted set out. **Numbers 9:19-22 (NIV)**

The words "set out" are so prevalent in this section of Scripture. When God commanded, the Israelites set out in obedience. If God did not command, the Israelites stayed encamped. Jesus said to His disciples in Matthew 16:24, "If anyone desires to come after Me, let him deny himself, and take up his cross and follow Me." As believers, we are commanded to follow Jesus. Many times we act as if God should be following *our* desires, *our* goals and *our* agendas. Jesus doesn't follow us. He commands us to be obedient and follow Him. We have the presence of the Lord dwelling within us. We can be led like the Israelites by watching for His lead and listening to His voice through the Word of God.

There are times He tell us to "go" and there are times He tells us "to be still and know that He is God." Following Jesus leads us to the tops of mountains as well as through the valley of Baca. But He is there, leading and guiding and loving us through it. We will learn to weep with Him as we see things that hurt His heart and rejoice with Him as we embrace the things that bring Him joy. One thing is for certain, He will never leave you or forsake you. Our tendency is to be tempted to leave Him, never the other way around.

Today, be sure that you spend time in His word, listening for His counsel and ask Him to help you follow His lead. Make His footsteps your pathway as you turn to Jesus throughout the day in obedience before you **set out**.

July 23

Today's Reading: Psalm 33-34; Acts 24
Today's Thoughts: Facing the Giants Today

David said to Saul, "Let no man's heart fail because of him; your servant will go and fight with this Philistine." And Saul said to David, "You are not able to go against this Philistine to fight with him; for you are a youth, and he a man of war from his youth." 1 Samuel 17:32-33

I remember learning the story of David and Goliath in Sunday school. As a child, I tried to picture the scene of this teenager going up against this scary, mean giant. How big was Goliath in comparison to David? Where did David get such courage to even think he could defeat him? All of the odds were against David. He had no experience, no armor, no weapon, and no one fought alongside him. Who was he kidding? But David had something far more powerful than physical stature, experience or weapons: he had faith in his God.

Do you ever feel like you are in the midst of a battle where the odds are stacked against you? Or maybe you see the giants, hear their taunts and decide to flee and hide out, as opposed to standing up to them, as the Israelites did before David arrived? Today, we are all in a battle and there is one clear enemy. Make no mistake about this fact: "your adversary the devil walks about like a roaring lion, seeking whom he may devour" (1 Peter 5:8). Satan will use whatever tactics he can to get to us, to frighten us, and to keep us from fighting back. He will play tricks on our minds so that we see giants that are not real. He will deceive us into believing that we have no ability to win. One of the best tools Satan has against us is **fear**.

Fear keeps us from facing the giants. Fear keeps us from ever achieving victory because we are afraid and run from the battle. Fear becomes our own trap of despair, depression and disillusionment. So often, the battle takes place in our own minds with giants conjured up in our thoughts, but we must understand that this is not God's will for us.

Where did David get his courage? He believed in his God, the God of Israel. He believed in all that he had been taught about God, and he had witnessed God's faithfulness and protection in his own life. And on this day, he was willing to put his faith to the test. David took a huge step of faith, depending completely on the Lord to save him. With one sling of one rock, the giant fell. For us, today is our day of victory. Today is the day to face the giants in our lives. In faith we step out, stop running, stop hiding and stop believing the lies of the enemy, Satan. In faith, we face the enemy head on and we claim victory in the name of our Savior Jesus Christ. He has fought the battle and won, on our behalf, but we need to step out, stand up and sling the rock. Victory is ours in Christ.

July 24

Today's Reading: Psalm 35-36; Acts 25
Today's Thoughts: Blessings of the Faithful

***"And Jacob called his sons and said, "Gather together, that I may tell you what shall befall you in the last days...."* Genesis 49:1**

Jacob was at the end of his life. He had the opportunity to tell his sons what would happen to each of them in their future before "he drew his feet up into the bed and breathed his last, and was gathered to his people" (Genesis 49:33). Jacob prophesied, or spoke forth the future of their tribes, to his sons, who are the heads of the twelve tribes of Israel. Genesis 49:28 says, "All these are the twelve tribes of Israel, and this is what their father spoke to them. And he blessed them; he blessed each one according to his own blessing." If you read the words that Jacob spoke, they were very insightful and discerning, but are his words really a blessing?

The definition of blessing is to speak well of someone, to impart good favor and to endow prosperity on someone. We use the word "blessing" as a way to make someone happy. Even in the Sermon on the Mount in Matthew 5, Jesus starts each sentence with "Blessed are they that..." (KJV). The word "blessed" is defined as "to be well off, happy, fortunate." Each person described in Matthew 5, from a worldly perspective, is not someone that seems highly favored or happy today. Who wants to "mourn, be poor in spirit, be meek or be persecuted?" In addition, if we read the words that Jacob spoke to some of his sons (like Reuben, Simeon and Levi), we would not define them as words of blessing. Personally, I would not have wanted to hear those last words from my father on his death bed.

So what are some other definitions for blessing as illustrated from Jacob's last words in the Old Testament? Hebrews 11:21 describes Jacob's very last act of blessing his children as his supreme act of faith. Jacob believed that God used his words to speak forth His truth concerning the future of his sons. For parents, blessing our children is an act of faith in which we trust them to God. We also bless our children by discerning God's talents and spiritual gifts and then, we release them to fulfill God's purpose in their lives. Even if the words are so truthful that they are interpreted as piercing the heart instead of blessing the heart, we have to remember that the truth sets us free. By embracing the truth and asking forgiveness, we are blessed by being made alright through the grace and mercy of the Lord. There is no greater blessing than the covering of Jesus' blood which sets us free from sin and cleanses us from all unrighteousness. How "blessed" is the man who trusts in Him (Psalm 84:12).

July 25

Today's Reading: Psalm 37-39; Acts 26
Today's Thoughts: A Weapon and a Tool

***And it happened, when our enemies heard that it was known to us, and that God had brought their plot to nothing, that all of us returned to the wall, everyone to his work. So it was, from that time on, that half of my servants worked at construction, while the other half held the spears, the shields, the bows, and wore armor; and the leaders were behind all the house of Judah. Those who built on the wall, and those who carried burdens, loaded themselves so that with one hand they worked at construction, and with the other held a weapon.* Nehemiah 4:15-17**

Nehemiah left Babylon to return to Jerusalem because he had heard that the walls around the city were still in rubble. Though others had already returned to rebuild the temple, no one had determined to repair the walls and gates. This news brought Nehemiah to tears, as he fasted and prayed fervent prayers to God. He returned to Jerusalem determined to fix the broken walls, but he needed help and support from those around him. Nehemiah's story is truly amazing in its purpose, preparedness, and planning. But even the best laid plans can come under threatening attacks—and his certainly did. Though they tried, those who were against the restoration of the walls and gates could not thwart Nehemiah's plans. His men kept working, even as they had to strap a sword to their side for protection.

Our enemy is always lurking about us, roaming and seeking whom he may devour. But remember this: he cannot thwart God's plans, even if he frightens us along the way. Nehemiah was aware of his enemy's schemes and he did not allow those tactics to intimidate him or his workers. They kept a hammer in one hand and weapon in the other. Despite obstacles, persecutions and negative influences, we need to keep pursuing the things of God. Our enemy will use every trick in the book to keep us from being effective, productive and fruitful; but once we are aware of his schemes, we can fight back. Do not let fear, intimidation, insecurity or guilt keep you from being successful. Pray, plan and purpose in your heart to get going and do not stop. Be strong and courageous, and ask the Lord to guide your every step.

July 26

Today's Readings: Psalm 40-42; Acts 27:1-26
Today's Thoughts: The Pit or the Rock

I waited patiently for the Lord; And He inclined to me, And heard my cry. He also brought me up out of a horrible pit, Out of the miry clay, And set my feet upon a rock, And established my steps.
Psalm 40:1-2

The house in which I grew up was built in the late 1800's and still had some of the relics from that period, such as an old cistern located in the backyard. I think at one time it had been a source of water. Since it was no longer used or needed, the opening was loosely capped with a rusty round lid. As a kid, I remember being in fear of accidentally falling into it. My imagination conjured up this image of a deep dark pit with a sinking bottom in which I would be forever trapped. I look back and think that maybe my fears were a little silly, but then I realize that sometimes as an adult I feel as if I wake up in that pit, dark and damp with no way out. Maybe a lot of us feel this way at times.

Reading the words of King David in this Psalm should give us all hope and encouragement. If David had such moments of despair, then we must know that our feelings of being trapped in a pit of miry clay are not unknown to our Lord. Jesus knew how David felt and He knows how we feel. The Lord hears our cries, the Lord inclines His ear to listen to us, and He will not only lift us up, but will lead us out. The key is to trust Him to do so. We must believe that the Lord is going to rescue us because we cannot do it in our own strength. Trying to climb our way out of the darkness is exhausting and depressing. And sometimes it is in that darkest moment, when we sense a hand reaching down, picking us up and setting our feet back on the rock. We did not need to try; we just needed to surrender our fears to the Lord.

Where are you today? Are you in the "horrible pit" or "upon a rock"? Regardless of the place you find yourself, just know that the Lord is right there with you. He is the only way out of the pit and He is the only one Who can establish our steps in the right direction. Pray today that you can trust Jesus for all your needs and that in trusting Him, you can wait patiently for His perfect timing.

July 27

Today's Reading: Psalm 43-45; Acts 27:27-44
Today's Thoughts: What If?

And a woman who has a husband who does not believe, if he is willing to live with her, let her not divorce him. For the unbelieving husband is sanctified by the wife, and the unbelieving wife is sanctified by the husband; otherwise your children would be unclean, but now they are holy. **1 Corinthians 7:13-14**

How many times do we, as adults, reflect upon our childhoods and wonder "what if?" What if my Christian mom would have left my non-Christian dad? Thoughts of my childhood bring a mixture of emotions. Growing up in a home with a father who did not know or believe in Jesus Christ was hard, even in the best of times. I felt the tensions, had fears, and though I did not understand everything, I knew we were in a battle. My mom fought one of those battles every Sunday, but somehow she persevered to get herself and two kids to church. Those mornings irritated and angered my dad. I now realize just how serious those battles were, for the warfare is not against flesh and blood but against spirits and principalities (Ephesians 6:12). But through His power and love, God gave my mother the strength to stay with my father. And because of the foundation my mother laid in the Lord, I learned to turn to God for help at a young age. God heard my prayers and I learned to trust in His protection, even in the midst of oppression.

Just recently, my mom and I talked about her relationship with my dad. I am so thankful she stayed with him. Paul says that the "unbelieving husband is sanctified by the wife" and that the "children would be unclean, but now they are holy." God honors our desire to keep the commitment we make with our spouses and God knows that the children need both parents together. Even if only one parent knows the Lord, God blesses the home. Besides, we never know when the Holy Spirit will change the heart of the unsaved parent. As a child, I prayed every night for my dad "to be saved," and though he never went to church, my mom remembers the night he prayed the sinner's prayer after watching Billy Graham. He died soon after. We may not understand why God allows certain situations to be so painful and hard for so long, but He honors our desire to please Him over ourselves. What if my mother had chosen to leave my father? Yes, my life would have been different but I am so thankful that my mother stayed.

God gave us His strength when we were weak. God gave us hope when we despaired. Because of those years, I learned to cry out to God and depend on Him at a young age. I learned invaluable lessons that have shaped my faith today. Despite your circumstances, no matter how dreadful they seem, God is with you. Turn to Him for the strength to make it.

July 28

Today's Reading: Psalm 46-48; Acts 28
Today's Thoughts: Watching the Clouds

And when he had said these things, as they were looking, he was taken up; and a cloud received him out of their sight. And while they were looking steadfastly into heaven as he went, behold, two men stood by them in white apparel; who also said, Ye men of Galilee, why stand ye looking into heaven? This Jesus, who was received up from you into heaven shall so come in like manner as ye beheld him going into heaven. **Acts 1:9-11**

The apostles had the awesome privilege of watching the Lord ascend into heaven. They gazed up to the heavens as He departed on the clouds! No wonder they all stood in awe and amazement. I am sure they dealt with thoughts such as "Where are you going? Are you coming back? Can I go too? When will we see you again?" And even, "I can't believe I am seeing this." But the angels had an answer to all these questions. They said, "....come back...." As a result, the apostles fully and completely believed that the Lord would be coming back in their life time. Time was short, the message needed to get out. They lived every day for the Lord. Think about it...would you live differently if you thought the Lord was coming back today?

I asked the Lord to impress upon me the immanence of His return. I want every day to count for Jesus. I want to be looking for Him, listening to Him and ready for Him. Often, when I look into the clouds, I pray to see Him. To live every day as if today was the day of His return has changed my priorities, my desires and my focus.

Pray for eternal eyes. Pray for an excitement for His return. Pray to live as if He is coming back today. And ask the Lord to remind you of His promise to return every time you gaze into the clouds.

July 29

Today's Reading: Psalm 49-50; Romans 1
Today's Thoughts: Would you have supported Jesus' ministry?

For in it the righteousness of God is revealed from faith to faith; as it is written, "The just shall live by faith." **Romans 1:17**

Would you have liked to have lived during the ministry years of Jesus? Do you think you would have sponsored His ministry? In haste I would answer, "Why…yes, I would have liked to have been one of the disciples." But after pondering these questions, I realized that His teachings would have been hard for me to understand. Firstly, none of his listeners or disciples had the indwelling Holy Spirit (John 7:39) so they could not understand many of Jesus' teachings. Also, the things Jesus did and said would not have been consistent with what I thought He should do or say. For example, He would heal someone, then tell them not to tell anyone. He would say things like: "the Son of Man has no place to lay his head," "go, sell your possessions and give to the poor," "unless you eat the flesh of the Son of Man and drink his blood, you have no life in you." People were confused and divided over Jesus. His ministry stumbled others because man tried to define who God should be according to their wisdom and their standards.

We are still struggling with that problem today. How many times have you wondered if a circumstance in your life was from God or Satan? And—think of your prayer life. Can you say it has a tendency to reflect the will of the Father or your will? We get confused because a lot of times our flesh pleads louder than our spirit. Even as believers, we have our own ideas of what God should do and how He should do it. And if Christians struggle with this, and we have the word of God and the Holy Spirit, think of what the people of the world are up against. The world wants to worship something that makes sense to them. The world wants to see, hear and do what feels good to them. That's not God's way. God is pleased when we walk by faith. God desires that we give thanks in all circumstances regardless of how we feel. The Lord wants us to persevere in faith instead of please our flesh.

I had a friend who died at age forty-two, leaving eight kids, from ages two to thirteen. Every part of me was telling God that I couldn't support His decision to let her die. My attitude reflected my struggle in that I was questioning God's ways over my friend's death. My family tried to help me get through this tough time. Even my kids reminded me that God's ways are not our ways. The righteous shall live by faith…no matter how it feels, or how weak we are, or how confusing the circumstances may be. We need to have an attitude of worship in our thoughts, our actions and our words. I want to sponsor Jesus here on earth regardless of what I think or how I feel because when I meet Him face to face, I want Him to sponsor me.

July 30

Today's Reading: Psalm 51-53; Romans 2
Your Thoughts on Today's Passage:

July 31

Today's Reading: Psalm 54-56; Romans 3
Today's Thoughts: His Most Precious Gift

***His lord said to him, 'Well done, good and faithful servant; you have been faithful over a few things, I will make you ruler over many things. Enter into the joy of your lord.'* Matthew 25:23**

My husband bought me a watch for my birthday. When I first received it, I cherished it and admired its beauty, not thinking too much its function. As time passed; however, I noticed that I was becoming careless with it. When talking on the phone, I would take it off, not noticing where I placed it. I would take it off by the sink and leave it there. One morning, I could not find the watch. I began to panic. Where did I leave it? It was amazing just how sentimental and valuable that watch became again.

As Christians, we sometimes treat our salvation the same way I treat my watch. At first, we cherish our salvation, knowing that it is a gift too precious to ever earn. We thank God for the privilege of knowing Him and we praise Him for His Spirit dwelling inside us. Then as time passes, we take our Christianity for granted, mostly because we get used to being a Christian.

But what if we could lose our salvation? How are we treating this most precious gift? Are we careless or slothful? Do you share it with others? Are we thanking the Lord for it still… every day? Our salvation is the most precious of all gifts. How are we treating it? These are sobering thoughts. Personally, I want to be found faithful with the gifts and talents He has given me. These gifts are too valuable to be taken for granted. When all is said and done, I want to hear, "Well done, good and faithful servant. You have been faithful over a few things, I will make you ruler over many things. Enter into the joy of your Lord!"

Today, ask the Lord to impress upon you where your heart is. Do you really value all He has done for you and appreciate all that you have? Ask Him to convict you of any careless behavior that is not pleasing to Him. At any time, you can ask the Lord to forgive your careless behavior regarding the privilege of knowing Him. Then, ask the Lord to return the joy of your salvation to lead others to Him. Jesus has given Himself to you as a gift; nothing is of greater value than that.

August 1

Today's Reading: Psalm 57-59; Romans 4
Today's Thoughts: Fear is Contagious

The officers shall speak further to the people, and say, 'What man is there who is fearful and fainthearted? Let him go and return to his house, lest the heart of his brethren faint like his heart.'
Deuteronomy 20:8

Today's verse is taken from a message that God gave Moses to give to the Israelites in preparation for their battles in the Promised Land. The Lord began preparing the people before they crossed the Jordan River to encourage them that He was with them and would fight for them. The Lord made a point of addressing the danger of fear in their hearts. He tells the "fearful and fainthearted" to stay out of the battles and to go home instead. Why? Because fear is contagious.

The message for us two-fold: we need to be aware of how our fears influence others and we need to be wise of those whose fears can influence us. Fear is a deadly weapon and a powerful tool used by the enemy. Unfortunately, it is easier said than done to rid ourselves of fear. Even if we can identify our fears, this does not mean we can overcome them on our own. The first step is to be honest with ourselves and with the Lord. When we bring our confessions to Him, the healing process begins. The next step is to read God's Word and believe what it says. Faith and fear cannot exist together. Faith is built on knowing and believing the Word of God.

>Does it still take time to change old habits? Yes.

>Will the Lord lead us to make certain changes? Yes.

>Are we responsible for following God's guidance? Yes.

>And, will the enemy continue to remind us of our fears? Yes!

But, the Truth sets us free, if we are truly willing to believe and apply it to our lives. Today, if you are struggling with fear, examine your closest influences and the fears they may be invoking. Ask for wisdom in how to handle them. It is hard to fight your own battles when others are bringing discouragement. Be honest with the Lord in your personal battles with fear and prepare your heart to take a stand against the lies of the enemy. Christians in the world today are in warfare from all sides but Jesus told us to be of good cheer for He has overcome the world (John 16:33). It is our faith that is the victory that overcomes all of the fears that the world brings (1 John 5:4). The Lord will go before you and fight for you, just as He did for the Israelites.

August 2

Today's Reading: Psalm 60-62; Romans 5
Today's Thoughts: Just Pick Up the Phone

If you remain in me and my words remain in you, ask whatever you wish, and it will be given you. John 15:7

My mother was over to visit yesterday and we talked about buying tickets for a stage play coming at Christmastime. We both were unsure of what kind of seats to buy. My mom said to me, "Call your aunt. She went to see that play last year and she will know." I hesitated to call her so my mom picked up the phone and called. When my aunt (my mom's sister) answered, my mom didn't even say, "Hello" but just asked the questions. My mom received the answers and hung up.

The Lord impressed upon me the difference between relationships. If I were to pick up the phone and call my aunt, I would have to go through a whole bunch of introductory comments first. I would have to say, "Hi, how are you? I haven't talked to you in a while. What's new?" and then, I could ask the questions. Because my mom talks to my aunt frequently, their relationship is different and their conversation just continues where they left off.

That is how it is with the Lord. When we develop this relationship with Him, we can just start talking and asking Him whatever is on our mind. However, if we haven't spoken to the Lord in awhile, then we sense the need to reintroduce ourselves to get on the same page before really sharing our hearts. God wants a close relationship with us. He is very near to us; He is in our mouths and in our hearts. The kind of relationship we choose to have is up to us. I have had both kinds of relationships with the Lord and I can tell you that you will have so much more peace when you are closer to Him. Today, ask the Lord to help you learn to draw close to Him. You will find that you don't even have to dial the phone to get His answers.

August 3

Today's Reading: Psalm 63-65; Romans 6
Today's Thoughts: Dead to Sin, Free to Live

Consider yourselves to be dead to sin, but alive to God in Christ Jesus. **Romans 6:11**

Even though you are dead to sin, sin's strong appeal will often cause you to struggle with feeling that you are more alive to sin than you are to Christ. But Romans 6:1-11 teaches us that what is true of the Lord Jesus Christ is true of us in terms of our relationship to sin and death. God the Father allowed His Son to "be sin" in order that all the sins of the world—past, present and future—would fall on Him (2 Corinthians 5:21). When He died on the cross, our sins were on Him. But when He rose from the grave, there was no sin on Him. When He ascended to the Father, He carried no sin on Him. And today, as He sits at the Father's right hand, there is no sin on Him. Since we are seated in the heavenlies in Christ, we too have died to sin.

Christ already died to sin, and because you are in Him, you have died to sin too. Sin is still strong and appealing, but your relationship with sin has ended. I've met many Christians who are still trying to die to sin, and their lives are miserable and fruitless as a result, because they are struggling to do something that has already been done. "For the law of the Spirit of life in Christ Jesus has set you free from the law of sin and of death" (Romans 8:2).

Romans 6:11 summarizes what we are to believe about our relationship to sin because of our position in Christ. It doesn't matter whether you feel dead to sin or not; you are to *consider* it so because it *is* so. People wrongly wonder, "What experience must I have in order for this to be true?" The only necessary experience is that of accepting Christ on the cross, which has already happened. But as long as we put our experience before our belief, we will never fully know the freedom that Christ purchased for us on the cross. When we choose to believe what is true about ourselves and sin, and walk forward on the basis of what we believe, our right relationship with sin will work out in our experience.

August 4

Today's Reading: Psalm 66-67; Romans 7
Today's Thoughts: Relationship Over Religion

Therefore, as through one man's offense judgment came to all men, resulting in condemnation, even so through one Man's righteous act the free gift came to all men, resulting in justification of life. For as by one man's disobedience many were made sinners, so also by one Man's obedience many will be made righteous.
Romans 7:18-19

Religion says that we must follow rules and laws. Relationship says something different. We are "to be" not "to do." Jesus just wants us to be with Him. You ask, "How can I be with someone I cannot see or touch?" Have you ever talked with someone over the phone and felt as though you were right there with them? And even with emails or letters, we can express and share love and support without touching or seeing the person on the other side. Why? Because we have a relationship with that person. We know them. We love them. We know they love us. Whatever the case may be, we have a personal connection. A true friendship does not require rules and guidelines but requires time, attention, forgiveness, communication, openness, unselfishness, honesty and most of all, love. Jesus wants more than just a relationship with us; He wants our friendship. Why is this relationship so important?

When Adam and Eve sinned, all of mankind was doomed to death. One sin brought down every human being who would ever be born. From that moment on, we would all be born of sin, and death would be our destination. One sin is all it took to separate us from God. When Jesus died for us, His one act of sacrifice on the cross was all it took to restore us back to God. Through one man, death entered; through one Man, life was restored. We had no choice in being born. And we had no choice in the condemnation of sin that is within each one of us. But we have a merciful God who by His grace has given us a choice today. We have a choice to believe in His Son. We have a choice to accept His free gift of eternal life. We have a choice for total and complete restoration back to our heavenly Father. We cannot change what happened through Adam, but Jesus changed the outcome that awaits each of us. And not only has Jesus given us life eternal, but also He has given us His peace. There is no need to work or to try to earn His favor. His grace promises us unmerited favor and mercies that are new each morning, not through religion but through a relationship with Him.

Spend some time today just *being* with Jesus. Talk to Him. Walk with Him. Sit at His feet. He will meet you no matter where you are. Stop working, stressing, trying, and striving. Think of Jesus as you would your best friend. He will help you get to know Him better. Pray to have the relationship with Jesus that He wants to have with you.

August 5

Today's Reading: Psalm 68-69; Romans 8:1-21
Today's Thoughts: Who is Greater Than My God?

1 Bless the Lord, O my soul! O Lord my God, You are very great: You are clothed with honor and majesty, 2 Who cover Yourself with light as with a garment, Who stretch out the heavens like a curtain. 3 He lays the beams of His upper chambers in the waters, Who makes the clouds His chariot, Who walks on the wings of the wind, 4 Who makes His angels spirits, His ministers a flame of fire. 5 You who laid the foundations of the earth, So that it should not be moved forever, 6 You covered it with the deep as with a garment; The waters stood above the mountains. 7 At Your rebuke they fled; At the voice of Your thunder they hastened away. 8 They went up over the mountains; They went down into the valleys, To the place which You founded for them. 9 You have set a boundary that they may not pass over, That they may not return to cover the earth. 19 He appointed the moon for seasons; The sun knows its going down. 20 You make darkness, and it is night, In which all the beasts of the forest creep about. 24 O Lord, how manifold are Your works! In wisdom You have made them all. The earth is full of Your possessions—Psalm 104:1-9, 19-20, 24

Take a moment and read these verses. Take even more time and read the entire Psalm 104. Sometimes we need to be reminded of God's awesome power and greatness. This Psalm captures the beauty of the Lord in poetic imagery. We can read the verses and almost visualize God's majesty. He is God and there is none other. He is worthy of all praise and worship.

Who covers Himself with light? God is light. Who stretches out the heavens like a curtain and lays the beams of His upper chambers in the waters? God. The same God moves upon the clouds as though they were His chariots and walks on the wings of the wind, because He made them. Who placed the waters in their place and put a border around them? God. The same God put the moon and the sun in their places, for seasons and for days, for light and for darkness. There is but one true God. Creator God. Sovereign Lord. Lord of heaven and earth.

And as if all of this is not enough, He created us in His image. He has a plan for each of us. He loves us beyond our understanding. And He wants to spend eternity with us. He has even prepared the place in heaven that awaits us. What do we have to do in return? Believe in His Son Jesus Christ, and receive the free gift of salvation. Take time today and praise the Lord, not just for all that He has done, but for just who He is.

August 6

Today's Reading: Psalm 70-71; Romans 8:22-29
Today's Thoughts: Our Only Hope

For we were saved in this hope, but hope that is seen is not hope; for why does one still hope for what he sees? But if we hope for what we do not see, we eagerly wait for it with perseverance.
Romans 8:24-25

Have you ever talked with someone who seemed to have no hope? They were beaten up, beaten down and hopeless. My heart breaks for those who have no hope in the future and have no idea how to take a step towards it. Maybe some of us know people who fit this description; maybe some of us have these same feelings ourselves. To a certain extent I think most people have times in life when it appears that all hope is lost. A sudden tragedy, a chronic illness, or just life in general can bring feelings of hopelessness.

But what is *hope*? To hope for something is to have an expectation and anticipation of getting it. In a way, hope is like faith because we cannot see hope; therefore, we must believe in what we cannot see. This is easier said than done, especially when we cannot see the sky for the clouds. What can we do in times like these? Our only "hope" is to turn to God's Word and ask Him for help. Romans 8:25 is a good verse to pray back to the Lord in asking Him to help us with our hope. A sample prayer might be:

> Dear Lord, I am struggling with hopelessness and I cannot see the light at the end of the tunnel. I need Your word to give me hope and I need Your presence to lift me up today. Your word says I am to hope for what I do not see. I do not see (_____) but I pray for the perseverance to eagerly wait for it. Help me to believe in the hope that only You can give so that I may see Your hand upon my life today. Thank You for saving me. Thank You for giving me hope. In Jesus name, Amen.

Try writing your own personal prayer today. Ask the Lord to help you find hope in Him. Explore other verses that speak to your heart and pray them back to God. He is listening and ready to help.

August 7

Today's Reading: Psalm 72-73; Romans 9:1-15
Today's Thoughts: Jesus Has Been Waiting For You

Here I am! I stand at the door and knock. If anyone hears my voice and opens the door, I will come in and eat with him, and he with me. **Revelation 3:20**

I recently received an email from someone who I love very much. She was hurting and questioning the reason for all the pain. The following is a letter I wrote back to her. Today, you may be in the same place or know someone who is struggling with the same issues. We pray that these words minister to your heart.

Jesus has been waiting for you your whole life. He knows exactly where you are and how you feel. He knows what it is to suffer and hurt. Hebrews 4:15 says that we have a great High Priest who understands our sufferings and pain because He, Himself, endured so much more. Our only hope is Jesus. He is the only one who can heal you. And He will heal you, that is why He died. He had to die or we would all be lost. His Holy Spirit is now moving on your heart, which is why you are asking these questions. Your questions are really directed towards Him. All He wants from you is your heart, nothing more. All you have to do is let Him in—you will have no peace until you do. The Bible has your answers but you will not understand what the Bible says until you have been saved by the very blood that was shed when Jesus died. "He was wounded for our transgressions, He was bruised for our iniquities—and by His stripes we are healed" (Isaiah 53:5).

He loves you and He wants you to be His child. You are precious in His sight. The Lord will give you answers but you must go through His Son, Jesus. John 14:6 says, "I am the way, the truth and the life. No one comes to the Father except through Me." Many people refuse to accept this truth but it is the truth. Jesus is the only way. I am going to give you the prayer to ask Him into your heart. Pray this prayer:

Dear Lord, I thank You for Jesus, whom You sent to the cross to die for me. I know, Lord, that I am a sinner. Please forgive me of my sins. Please cleanse me of them all and make me new today. Jesus, I ask You to come into my heart and save me. Please heal me and make me whole. Please fill me with Your Holy Spirit that I may understand Your Word and live a new life in You. Thank You for saving me. Thank You for loving me. Help me to live for You. In Jesus name I pray. Amen.

August 8

Today's Reading: Psalm 74-76; Romans 9:16-33
Today's Thoughts: Wisdom with Self Control

Then Joseph could not restrain himself before all those who stood by him, and he cried out, "Make everyone go out from me!" So no one stood with him while Joseph made himself known to his brothers. **Genesis 45:1**

Personally, I can't believe Joseph could keep his secret this long. Joseph had God-given self-control. Can you imagine how difficult it must have been to not reveal yourself to your brothers? Joseph handled this situation with such wisdom.

Wisdom and self-control are not natural traits. They have to be developed by and in a person. Wisdom is given by God and self-control is the last attribute listed in the fruits of the Spirit. It is through life experiences that we gain wisdom and it is through intense trials that we learn self-control, However, experience and trials do not necessarily mean that people become wise and self-controlled.

God desires to develop both of these traits in His children. When God can trust you, He entrusts you with greater gifts and callings. Obviously, God tests our hearts through a variety of trials in life. How we react and respond to those trials shows us where we place our trust. How often do we want to look to man or impulsively react to what we think is best? Patience, endurance, waiting and resting are not eagerly sought characteristics today. Thank the Lord that, in Christianity, we are under the covenant of grace which means that we can take the test over and over until we pass.

A person with wisdom and self-control blesses God and others, as well as himself. Joseph is a great role model for each of us. God was so gracious to give us an example of a man who endured years of trials and abuse but overcame the circumstances to be entrusted with governing Egypt. Ask the Lord to show you what's missing in your walk with Him. Then, ask for wisdom and pray for self-control.

August 9

Today's Reading: Psalm 77-78; Romans 10
Today's Thoughts: Talk About Jesus

How, then, can they call on the one they have not believed in? And how can they believe in the one of whom they have not heard? And how can they hear without someone preaching to them? And how can they preach unless they are sent? As it is written, "How beautiful are the feet of those who bring good news!" But not all the Israelites accepted the good news. For Isaiah says, "Lord, who has believed our message?" Consequently, faith comes from hearing the message, and the message is heard through the word of Christ. Romans 10:14-17

There are a lot of misconceptions about evangelism. First, evangelism means to share the good news of Jesus Christ. It is as easy as sharing your story (or testimony) of how Jesus Christ changed your life. No one can argue with your own personal story, and you do not need to know any Scripture verses to share your story.

How do I witness or tell someone else about Jesus? Well, there are three things that everyone loves to talk about: themselves, their work and their kids. I start asking tons of questions about each of these three things, to let them talk. After they have shared about themselves, they usually ask me something about myself. Before answering, I pray silently to the Lord for the right sentence to direct the conversation to Christ. I want to use all that the person has shared in a way to lead the conversation to the Lord. God has always honored my desire to witness. He has always given me the words and I have seen the fruit. I have seen many people (complete strangers) come to Christ in a variety of places, from hospital beds to airplanes, and even on the phone to an operator in another state who was helping me make a hotel reservation!

People need Jesus. I believe that so firmly that I share Jesus with everyone I can. Jesus wants people to come to Him. We are the bridge between Jesus and the lost (through the Holy Spirit). A long time ago, I learned that **it is better to say the WRONG thing than to say NOTHING at all.** The Holy Spirit will use anything to minister to a heart, but the words have to be spoken for Him to use them. The Holy Spirit needs two things from each of us: a face and a voice. Paul said, "faith comes from **hearing** the message" and how will anyone hear without someone telling them? (Romans 10:17)

Tell the world about Jesus, starting with one person at a time. Be sure you ask them, "Do you want to pray with me right now to come to know Jesus?" You will be shocked by how many say, "Yes, I do." And remember, **God blesses obedience, not results.** Be obedient to share the gospel and leave the results to God.

August 10

Today's Reading: Psalm 79-80; Romans 11:1-18
Today's Thoughts: Do You Know This Jesus?

"The Spirit of the Lord God is upon Me, Because the Lord has anointed Me To preach good tidings to the poor; He has sent Me to heal the brokenhearted, To proclaim liberty to the captives, And the opening of the prison to those who are bound" Isaiah 61:1

I saw a dear friend the other day. We "just happened" to be parked next to each other on a street I have never parked on before. I have to say that the strangest circumstances and the oddest timing led me to that place. But I knew when I left her, God had orchestrated our meeting.

My friend told me a tragic story about a person I have watched grow in the Lord and have frequently prayed for. Hearing the details, I realized that my heart was grieving and crying out to God once again for this precious person. This world is sickened with sin. The wages of sin is death and this world will die.

When you face circumstances that taint the very face of God, it is difficult to know how to think, let alone pray. When evil seems to prevail over the innocent, we wonder and question and ponder if we really can trust God. We question where He is in those times. We question what He meant when He said He would never leave us or forsake us. Are God's definitions different? Do we have to spiritualize everything that happens on earth, waiting to be vindicated only in heaven?

No and no. The Christianity that is represented today in most people is not the Jesus that I worship. I am sick and tired of Christians who quote verses to the hurting just to make themselves feel better. My Jesus can give you Living Water that heals every hurt at the depths of your broken and abused soul. We should not settle for what other Christians have to offer. We need Jesus. We need everything He has to offer. We need Jesus to reach down into the depths of our hearts to bring healing and life and liberty to the captive. He has come to set you free; do not settle for anything less. If you need help, let us know. We are here for you: the abused, the lonely, the confused, the disillusioned, the helpless.

You cry out to Him. You be honest. He will deliver you. Do not settle for anything less than a miracle that transforms your life so powerfully that you know Jesus Himself is holding your heart. That's Jesus' Christianity.

Do you know the real Jesus? Do you feel Him as you read His words? Are you empowered with supernatural eyes that see beyond the physical to experience the spiritual? Please do not settle for anything less. God has so much more for you.

August 11

Today's Reading: Psalm 81-83; Romans 11:19-36
Today's Thoughts: Ministry

Therefore, since we have this ministry, as we received mercy, we do not lose heart. **2 Corinthians 4:1**

Ministry is an interesting word. What defines a ministry? Is it a job in a church, a job outside of a church, or is everything in the Christian life somehow a form of "ministry"? Regardless of where or how ministry is conducted, one thing is true—

it is a tough job. If we are not careful, we can become very weary and discouraged in doing the work of the ministry.

I have noticed in my own experience that I need to be reminded repeatedly that God does the work. I just need to be faithful to do what He has asked of me today, depending on Him for everything. Somewhere during the busyness, I can forget what dependency feels like and try to make sense of ministry in my own flesh instead of relying on His Spirit. A good saying is: "Where God's finger points, there God's hand will make a way."

For those of you who are serving in a ministerial role, pray for the Lord to give you verses to anchor your ministry goals. While reading the Scriptures, look for a verse that expresses how you feel or your heart's desire. Keep searching the Scriptures and praying until you find at least one verse or one story that you can pray back to God and meditate on. Those verses will give you the strength to get through the trials as well as the focus to finish what you started. God will bring to your memory the Scripture throughout the days ahead when the times get difficult. God wants us to count the cost and be responsible in whatever service we do, but God takes us as His responsibility to get us through. And remember, we are all dealing with the same thoughts and struggles. You are not alone when serving the Lord.

So we must not get tired of doing good for we will reap at the proper time if we don't give up. **Galatians 6:9**

August 12

Today's Reading: Psalm 84-86; Romans 12
Today's Thoughts: A Shepherd, Not A Hireling

***But a hireling, he who is not the shepherd, one who does not own the sheep, sees the wolf coming and leaves the sheep and flees; and the wolf catches the sheep and scatters them. The hireling flees because he is a hireling and does not care about the sheep.
John 10:12-13***

Have you ever left something or someone in the hands of another person? Is it difficult to leave your child with a babysitter; or your home with a house sitter? For those of us who have left our "valuables" in the hands of another person, the feelings can be unsettling at times, regardless of how much we may trust the person. We know that their care and concerns are just not the same as ours. Why? Because we know that person is a "hireling."

Jesus uses this same illustration to show His care and concern for us. No one loves and cares for us more than Jesus. He is our Good Shepherd and we are His sheep. Jesus looks at each one of us as His very own precious little lamb. He treasures us so much that He would lay down His life for us. He would do anything to save us from harm or death. And that is exactly what He did: He gave His life for each one of us.

Take comfort today in knowing that you have a Good Shepherd watching over you. He watches you come in and go out. He knows your every move because He cares about your safety. If you should find yourself straying from Him, maybe venturing out too far, just listen for the sound of His voice. As a Christian, you are His lamb and part of His sheepfold. You will recognize His voice, just as a sheep recognizes its shepherd; and you will hear Him calling you back. Jesus never leaves us in the hands of someone else because He knows that no one will watch over His valuables more than Himself. Trust in your Good Shepherd and know that He will never leave you.

August 13

Today's Reading: Psalm 87-88; Romans 13
Today's Thoughts: Knowing Christ

But the Jews did not believe concerning him, that he had been blind and received his sight, until they called the parents of him who had received his sight. **John 9:18**

To open the eyes of the blind was a clear sign that Jesus was the Messiah. The religious leaders knew that no one could open the eyes of the blind. However, they did not believe that the man had really received his sight until his parents confirmed the fact that he had been born blind. The Jews choose to not believe that Jesus healed his sight.

Today, Jesus is still performing miracles as He gently touches us and others. He is still answering prayers on a daily basis. Nevertheless, like the Pharisees, we do not always appreciate the work of God because of the hardness of our own hearts. We personally want to tell God what He can and cannot do. We want to challenge the miracles and call them "coincidences" or deny that a miracle happened at all. We talk Christianity and try to exclude the "Christ."

I had a man say to me once that he tried "that Christianity stuff and it did not work". He went on to tell me that he read the Bible, prayed and did good works within the church but it was only a phase. I then asked him the following questions: "Have you ever gotten down on your knees, begging God to cleanse you from your sins? Have you ever told the Lord that you only desired to live for Him, to serve Him and to follow Him wherever He wanted? Have you ever had such a hunger and thirst to read His words that He placed within His book just to get to know Him better? Did you ever want Jesus more than you wanted anything else and desired to please Him in everything you did?" He clearly and quickly answered, "NO." And I said, "Then I don't think that you ever came to know Jesus, and without Jesus, there is no Christianity." Christianity without Christ is just a lot of wasted time and effort, doing some good things that do not really matter. Jesus does not need us to do anything. He wants us to come to Him, listen to Him and enjoy His fellowship.

There are many reasons why we may want to be religious. God does not want our actions; He wants our hearts. When our hearts align with His, our actions will follow to please Him. No one has to pretend or try to act like a Christian. As you believe in Him, He will change you. Pray that your heart will stay soft to recognize the works of God in your life. Pray that the miracles of Jesus make a difference in your day and affect the lives of others. Pray that you see Him moving in you, touching you and healing you from the depths of your heart today. Jesus wants to gently and quietly love on you. God does not need you to *play* Christianity. God wants you to *know* Christ.

August 14

Today's Reading: Psalm 89-90; Romans 14
Today's Thoughts: Why Do We Suffer?

Yet it was the Lord's will to crush him and cause him to suffer, and though the Lord makes his life a guilt offering, he will see his offspring and prolong his days, and the will of the Lord will prosper in his hand. After the suffering of his soul, he will see the light of life and be satisfied; by his knowledge my righteous servant will justify many, and he will bear their iniquities. Isaiah 53:10-11

We struggle when we think that we will have to experience "trials of various kinds." Various kinds of trials invariably come with suffering. It does not take much effort to see that there is so much suffering, but we wonder: Why do Christians have to experience it when Jesus came to give us life abundantly? We frequently forget that the abundant life starts here, amidst all the suffering. We all know that there are more ways to suffer than just physically. Mental, emotional and circumstantial events can all bring aspects of suffering. At any given time, we could be faced with a serious trial by any one of these means. Why would God allow that to happen?

The word Christian means "little Christ." As Christians, we are not greater than our Master and His call on earth was to suffer. The NIV says that *He was despised and rejected by men, a man of sorrows, and familiar with suffering. Like one from whom men hide their faces he was despised, and we esteemed him not* (Isaiah 53:3). Jesus was filled with all the fruit of the Spirit (love, joy, peace, patience, goodness, kindness, faithfulness, gentleness and self-control) while still suffering and in sorrow. That is the goal of the Christian's life. Despite all the emotional, mental, physical, circumstantial torment, we have love, joy, peace, patience—through it all.

The rain falls on the just and unjust alike. It falls on the saved and the unsaved. We are not spared from trials because we are Christians. But it is through the everyday issues of life that we are changed into Christ-likeness because of them. Jesus is our example as well as our intercessor who knows and understands whatever we are facing. Turn to face Him, knowing that the things of earth will grow strangely dim in the light of His glory and grace.

August 15

Today's Reading: Psalm 91-93; Romans 15:1-13
Today's Thoughts: Under the Shelter of His Wings

And He shall cover you with His feathers. And under His wings you shall take refuge; His truth shall be your a shield and buckler. Psalm 91:4

Some summers ago, our family purchased a 6-week old male cockatiel. But it was around a month ago that I started noticing behavioral changes in Mercy as he began to softly sing songs, slowly rocking back and forth, while hiding in dark, quiet and soft places (like our sock drawer). He seemed to be nesting. Was Mercy having an identity crises or was our male bird "a girl"? One day, the mystery was solved as Mercy laid an egg. She actually laid a total of 3 eggs during 1 week period of time. But with the first egg, Mercy took on the role of a mother. She would stay in the corner of the cage to protect the eggs. She did not want to get her eggs dirty so she would not leave any droppings while in the cage. Mercy waited until we lifted her off the eggs and got her out of the cage. She stopped chirping to not give away her position. Our little bird suddenly and instinctively changed. It was amazing.

After watching how Mercy sheltered her eggs, I began to understand why God uses a bird's wings to describe Himself to us. Mercy would skillfully gather her rolling eggs and gently sit on them. However, her little body was not big enough to cover all three eggs. So, she would then tilt her body slightly forward and gracefully expand her wings to keep the three eggs protected from injury and to keep the eggs very warm. She was completely determined and resolved to stay in this position to keep the eggs under her wings. Mercy would even sleep in this position as her head would fall forward and rest on the floor of the cage. Only if I lifted her off the eggs to eat, drink and exercise, would Mercy leave her eggs.

If this little bird would do all this for empty, unfertilized eggs, how much more does our Heavenly Father protect and comfort us? The Lord's biggest difficulty with us must be that His little eggs have feet to get up and leave the nest! I suddenly realized that He honors our unwillingness to not stay under the shelter of His wings. It is not best for us but we have a choice.

How many of us need to return to His nest? Do you need to find a place of rest and warmth by the counsel and comfort of His Spirit? God is faithful and during this difficult season in our economy, know that He never leaves you nor forsakes you. Rest in His nest as you open His Word and seek Him for wisdom and discernment. He will provide for us and protect us. It is God's heart desire to gather us as chicks.

August 16

Today's Reading: Psalm 94-96; Romans 15:14-33
Today's Thoughts: A Man who Found his Home

***For what will it profit a man if he gains the whole world, and loses his own soul?* Mark 8:36**

Several years ago, I had asked my neighbor if she knew of someone who did handyman work. She immediately replied, "Use Larry. He isa Christian and homeless. He only has a car and a beeper." I had seen this man frequently in the neighborhood. He was very clean-cut and polite. While speaking to him, I knew that the Lord wanted me to ask him about his home life. I said to the Lord (in my mind), "But Lord he doesn't have a home." I heard again, "His home in Heaven." So, I asked Larry about his home life and that sparked a long response as to what happened to him. Larry told me that he used to be a multimillionaire who lived in a 32-room mansion. At first, I was not sure if I believed him, but my neighbor confirmed that his story was true. She knew Larry during those days. Through a variety of circumstances, he became homeless. Now he lives in his car, doing handiwork jobs for a living and ministering to the homeless. He said to me, "I have lived in the best addresses in Rancho Santa Fe but I'm no longer concerned about those addresses. I constantly keep my mind focused on my next Home address." He said that many people are living in homes but they are truly homeless because they are empty inside. They have no peace. He quoted me his favorite verses that he claimed on a daily basis:

"*I am not saying this because I am in need, for I have learned to be content in whatever the circumstances may be. I know what it is to be in need, and I know what it is to have plenty. I have learned the secret of being content in any, and every situation, whether well-fed or hungry, whether living in plenty or in want*" (Philippians 4:11-12 paraphrased).

After we finished talking with each other, I started meditating on those verses. I noticed that "I *have learned"* is mentioned twice. Learning to be content is a choice. Larry was not immune to fear, loneliness, worry or doubts. Larry had troubles but he trusted in Jesus' words. We are only passing through earth in these temporary tents to get to our eternal home in heaven. The world's promises do not bring peace, but Jesus brings peace. Remembering that we have a reservation in heaven keeps us focused on what matters most.

The fruit of Larry's homeless life has led many homeless and hopeless people to Christ. Larry learned to live for what matters most. Larry is now at his new permanent address as he went home to be with the Lord in his early 60s. Larry lived the second half of his life with nothing but a car and beeper; however, Larry gained everything because he lived his life here for his eternal home.

August 17

Today's Reading: Psalm 97-99; Romans 16
Today's Thoughts: A Gentle Answer

A gentle answer turns away wrath, but a harsh word stirs up anger.
Proverbs 15:1

If a gentle answer turns away wrath, why do we choose to say harsh words? We hear harsh words everywhere we go. People seem to be so angry these days. The southern California freeways are notoriously known for drivers having "road rage." I do not even want to know what kinds of words are being spoken in the cars of others while driving.

Out of the abundance of the heart, the mouth speaks. Our mouths are speaking right from our hearts. The difficulty with having a "gentle answer" is that the character trait of self-control has to be developed first. If we look at the fruits of the Spirit, the last one is self-control. It starts with love and ends with self-control. We all have the ability to speak gentle and kind words to each other when we want to be kind and loving. However, we do not want to be gentle and kind when we have been criticized or mistreated. These situations require self-control so we can watch our words and answer with gentleness and kindness.

I pray for self-control daily. Not only do I not want to regret my words, but also I really want my heart to reflect the Lord. I see the Lord's reflection in me when I hear the words that come from my heart. Listen to yourself today. What kind of words do you use? What kind of tone or attitude do you have? Are your words edifying and encouraging or destructive and negative? Then, when you get angry, evaluate your words again. God wants us to have peace: He left it and He gave it. Anger is not a sin but stirring anger up is. Pray for the Lord to take hold of your heart to help you to hold your words. Pray for self-control of your tongue.

August 18

Today's Reading: Psalm 100-102; 1 Corinthians 1
Today's Thoughts: Thanks be to God

Make a joyful shout to the LORD, all you lands!
Serve the LORD with gladness;
Come before His presence with singing.
Know that the LORD, He is God; It is He who has made us, and not we ourselves;
We are His people and the sheep of His pasture.
Enter into His gates with thanksgiving,
And into His courts with praise.
Be thankful to Him, and bless His name.
For the LORD is good;
His mercy is everlasting,
And His truth endures to all generations. **Psalm 100**

Let us all be thankful for our freedom and for our founding fathers who laid the God-fearing foundation for this great nation. But above everything else, let us not forget the One who made us. We are God's people—we are His sheep and this nation is still His pasture. Take time today to give Him praise and thanksgiving for the true freedom that we have in His Son Jesus Christ. Make a joyful noise to the Lord! Pray for our nation today. As Christians, we should all be greatly concerned at how our lawmakers are turning away from God in their choices and decisions. God's mercy is everlasting but there is a day of judgment coming. Pray that our nation will turn back to the Lord. One by one, as Christians, we can make a difference. Let's start today.

August 19

Today's Reading: Psalm 103-104; 1 Corinthians 2
Today's Thoughts: Our Father Helps Us

I was with you in weakness, in fear, and in much trembling.
1 Corinthians 2:3

My children come to me the quickest and the loudest when they are weak, fearful or trembling. Someone knocked them down, something scared them or something just does not "seem fair." As they bare their concerns and worries to my listening ear, they regain composure to walk right back into the same exact situation. Usually nothing has changed, neither the circumstances nor the playmates, but by expressing those concerns to me, they were changed.

We need to remember that God is our heavenly Father. He is waiting for us to come to Him and bare our souls. It is in those times of weakness, fear and trembling that we need to learn to talk to Him. We need His strength to get back out there again. I pray for this frequently. Why? Because, like Paul, I have found in my walk with Christ that, "when we are weak then He is strong" (2 Corinthians 12:10). God's strength is made perfect, or complete, in weakness. In times of helplessness and panic, I find hope and peace only when giving those burdens to Christ. And after sensing His strength and power flowing through that weakness, fear and trembling, I know without a doubt that the living God alone has heard my cry. The situation or circumstance may not change but I change and when I change, He gets the glory.

Today, give God whatever situation is making you feel weak, fearful and helpless. Ask Him to fill you with His power and wisdom. Ask Him to work a miracle and change the situation, or ask Him to just change you—which also would be a miracle. Record it and watch what God does.

Pray about it: Oh Lord, I pray that You may increase and I may decrease (John 3:30). I pray that your power may be perfected in my weakness, for I know that when I am weak, You are strong. I want to draw from Your strength today. Use me for Your glory.

August 20

Today's Reading: Psalm 105-106; 1 Corinthians 3
Today's Thoughts: Constantly changing

For I will not contend forever, Nor will I always be angry; For the spirit would fail before Me, And the souls which I have made. Isaiah 57:16

Christianity is like redecorating an older home. When you change the look of one room, suddenly you realize that the rest of the house does not look as good any more. The Lord does the same with us. He starts working on one area of our character and then suddenly we realize that a lot more change is needed.

How do we not get overwhelmed by it all? How do we prevent ourselves from getting discouraged? There are times, after I have been confronted with an issue, that I have wondered how I got along all these years without realizing that change was needed. Other times, after God addresses something, I have wondered why God has decided to work on these aspects of my character now. The saying, "He never gives up on you," is so true. Whether good or bad, the Lord takes you as His own and will make you like His own. He will not give up, even when we ask Him.

The only way to faithfully get through the refining process every day is by understanding the character of God. He is a God of love, patience, kindness and gentleness. He will never lead us where He cannot keep us. But for us to receive more of His blessings, He needs to make sure our own character can handle them. The Lord disciplines those He loves and desires for each of us to live in His abundant blessings. However, there is a level of responsibility with each blessing, and that requires change.

If you find yourself in the fiery furnace right now, find promises on the love of God. Think on these things. You will find yourself yielding to Him, with less resistance and you will get through just fine.

August 21

Today's Reading: Psalm 107-109; 1 Corinthians 4
Today's Thoughts: The Power of Imitation

"Therefore I urge you, imitate me." **1 Corinthians 4:16**

We are masters at imitation. Whether we are copying someone else or they are copying us, we are continually being influenced and adapting our personality to the conformity of others. We may not realize that the things we do or wear or say are a result of imitating someone or something in our culture.

Paul understands these issues when he says to the Corinthians "imitate me." They were living *in* the world and enjoying being *of* the world. Paul knew that they needed a living example to follow. He became their example, a real person from whom to watch and to learn. Paul also knew that their only hope was in Jesus. He was the One they must ultimately learn to imitate.

Second Corinthians 4:18 says that, "we do not look at the things which are seen, but at the things which are not seen. For the things which are seen are temporary, but the things which are not seen are eternal." Our focus should only be on Jesus Christ our Savior. This world and all that is in it will someday be gone. Where will you be when this happens? Do you choose your friends wisely, those who set the example in Christ as Paul did? Or are you influenced more by the world's temporary offerings, instead of Christ's eternal promises? How accountable are you to set godly examples for those who are imitating you? Ask the Lord to help you become the role model and example that He wants you to be for others. His Holy Spirit will guide you in this role.

August 22

Today's Reading: Psalm 110-112; 1 Corinthians 5
Today's Thoughts: To Stay or To Go

I do not pray that You should take them out of the world, but that You should keep them from the evil one. **John 17:15**

After reading the prayer of Jesus, I realize how far I am from understanding the truth of His message. Jesus prayed that we not be taken out of the world but that the Father would keep us from the evil one. There are times when some of us want to be taken out of this world as soon as possible just to be in heaven with Jesus. When you love someone, you want to be with them, and as Christians, our hearts yearn to be closer to Jesus. Other times, life here on earth is so hard and such a struggle that we want to be taken out just to get out of here. Admit it; we all have felt this way at times. But these feelings are no surprise to the Lord.

Jesus' prayer to keep us in the world was truly based on love. He has provided everything we need for life and godliness, despite our circumstances. He has given us creation to see His handiwork and understand His unlimited abilities. He has given us His Word that sanctifies us and sets us apart for Him. And He has given us His own testimony as Jesus, Son of God, who walked on this earth and lived in this world. He prayed that the Father would not take us out of this world but to keep us from the evil one. Jesus prayed that we be made *one* with Him while remaining on earth. At times, it seems impossible to walk against the pattern of this world. We can take comfort in knowing that Jesus understands what we are going through. Our only hope is to depend on Him, which is exactly what He wants in the first place.

Where are you today? Do you secretly desire at times to be taken to heaven? Only you know the reasons why you feel this way but know today that Jesus has a plan for you being right here, right now. He loves you and has a plan for your life. Ask the Lord to give you strength in times of trouble, peace in times of struggle and hope when the world seems to be hopeless. Only in Jesus can we stay in this world and experience heaven on earth.

August 23

Today's Reading: Psalm 113-115; 1 Corinthians 6
Today's Thoughts: Such Were Some of You

Or do you not know that the unrighteous will not inherit the kingdom of God? Do not be deceived; neither fornicators, nor idolaters, nor adulterers, nor effeminate, nor homosexuals, nor thieves, nor the covetous, nor drunkards, nor revilers, nor swindlers, will inherit the kingdom of God. Such were some of you; but you were washed, but you were sanctified, but you were justified in the name of the Lord Jesus Christ and the Spirit of our God. **1 Corinthians 6:9-10**

We know that the unrighteous will not inherit the kingdom of God and most of us would agree with Paul's list of those people. But to then read, "Such were some of you," should bring us great comfort. God reaches out to all kinds of sinners. No sin or sinner is beyond the help of God or His hand of mercy. God reaches out in love to save us from any and all sins. Some Christians believe that certain sins could exclude others from receiving God's grace. Is there such a thing? Could our sin go too far or be repeated too much for God to accept us?

Let us answer that question by using Romans 1. Romans1 describes the person who knows God but chooses not to honor Him (verse 21), exchanging the truth of God for a lie (verse 25). As a result, this person is given over to his unnatural lusts, ending in the sin of homosexuality. The sin of homosexuality did not exclude the person from having a relationship with God. The choice to reject God came first, which led to the one and only unpardonable sin, rejecting and rebelling against the Holy Spirit. In this case, homosexuality becomes a symptom of the disease of rejection. If a homosexual has never heard of the truth of Christ, his sexuality will not exclude him from a new life in Christ. The attitude of the heart is greater than the behavior of the flesh. God judges the attitude of the heart. We are to represent in word and action the message of salvation to everyone. No sin is too great for the God of love to penetrate. No act is too deeply depraved for the blood of Christ to cover. All sinners (all of us) need a savior.

As we accept Christ as our Savior and Lord, He washes us from those sins and makes those sins as white as snow. Next, He sanctifies us (or set us apart from those sins) to be used for His service on this earth. Finally, He justifies us before the Father "just as if I" never sinned. We can now live in this sin-filled world with these sinful bodies and have a relationship with the Holy God being led by His Holy Spirit.

When we come to Christ, the struggle of sin does not go away but if we choose to love Jesus more than anything else, He will get the glory despite our struggle.

.August 24

Today's Reading: Psalm 116-118; 1 Corinthians 7:1-19
Today's Thoughts: In the Morning

*My voice You shall hear in the morning, O L*ORD*; In the morning I will direct it to You, And I will look up.* **Psalm 5:3**

Have you ever noticed the difference between waking up with an alarm clock versus sleeping until you naturally wake up? When it comes to doing morning devotions, there is a huge difference between the two. Personally, I love to sleep in. But once I am up, I do not slow down; I make the most of my day. I rarely sit down unless I am eating or in a meeting. So I am thankful to the Lord that I have no trouble falling asleep and staying asleep. It is the *waking up* that gives me the most trouble.

For years, I set aside a "quiet time," to just be quiet to pray and to read my Bible during the day. But because of my personality and because of the activities of the day competing for my attention (like work, the kids and the phone), I became too distracted to concentrate on the things of God. I would read verses like Psalm 5:3 and know deep within my heart that I needed to get up before the day started. It took a long time and a lot of prayer to get me into the habit and pattern of waking up early. It is only in the early mornings that I am in a better place to hear from the Lord because there is nothing competing against His words. During this quiet time, I can enjoy the freedom from the noise of the day. The house is quiet and peaceful in the pre-dawn hours.

It is funny that if I was going on vacation or needed to catch a plane, I could more easily get out of bed to the buzzing noise of an alarm clock. But how come with the things of God, we resist so much? Personally, I believe it is because we resist the words 'discipline' and 'dependency'. We need to discipline our bodies to wake up early and we do it to show our dependency on needing the Lord to help us. God loves us and has so much to give to us. His ways are of peace. For me, I have learned that an extra hour in bed cannot compare with the peace I receive spending quiet time with Him first thing; for Jesus says, "But seek first the kingdom of God and His righteousness and all these things shall be added unto you" (Matthew 6:33).

My encouragement to you is to try it. Set your alarm half an hour earlier than you normally get up and make a commitment to do it every day for a week. And give that half hour to the Lord, ask Him to wake you up to His words and to give you the peace that will carry you throughout your day. When you make this change in your daily habits, God will bless you.

August 25

Today's Reading: Psalm 119:1-88; 1 Corinthians 7:20-40
Today's Thoughts: Our Witness to Our Faith

Therefore, having obtained help from God, to this day I stand, witnessing both to small and great, saying no other things than those which the prophets and Moses said would come—that the Christ would suffer, that He would be the first to rise from the dead, and would proclaim light to the Jewish people and to the Gentiles." **Acts 26:22-23**

My husband has been working with a client who was born and raised Buddhist. She is vice president of a large publishing company. She is smart, tenacious, tough and lost. In working with her over the past few months, my husband has made his faith very clear to her. As the project comes to a close, it seems that their conversations on faith have increased even more. She is open to talking with him about the subject. So often, we find ourselves in situations where we either get that one shot to witness, or we get shot down before even starting. In this case, my husband and this high-powered executive keep debating the issues. These conversations have brought her to a point of admitting how little she truly knows about her own religion. He has been asking her to tell him exactly what she believes. How does it work? And most importantly, he asked her a simple question, "Do you really believe in it?"

Jesus asks this question to those around Him, "Do you believe in the Son of God?" Many people find a place in organized religion but have no idea of what or whom they really believe in. We are born and raised in religious traditions. We like the routines, the fellowship, and the comfort such activities bring to our lives. Since when did Jesus use the word "comfort" to describe our walk with Him? Yes, He gives us comfort and peace through His Holy Spirit. But Jesus was clear in saying that He came with a sword, dividing friends and families (Matthew 10:34-35). Placing anyone or anything, including church traditions, above Jesus Christ will bring division between people who choose Him against those who choose their religion.

Many people do not know what to do or what to say at these times. What do we say? How much do we say? When should we speak about our faith? And on and on we wonder and ponder. I like what Paul says in today's verses. First, he says that he gets "help from God." The Holy Spirit will impress, lead and direct us in what to do. Next, Paul knows he must "stand." Sometimes we need to stand, hold our position, be honest, but not start debates unless we are led to do so. Then, Paul says he witnessed "both to small and great." We must seize the opportunity when we know it is time, regardless of who it is, how many there are, or whatever the circumstances. Sometimes we need to not think too much and just let the Lord work through us.

August 26

Today's Reading: Psalm 119:89-176; 1 Corinthians 8
Today's Thoughts: Unconditional Promises

For the children of Israel walked forty years in the wilderness, till all the people who were men of war, who came out of Egypt, were consumed, because they did not obey the voice of the Lord--to whom the Lord swore that He would not show them the land which the Lord had sworn to their fathers that He would give us, "a land flowing with milk and honey." Then Joshua circumcised their sons whom He raised up in their place; for they were uncircumcised, because they had not been circumcised on the way. So it was, when they had finished circumcising all the people, that they stayed in their places in the camp till they were healed. Then the Lord said to Joshua, "This day I have rolled away the reproach of Egypt from you." Therefore the name of the place is called Gilgal to this day.   Joshua 5:6-9

God's promises are not conditional. He will fulfill every promise, but in His way and in His time and to whom He wills. God heard the cries of the Israelites when they were slaves to Pharaoh. God had promised Abraham that after 400 years of bondage, He would rescue his descendents. That time came and God fulfilled His promise by sending Moses. Despite the Egyptians hard hearts and selfishness, God had His way and the Israelites left Egypt. Through God's outstretched arm and mighty powers, God brought plagues and parted the Red Sea. Nothing is impossible for the Lord. However, the Israelites' hearts were hardened too. After different trials and tests in the desert, it was obvious that their prayer for freedom had not been about the freedom to worship God, but for freedom from difficult circumstances. Once the circumstances became difficult in the desert, these same people chose to forsake their God who had rescued them.

Because of Moses' intercession, the next generation would be the ones who would receive the promises of God and enter into the land flowing with milk and honey. Their willingness to obey and attitude mattered to God. They needed to be obedient to His commands; however, God did not remind them about obedience to the covenant of circumcision until after they crossed the Jordan River. So you see that God's promises were not predicated upon their flesh but on His promises. However, in time—God's time, God would make sure that they kept up their end of the deal. We find ourselves in the same kind of relationship with the Lord today. We can accept Christ and receive His Holy Spirit regardless of our behaviors. He will empower us and use us to fulfill all that He has for us. But as time goes by, He continually reminds us and asks us to keep in line with our part of the relationship. Our willingness and attitude matter first and then our obedience proves if we take His promises seriously.

August 27

Today's Reading: Psalm 120-122; 1 Corinthians 9
Today's Thoughts: Preparing to Run

Do you not know that those who run in a race all run, but one receives the prize? Run in such a way that you may obtain it.
1 Corinthians 9:24

How often do we feel as though we are running a race? Life seems to be one big race. We run from place to place to get things done, just to accomplish one more task. But what about the race of life? Some of us run from parts of our lives, some run from the past, some run towards the future, and some just run in circles. Do we ever win this race? Do we even see the finish line? What prize do we expect to receive?

Let us examine the basic components of running a physical race to help better understand the spiritual race. The first step usually involves some sort of training, which depends upon the type and length of race. Good equipment and accessories are integral parts of preparation and participation, such as clothing, shoes, and personal timing devices. Because slight dehydration can jeopardize the heartiest competitor, nutritional supplementations and liquid replacements are critical for endurance and stamina. And a final key component for any serious participant involves their own personal desire and commitment to complete the race, despite hardships and pain. One should not even consider entering a race without some of these basic necessaries.

In Hebrews 12:1, Paul says that we should "run with endurance the race that is set before us." This race is not about our physical lives but our spiritual lives for Christ. As Christians, we should be racing towards "the goal for the prize of the upward call of God in Christ Jesus" (Philippians 3:14), not racing for the things of this world but for the eternal prize in God's kingdom.

So how do we successfully compete in this spiritual race? Our training must involve a progressive and disciplined program that equips us with God's wisdom, knowledge and discernment. A Christian church teaching God's Word is an essential component. Other tools include reading the Bible daily accompanied with prayer. A good church, Bible reading and prayer will be used heavily during the race. These necessities are best when supplemented with Bible studies, prayer groups, and volunteer services. The most critical component that CANNOT be compromised involves our sincere desire to surrender our heart and everything in it to the Lord Jesus Christ. If we commit to these basic steps, our race will be run with the assurance of a strong finish to this earthly life.

August 28

Today's Reading: Psalm 123-125; 1 Corinthians 10:1-18
Today's Thoughts: Whom Are You Seeking?

Jesus said to her, "Woman, why are you weeping? Whom are you seeking?" John 20:15

The first person Jesus appeared to after His death was to Mary Magdalene. She had already told the disciples that Jesus' body was missing. Peter and John ran to see. Peter did see but John saw and believed. The word "believe" means that he put his trust in God for what had happened. None of them had yet understood the resurrection. He just believed that God was responsible for the missing body of Jesus.

While the disciples went home, Mary stayed at the tomb. She wept and stooped down to look into the tomb. Two angels in white were seated there and they spoke with her. She continued crying and questioning and seeking Jesus, despite the actual presence of angels. She would only be satisfied with Jesus Himself. Jesus stood behind her, but it wasn't until He called her by name that she turned and clung to Him. I can imagine Mary was thinking, "You got away from me once and I am not letting You go again." Once again, Mary went to the disciples, but this time with a different message. She had sought Jesus and He found her.

The first person Jesus revealed Himself to as "the Christ" was the Samaritan woman (John 4). And the first person Jesus revealed Himself to after His death was Mary Magdalene. If you, male or female, are seeking Jesus, He will reveal Himself to you. He shows no partiality or favoritism. We are all given the same Holy Spirit, and one day, we each will behold our Lord face to face. But as of today His promise stands that if you seek Him, you will find Him if you search for Him with all your heart. Whom are you seeking?

August 29

Today's Reading: Psalm 126-128; 1 Corinthians 10:19-33
Today's Thoughts: The Power of God's Voice

"Then you came near and stood at the foot of the mountain, and the mountain burned with fire to the midst of heaven, with darkness, cloud, and thick darkness. And the LORD spoke to you out of the midst of the fire. You heard the sound of the words, but saw no form; you only heard a voice." **Deuteronomy 4:11-12**

In the book of Exodus, the Israelites begin their journey from Egypt, across the Red Sea, heading to the Promise Land. Along the way, they stopped at the base of Mount Sinai and had an experience of a lifetime. It was at this place that the Lord appeared to them for the first time, and they could not handle what they saw. Moses recounts this experience in the book of Deuteronomy and reminds them of what happened on that amazing day. The Lord appeared in smoke, fire and thick darkness. There was thundering and lightning in an awesome display of God's power on the top of that mountain. And, they were scared to death! They told Moses that they did not want to hear from God, just from Moses. He could speak on behalf of God and they would just listen to Moses. The Lord's voice was too much for them to handle.

For us, maybe the Lord's voice is a bit too much to handle at times. Sometimes, His voice seems to be loud and consuming—and scary. Maybe the dark clouds that come into our lives are evidence that the Lord is trying to speak to us. Instead of being afraid, we need to be alert to what God is trying to say to us. Instead of hiding, we need to start seeking Him even more. Our God is a consuming fire, not to our harm or destruction, but to our reverence and worship of Him. He is God. He is a jealous God, and He does not want to share us with other gods. His words from the top of mountain warned the people not to go out and make Him into a molded image, nor to worship any other images of gods created by their hands. His words today are the same to us. The Lord wants us to know Him personally, to love Him with all that we are, and to fear Him in respect and honor. Why? Because He wants to bless, protect and lead us in His ways. We are the safest and the most peaceful when we are obeying His words and willing to listen to His voice, even when it seems too loud and scary to handle.

August 30

Today's Reading: Psalm 129-131; 1 Corinthians 11:1-16
Your Thoughts on Today's Passage:

August 31

Today's Reading: Psalm 132-134; 1 Corinthians 11:17-34
Today's Thoughts: Exercise Your Faith

After these things the word of the Lord came to Abram in a vision, saying, "Do not be afraid, Abram. I am your shield, your exceedingly great reward." But Abram said, "Lord God, what will You give me, seeing I go childless, and the heir of my house is Eliezer of Damascus?" Then Abram said, "Look, You have given me no offspring; indeed one born in my house is my heir!" And behold, the word of the Lord came to him, saying, "This one shall not be your heir, but one who will come from your own body shall be your heir." Then He brought him outside and said, "Look now toward heaven, and count the stars if you are able to number them." And He said to him, "So shall your descendants be." And he believed in the Lord, and He accounted it to him for righteousness. **Genesis 15:1-6**

When God gave Abram the promise that his descendants would be as countless as the stars in the heavens, Abram and Sarai were old and had no children. This promise was still several years away from fulfillment even as the Lord made this covenant with Abram. But despite the circumstances of the current situation, Abram "believed in the Lord" and all that He was telling him. God accounted him as righteous because of his great faith.

It takes incredible faith to believe in a God you have never seen. It takes even more faith to believe in His promises that will directly impact your life. Great faith comes from our resolve to believe in the Lord and who He is, not what we can prove or intellectualize about Him. As a matter of fact, faith can be downright silly in man's eyes, even foolish. Instead of being self-conscious or embarrassed we should all pray to be more foolish in our faith. Why not expect God to move mountains in our lives?

Where is your faith today? Faith is meant to be active, not passive. If we do not keep exercising our faith, we get weak and can be tossed to and fro when the enemy comes to strike. Get into the Word of God on a regular basis and ask the Lord to increase your faith. When you get that sense within your spirit that you need to move, do it...step out and go for it. Sometimes when the fear is the greatest, you have to ask yourself: who is trying to stop you and *why*? Fear is not of the Lord. So why is the enemy causing you to fear? He does not want you to succeed. Faith brings success.

September 1

Today's Reading: Psalm 135-136; 1 Corinthians 12
Today's Thoughts: Jesus Lives

They asked her, "Woman, why are you crying?" "They have taken my Lord away," she said, "and I don't know where they have put him." At this, she turned around and saw Jesus standing there, but she did not realize that it was Jesus. "Woman," he said, "why are you crying? Who is it you are looking for?" Thinking he was the gardener, she said, "Sir, if you have carried him away, tell me where you have put him, and I will get him." Jesus said to her, "Mary." She turned toward him and cried out in Aramaic, "Rabboni!" (which means Teacher) John 20:13-16

Can anyone take the Lord away from us? Mary thought so. Her heart was broken, her Lord was dead and now she thought His body had been taken away. Mary cried deeply, thinking that someone could have taken the only part left of Him away from her. At this point, she did not even notice that angels were speaking to her. She just wanted Jesus back. Even when Jesus was standing (literally) behind her, Mary thought He was the gardener. In her loyalty and despite her depression, Mary was willing to do whatever it would take to "get Him" back.

Her devotion is sincere but how can she serve a dead God? What kind of Lord would we have if others could change His position or steal His body? What kind of God would we serve if we have the ability to lose or misplace Him? That kind of god puts the burden on us, regardless of how devoted or sincere we may be. Thank the Lord Jesus that He arose from the grave. We do not worship a dead god but a living Lord. We worship a God who has placed all the burdens on Himself. We do not have to find Him because He never loses us. And God promises that if we seek Him, we will find Him as He comes to us. We are never lost from His sight. We may not see Him but He is always there, even sometimes from behind.

I love the verse in Isaiah 30:21 that says, "Your ears shall hear a word behind you, saying, 'This *is* the way, walk in it,' Whenever you turn to the right hand Or whenever you turn to the left." God does lead us from behind at times and we can feel that we have lost Him. But like Mary in John 20:16, when Jesus calls your name even from behind, you will recognize His voice. Remember to keep seeking Him and desiring to be with Him even if He seems to be missing. You have not lost Him because He can never lose you. He conquered death on a cross and rose again. Jesus is the Resurrection and the Life and He is always with us, leading us in every step.

September 2

Today's Readings: Psalm 137-139; 1 Corinthians 13
Today's Thoughts: You are Important to God

How precious also are Your thoughts to me, O God! How great is the sum of them! If I should count them, they would be more in number than the sand; When I awake, I am still with You. **Psalm 139:17-18**

How many of us really believe that the God of the universe is thinking kind thoughts about us? Doesn't He have more important things to think about? How can He really care about us and be all that this verse says? The key point for us is to realize that God considers us to be very *important to Him*. If we stop for a moment and consider some of the basics of what the Bible teaches, we may better understand the truth of these verses.

Over and over, the Bible is filled with verses of truth that describe His love for us. John 3:16 tells us that, "God so loved the world that He gave His only begotten Son..." for us. Jesus came to earth, fully man and fully God, to give His life for our sins so that we may know Him personally. As hard as it may be to believe, God loves us so much that all He wants is to be with us; the God of the universe wants a personal relationship with you and me. That is why He thinks about us so much. That is why He never leaves us, even as we sleep. He wants us to know that He is still with us when we wake up.

My dog wakes me up each morning, sometimes earlier than I like. Now, my dog does not have to be let out, he has his own doggy door. He is not hungry or thirsty. He just wants to be with me. He wants to be petted and loved on, even for a few minutes until I am fully awake and then he goes back to sleep. As I wake up, I know that the Lord wants me up too. He has been thinking about me all night, watching over me and waiting on me to get up and be with Him. Mornings spent with the Lord are precious, and it is in those moments that He lets me know just how much He loves me. The Lord does not need anything from me, but He wants my attention and my love, and He gives me so much more in return.

Today, stop and give some time and attention to the Lord. Tell Him you love Him. Thank Him for His thoughts of you and for always being with you. And when in doubt, read the full chapter of Psalm 139...you will be blessed.

September 3

Today's Reading: Psalm 140-142; 1 Corinthians 14:1-20
Today's Thoughts: Does Your Heart Burn?

Then their eyes were opened and they knew Him; and He vanished from their sight. And they said to one another, "Did not our heart burn within us while He talked with us on the road, and while He opened the Scriptures to us?" Luke 24:31-32

One of my favorite stories in the New Testament is the story of the two on the road to Emmaus. They had heard the account from the women who had gone to the tomb and found it empty. Jesus was gone. His body was not in the tomb. All of their hopes for a king were gone as well. But as they walked and discussed all that had happened, Jesus showed up. Jesus began to question them about what had happened, as they were completely unaware of Whom they were talking to. After breaking bread with them, Jesus vanished and then they knew. Their eyes had been opened to the spiritual realm and they got it. Their hearts burned within them as Jesus talked to them. They understood His Word. The Scriptures came alive to them, just as, "the Word became flesh and dwelt among them" (John 1:14).

Has your heart ever burned within you as if Jesus were speaking to you personally? Jesus left us His Word and His Holy Spirit. Through the Holy Spirit living within our hearts, we can discern spiritual things, and it is only through the Holy Spirit that we can comprehend the Word of God. When we fall in love with His Word, we have fallen in love with Him. We find ourselves unable to get enough of His presence. We just want more of Jesus. It is then that we begin to have our eyes truly opened to know Him personally.

Praise God today that we do not have to hope for an encounter with Jesus. If we have accepted Him into our hearts as our Savior, then we have His Spirit living within us. Through reading and studying the Bible, we can understand His ways and grow in the knowledge of Him. Our lives will change, day by day, from glory to glory. Read Luke chapter 24 today as part of a further study in the Word. Ask the Lord to open your mind to understand the Scriptures. Maybe you will find your heart burning within you, before you are done.

September 4

Today's Reading: Psalm 137-139; 1 Corinthians 13
Today's Thoughts: Admit Your Fears

Whenever I am afraid, I will trust in You. Psalm 56:3

Is it hard for you to admit that you are afraid or fearful of something? Think of the circumstances and people that occupy your mind during the day. Are your thoughts filled with peace or fear? It is difficult to admit as Christians that we are fearful. For some reason we believe that the presence of fear disqualifies the strength of our faith. But it is foolish to believe that Christians are not fearful. Every stage we enter in our walk with the Lord has to shake our faith in order for us to understand the seriousness of the choices at hand. I love the Word of God, the Lord allows us to be honest about our fears and then helps us work through them as we learn from the examples of others as described in His Word.

When reading Psalm 56, we learn that David wrote this Psalm. From what we know about David, we know a boy who could slay a lion and a bear to protect his sheep. Then we know him as a man who was a mighty warrior and wise king. And yet, God allows us to see into this man's heart and we find that David was a lot like us, struggling with fears at times. But what makes David, "a man after God's own heart," was that he knew what to do about his fears. Those fears did not separate him from the Lord or hinder his faith. David had faith amidst the fears. How?

First, David made a conscious effort in prayer to admit to God that he was fearful. As believers, we often tend to not include the Lord in our fears, knowing that we shouldn't be "fearing" anything but Him. Not David, he understood that the Lord sees and knows all things about us. We can't hide from the Lord, so why pretend He doesn't know how we are feeling?

Next, David confidently told the Lord that he chose to trust in Him about those fearful things. So now, think of those things again that are fear inducing in your life. Talk to the Lord about them. Tell Him everything you are thinking. Now, tell the Lord that you choose to trust Him with those very things. You might have to tell Him over and over, but in time, you will train yourself to be like David...a man after God's own heart.

September 5

Today's Reading: Psalm 146-147; 1 Corinthians 15:1-28
Today's Thoughts: Hold Fast

***Moreover, brethren, I declare to you the gospel that I preached to you, which also, you received and in which you stand, by which also you are saved, if you hold fast that word which I preached to you--unless you believed in vain.* 1 Corinthians 15:1-2**

One day, I spent time with a patient who had just been diagnosed the day before with breast cancer that had spread to her brain, liver and lungs. She was a young woman in her forties. I met her on the Oncology Unit in the hospital where doctors were just beginning a treatment plan. This patient seemed to have a peace. She spoke to me about her signs and symptoms that led up to this diagnosis; she was honest with her emotions and heart's desires to get better. When I had asked her what her source of strength was during this time, she told me that over 300 people were praying for her and that she had wonderful family members and friends to see her through this difficult time. As the conversation continued, I asked her if she had the assurance of her salvation. I asked her if she had ever accepted Jesus Christ as her personal Lord and Savior. Her answer was quick and confident as she said, "I don't need to do that because my father is a minister. I have the church's support." My heart was grieved for there aren't any grandchildren in heaven. God wants to adopt sons and daughters into His Kingdom. God desires to personally raise each one of us, to call us each by name and for His kids to call back to Him, "Daddy," knowing that His arm isn't too short too save or ear too dull to hear the cry of His own children.

Becoming a child of God and receiving the assurance that you are His child is so easy. However, we resist so hard. In 1 Corinthians 15, Paul lays down the basics of salvation again. A child of God accepts that Jesus came to earth, fully God and fully man, and that He died on a cross for our sins, He was buried and rose from the dead three days later. Jesus is the way, the truth and the life and no one comes to the Father but by Him. His death on the cross allowed sinful man to have a relationship with a Holy God. His resurrection gives us the power to live this earthly life for His purposes, instead of our own. Romans 10:9-13 says, "that if you confess with your mouth the Lord Jesus and believe in your heart that God has raised Him from the dead, you will be saved. For with the heart one believes unto righteousness, and with the mouth, confession is made unto salvation." If you want to be sure that you have come to know the Lord Jesus in a personal way, pray this prayer: *Lord Jesus, I want to know You personally. Thank You for dying on the cross for my sins and forgiving me. I want you to come into my heart and take control of my life. I want to live for You as I believe You died for me. Thank You for hearing my prayer. I want Your Spirit to fill me.*

September 6

Today's Reading: Psalm 148-150; 1 Corinthians 15:29-58
Today's Thoughts: Praise Him and Find Victory

Praise Him for His might acts; Praise Him according to His excellent greatness! Let everything that has breath praise the Lord. Praise the Lord! **Psalm 150:2, 6**

As we meet and talk with women, we often hear of their struggles, especially in their own thoughts. They feel unworthy to be serving the Lord. Thoughts like, "Who are you to be doing this?" or "Why are you working on this—someone else could do it better?" are common components of mental attacks. These thoughts then trigger certain emotions and the next thing for women is a pit of depression and hopelessness, just what the enemy had planned. The enemy wants you to feel so condemned and depressed that you quit or are too demoralized to minister effectively and powerfully.

Some Christians recognize that these types of thoughts are not of the Lord but they do not know how to stop them. Some of us quote verses, rebuke the devil, and attempt to reason our way out of it with logical comebacks, but still have no real victory. Instead of fighting back, we begin to justify the bad thoughts, thinking that maybe these self-condemning behaviors are from the Lord to keep us humble. Wrong!

Jesus came to give us life to the fullest. He wants us to share in His joy, not wallow in our flesh. We are sinners. We are unworthy to do anything to please God. We are to love Him and to worship Him. Our faith is the only way to please God. It is not for us to dwell on our unworthiness but to praise Him for His willingness to save us.

How can we change this process? Start praising the Lord! Put on praise music, read the psalms aloud (as if you wrote them to the Lord) and worship the Lord. The key is to get our eyes off ourselves and focus on Him. He is worthy and by His blood that cleanses us, we can do all things through Christ, even change those negative thoughts.

September 7

Today's Reading: Proverbs 1-2; 1 Corinthians 16
Today's Thoughts: Looking for God's Wisdom

So that you incline your ear to wisdom, and apply your heart to understanding; yes, if you cry out for discernment, and lift up your voice for understanding, if you seek her as silver, and search for her as for hidden treasures; then you will understand the fear of the Lord, and find the knowledge of God. For the Lord gives wisdom; from His mouth come knowledge and understanding; He stores up sound wisdom for the upright; He is a shield to those who walk uprightly; He guards the paths of justice, and preserves the way of His saints. Then you will understand righteousness and justice, equity and every good path. **Proverbs 2:2-9**

As King David was facing his death, he appointed his son, Solomon, as the new king of Israel. In the beginning of King Solomon's reign, the Lord spoke to Solomon in a dream and said: "Ask! What shall I give you?" (1 Kings 3: 5) His reply greatly pleased the Lord, as he asked for wisdom and discernment so that he could rightly judge God's people. The Lord gave Solomon such an abundance of wisdom that he became famous throughout Israel and other lands, attracting the attention of many who came just to hear his wisdom. His own people feared and respected Solomon because of the power of his words.

The Proverbs are filled with Solomon's nuggets of wisdom, which are still applicable for our lives today. However, this man of wisdom begins this book of wisdom by instructing us to seek for wisdom, knowledge and understanding. Solomon tells us to "seek her as silver, and search for her as for hidden treasures." He obviously was thankful that he had the right answer when God asked him years earlier, "What shall I give you?"

We need to get in the habit of asking the Lord for wisdom, knowledge and understanding every day. The New Testament writer, James, also instructs us to ask for wisdom as he says in Chapter 1 verse 5, "if any of you lacks wisdom, let him ask God who gives to all liberally and without reproach and it will be given to him." In other ways, anyone can ask for wisdom at any time and the Lord will give it liberally regardless of our shortcomings and faults. Asking for wisdom is a gift as much as begin the recipient of the wisdom that He gives. There are no prerequisites or strings attached. There will be tangible fruit in your life as you begin to understand things from a different perspective. Certain habits will break and your own words will inspire you as they bring edification and counsel to others.

Oh Lord, please give me wisdom, knowledge and understanding today so I can think Your thoughts as You direct my words, hands and feet.

September 8

Today's Reading: Proverbs 3-5; 2 Corinthians 1
Today's Thoughts: Direct My Paths

Trust in the Lord with all your heart, and lean not on your own understanding; In all your ways acknowledge Him, and He shall direct your paths. **Proverbs 3:5-6**

My mom and dad are dealing with a huge plumbing problem. The main pipe that carries their hot water from the garage to the kitchen broke underneath the cement slab. They knew something was wrong when their tile floor was warming their feet. Because we have earthquake building codes, the plumbers cannot fix the problem by digging down, but only by rerouting the pipe through the ceilings. This has caused a major upheaval to their home. The laundry room, downstairs bathroom, family room, kitchen, dining room and even the closet where they store all the Christmas decorations have been affected. All of their personal belongings are out of the cabinets and closets, and are on the couches, tables and floor. Then there are the new large holes in the ceilings and closets with dust from the holes everywhere. The plumbers have tried to be considerate by placing plastic over their personal belongings and a walkway runner on the rugs, but with all the work and workers, it is impossible to keep the décor a priority.

Well, my mother has handled it very well, all things considered but...there are times she finds herself overwhelmed with the situation. She has continued to ask the Lord to help her and to remain peaceful through it all (including the cold showers for days). God has been so faithful to hear her prayers. The other night, she went to bed praying about the pipes and when she woke up, she heard this in her mind:

> Trust in the Lord with all your heart and lean not on your own understanding. In all your ways, acknowledge Him, And He shall **redirect your pipes.** Proverbs 3:5-6
> *(bold words are paraphrased)*

The Word of God is active and living. If we put His Word into our minds and meditate on it, God has the ability to bring it back in each and every situation to maintain peace in your life. God is so good to us. He hears every prayer and cares about every circumstance. Trust in Him and He will direct your paths, and even your pipes, if you include Him.

September 9

Today's Reading: Proverbs 6-7; 2 Corinthians 2
Today's Thoughts: Aroma of Christ

But thanks be to God, who always leads us in triumphal procession in Christ and through us spreads everywhere the fragrance of the knowledge of him. For we are to God the aroma of Christ among those who are being saved and those who are perishing. **2 Corinthians 2:14-15**

There are two words in these verses that should catch our attention, the words "fragrance" and "aroma." Isn't it interesting that Paul uses words that relate to our sense of smell to describe how we should be seen by others (or *smelled* by others)? Most of us notice when a person passes by wearing cologne or perfume, especially when it is a strong scent. There are certain fragrances that evoke memories in us. The smell of spices and evergreen are most common at Christmas. Floral smells hint of springtime. And even someone's scent (good or bad) stirs emotions within us if we relate that fragrance to a particular time or event in our lives.

As Christians, Paul is saying that we, too, have a scent about us —"the aroma of Christ." The origin of the word "aroma" brings us back to Old Testament times when the sacrificial offerings of burning animal flesh were common. Our lives as Christians are offered as that living sacrifice, holy and acceptable to God (Romans 12:1). As we live for Christ (and are willing to die to ourselves), we become a sweet-smelling aroma to God. Not by our good works, but because of the offerings and sacrifices we are willing to make in our daily lives. But we live in a world of indulgence to self, not sacrifice. How can we be the fragrance of Christ in the world today?

The first step is in being aware of who you are in Christ. As a believer, you represent Him wherever you go and whatever you do. What does that look like in real life? Maybe by forgiving someone who has hurt you. Maybe by giving up something you want just to bless someone else. The root of sacrifice and unselfishness goes back to love. It is the love of Christ in us and through us that leaves that sweet aroma on others.

Take time today and think about the fragrance you are wearing. Put it in the perspective of being an offering and a sacrifice unto the Lord for the sake of others. What better role model for us than the apostle Paul? Of course, the ultimate sacrifice was given through God's only Son, Jesus Christ, and it is now up to us to make that sacrifice known to the world. Ask the Lord today to help you be the aroma of Christ everywhere you go. May you leave a scent of love on everyone that you pass.

September 10

Today's Reading: Proverbs 8-9; 2 Corinthians 3
Today's Thoughts: The Unveiling

But we all, with unveiled face, beholding as in a mirror the glory of the Lord, are being transformed into the same image from glory to glory, just as by the Spirit of the Lord. **2 Corinthians 3:18**

In the Old Testament book of Exodus, Moses describes the building of the tabernacle. This tabernacle is to be built on earth as a copy of heavenly things, a place where God can dwell with His people. Its design and construction are described in great detail in Exodus Chapters 26 through 28, then again in Chapters 34 through 38. One of its most beautiful pieces was a veil that separated the Holy of Holies from the Sanctuary. The veil represented man's separation from God; man at that time had no real fellowship with God. Only Moses could enter into such personal fellowship with God. He would come in such intimate contact with God that even his face shone with the glory of the Lord. Moses would use a veil to cover his face. 2 Corinthians 3:13 says that Moses used the veil to keep the people from seeing what was passing away, but the next verse states that their minds were blinded. It was as though the minds of the people had their own veil that kept them from really seeing the truth. Years later when Jesus died on the cross, this veil, still used to separate the two sections of the temple, was torn in two. Jesus died so that we may have personal fellowship with God. No longer was a veil needed, because there was no longer a need for separation. The only veil left is the one that we keep, over those areas we have not yet surrendered to God.

Do you have a veil covering part of your heart or mind? As Christians, we might be quick to say, "Of course not, I love Jesus and I know His Spirit lives within me." I am one of those Christians who not only would say it, but also would believe it with all my being. How can I be veiled in any area when I so desperately want all that the Lord has for me? How can I be blinded when all I want is to see with spiritual eyes? But the closer I walk with the Lord, the more I understand how blind I really am. A veil does not have to be thick. The veil in the temple was about three feet thick. I have learned that the veils that cover my heart and mind can be quite thin, enough to let me see through but not enough to let me see clearly. The problem comes when I think I am seeing clearly but am really only seeing through a fog. Then the Lord begins to show me things so crystal clear that I can barely look at them. The Lord begins to unveil parts of me.

We must go through this process with the Lord if we want to continue to grow in our walk with Him. Breakthroughs come and strongholds are broken when we get through these types of trials.

September 11

Today's Reading: Proverbs 10-12; 2 Corinthians 4
Today's Thoughts: To Believe is to Live

Therefore I said to you that you will die in your sins; for if you do not believe that I am He, you will die in your sins." John 8:24

The Pharisees were always trying to catch Jesus off guard. Regardless of what Jesus was teaching, they would test Him. The Pharisees were the religious leaders and teachers of the day. They knew the Scriptures better than anyone and were always ready to tell others how to live by them. They were superior in their righteousness, or so they thought. They enjoyed their status, their power and the honor given to them by man. Jesus however, created problems for them in His teachings. The people wanted to hear more of what Jesus had to say. The Pharisees were threatened as they saw their power diminishing, as more and more people followed Jesus. How sad to realize that the very One whom they had been waiting for their whole lives was in their midst and they did not recognize Him!

The people who denied Jesus would die in their sins because they refused to believe that He is the Son of God. When all is said and done, the verse above sums up the truth. If we do not believe in Jesus, we will die in our sins. Sadly, there are still Pharisees in our world today who think that by knowledge and good works they have the answers. There are those who can recite Scripture to justify any argument or prove any point and are always instructing others on how to live godly lives. Then, there are those who truly try hard to do what is right. They volunteer for ministry services, give freely to others in need and are kind-hearted to everyone. On the outside, both types look very godly and religious. But what about on the inside?

Jesus is not impressed with our "good" works. Our best intentions mean nothing in the kingdom of God. We are born of sin, live sinful lives and will die in our sins, unless we do one thing—**believe in Jesus**. It is not the kind of belief that acknowledges His existence or His role. James 2:18 tells us that, "even the demons believe and tremble." The word "believe" means to adhere to, rely on, trust in, and depend upon. Do you believe in Jesus Christ according to His definition of the word? Do you absolutely depend upon Him for your salvation? Do you know that there is nothing you can add to your salvation and that there is no good work that you can do for eternal security? The only requirement is faith. The only One who can cleanse us from our sins is Jesus Christ, and the only way we can be cleansed is by believing in Him alone. Today, you can know Him. *Dear Lord, I want to know Your Son Jesus. I know that I am a sinner who cannot do enough good works to earn my salvation. I am lost without You. I want You to be my Lord and Savior forever more.*

September 12

Today's Readings: Proverbs 13-15; 2 Corinthians 5
Today's Thoughts: A Slow Day

He who is slow to wrath has great understanding, But he who is impulsive exalts folly. **Proverbs 14:29**

Have you ever had one of those days where you started out groggy, as if in slow motion? I woke up tired this morning. As I sat in my chair and began to read my online Bible, I was even more tired. Coffee did not help. I felt like I was swimming against the current. My insides were trying but my reflexes were not responding. I was having a hard time comprehending and focusing on what I was reading. Even my prayers were challenging, as my mind kept wandering. The day seemed to flow with the same kind of beat. Traffic seemed extra slow and congested. Red lights were longer than normal. And my computer...the little hour glass icon next to the cursor became a great nuisance as I tried to double click through files and pages. My irritation and frustration grew at top speed while everything else around me crawled at a snail's pace. As my day is fast coming to a close, I am still tired and still in slow motion. What does the Lord want me to learn from this kind of day?

Several verses in the Bible include the word "slow," but are often followed by the words "to anger." Why? Because we tend to get angry very quickly. Also, when we act on impulse, we usually act foolishly. The more anxious we get at times, the more likely we are to behave rashly. The end result is not good, not to mention who gets hurt in the middle. We must learn the self-control to be "slow to wrath" and slow to anger. We must learn to recognize the triggers in our environment and circumstances that tempt us to lose control and react in the heat of the moment. For me and my day, the Lord reminded me of these lessons.

Since I woke up in slow motion, He meant for me to go with His flow, not mine. Today, God wanted me to slow down, sit longer at lights, and even wait on my computer to respond. Instead of impulsively getting upset, He wanted me to think about Him during those moments, talk to Him and let Him talk back. Instead of talking to the Lord, I was wondering just when that red light was going to turn green.

Today, and any day, when you are sitting in traffic, at a long light or waiting for service somewhere, stop and talk to the Lord. God has set your day in motion for His purposes. It is okay to go slow, especially when circumstances pull at our emotions. If you go with His flow and follow His tempo, you will end your day with peace and joy, instead of regrets and fatigue.

September 13

Today's Reading: Proverbs 16-18; 2 Corinthians 6
Today's Thoughts: God Has a Way of Balancing Things

***When the Lord saw that Leah was unloved, He opened her womb; but Rachel was barren.* Genesis 29:31**

God has a way of balancing the scales. I have seen it with money, with time and, as this verse says, with love. Leah was "unloved" but Rachel was adored by Jacob. Leah had children and Rachel was barren. Not all things are that cut and dry spiritually, but all things are in the hands of the Lord. He is the One who opens and closes wombs. He is the One who determines the day of our birth and knows the day of our death. Moreover, even though Jacob loved Rachel more than he loved Leah, Leah was married to Jacob years longer and Leah was buried with Jacob. God does not do these things to punish us or hurt us: but to balance us.

While there are blessings, there are also burdens. That balances us. If life here were perfect, would anyone want to go to heaven? Think about it: we all want to go to heaven but no one wants to die, even though life is hard here. You may not be getting something you have been praying for—a long time. It might be something that others get so easily, but for you, you are begging God for answers, yet without seeing it. If this is you, look at the circumstances differently.

Instead of begging God for what you do not have, start thanking Him for all you *do* have. Instead of feeling badly about what you are missing on earth, start asking Him to give you a heart to have many things in heaven. We are only passing through here but heaven is for eternity. God knows what is best to keep you balanced. Start praising Him. Start praying to Him about pleasing His heart instead of asking Him to please yours. He loves you. Try seeking His face instead of His hand, and you will receive His peace that surpasses everything else.

September 14

Today's Reading: Proverbs 19-21; 2 Corinthians 7
Today's Thoughts: Choose to Love

These things I command you, that you love one another.
John 15:17

As a child, one of my favorite words was "why." I know it drove my parents crazy when I used it repeatedly. *Why* do I have to brush my teeth? *Why* do I need to eat those green things? The answer was always the same, "Because I told you so." There are times when we must submit and do as we are told. Children have little choice in the matter, but as adults we can choose our own course. Jesus tells us to "love one another." And even in this command, we have a choice. Have you ever found yourself wondering *why* you have to love that person who seems so undeserving? The answer is because Jesus says so.

Love is so important to Jesus that He summed up all Ten Commandments into two: love the Lord with all your heart, soul, mind and strength, and love your neighbor as yourself. The Apostle Paul says in First Corinthians 13 that nothing we do really matters without love. But in today's world, we are confused about love. Movies and television lure us into fantasies about "true" love using sex, guilt and manipulation as tools to obtain love. When we are hurt by someone who says they love us, we begin to wonder if love really exists at all. Is love a feeling or is it an attitude or a behavior? Jesus knew exactly why He had to command us to love each other. Because we would not do it based upon our own feelings. We are human and our nature is sinful.

The only way to keep this commandment is to pray. Pray for God's love to fill your heart. Pray that you will love your enemies as well as your neighbor (the good and bad). Pray that you will love Jesus more everyday. It is only when we stop asking "why" and start asking "how" that we can truly know what love is.

September 15

Today's Reading: Proverbs 22-24; 2 Corinthians 8
Today's Thoughts: Small Adjustments for Big Breakthroughs

"Do not be overly righteous, nor be overly wise: why should you destroy yourself?" Ecclesiastes 7:16

I tend to be the main point person when dealing with the retreats that we attend. So it's up to me to do mostly everything from communicating with the church to finding out the specifics of the retreat. Well, last weekend's retreat was no different. When Bobbye asked me what she should bring, I pulled out the list and started reading it to her: sleeping bag, flashlight, pillow, towels, toiletries. She listened intently and asked, "Is this a camp ground? Are we going camping this weekend?" I said, "No, it's a retreat center," without giving it another thought.

The weekend came and we were greeted with willing arms to help us carry our things. As I popped open my trunk, the pillows, towels, extra blankets and sleeping bags were right on top. One of the girls said, "You do not need any of that." Confused, I rechecked the packing list just to find out that I had stapled my 7th grade daughter's retreat list to the church's information. I had been following the wrong list.

How can you be so prepared and organized while being so radically off base? It took a lot of extra time, thought and effort to pack things that I didn't need. I had never questioned the list, trusting in my own skill set.

This was a small mistake in the big picture of things but it made me examine other things in my life. How many times am I going out of my way to make things harder on myself, absolutely convinced that I am right? I might be heading the right way and even able to get to the right place but am I prepared with the right things when I arrive? Am I making the trip of life more difficult in the process? Am I willing to listen to others who are personally involved? Am I spending more time doing or am I really thinking things through? These were great questions for me to ponder while away in the mountains. Sometimes it is not about making big changes but making small adjustments that can lead to huge breakthroughs.

Everything considered, the retreat was amazing! Women got saved and clearly women changed. The testimonies absolutely reflected that God is still in the business of setting captives free. Despite packing the wrong kind of things, we really had the right kind of time.

September 16

Today's Reading: Proverbs 25-26; 2 Corinthians 9
Today's Thoughts: An Everyday Retreat

And be kind to one another, tenderhearted, forgiving one another, just as God in Christ forgave you. **Ephesians 4:32**

Have you ever attended a weekend church retreat or fellowship outing? It is usually a great time of fellowship filled with a lot of love and laughter. It is an opportunity to meet new friends, share hidden struggles and spend time together. But why do we often wait until we attend such an event to reach out and enjoy others? Why can't we get along like that all the time? In one weekend, we can find out more about a person than after a whole year in church together. What happens at a retreat that causes our barriers to come down and our kind, loving, accepting personalities to come out? There are many burdens carried by people who attend a retreat, but somehow those burdens become a lot lighter by the end. The answer is God.

A retreat is the one place that we incorporate the Lord in everything we do. We sing to the Lord, we learn about the Lord, we talk to others about the Lord, we pray together to the Lord, and we are at a facility that is dedicated to the Lord. It is all about the Lord. We live out Matthew 22:37-40, "Jesus said unto him, 'You shall love the Lord thy God with all of your heart, and with all of your soul, and with all of your mind. This is the first and greatest commandment. And the second is like it, you shall love your neighbor as yourself.'"

We need to pray that we can understand this kind of love and kindness all year round. We get busy and focused on our own struggles, trying so hard to make it through the day that we forget others are in need too. But what we are really forgetting is a Who—the Lord. We need to include the Lord in the busyness of our days so that we can represent Him to others. Pray that you can be used by the Lord and that you can incorporate Him in everything you do today. You just might be able to help someone along the way. And, keep going to those retreats—we all need those special weekends with the Lord and with others.

September 17

Today's Reading: Proverbs 27-29; 2 Corinthians 10
Today's Thoughts: Winning the Battle

"For the weapons of our warfare are not carnal but mighty in God for pulling down strongholds, casting down arguments and every high thing that exalts itself against the knowledge of God, bringing every thought into captivity to the obedience of Christ" 2 Corinthians 10:4-5

Did you know that there is an ongoing battle for your mind? Many of us realize that there is a battle taking place *in* our minds, but the real battle is *for control of* our minds. This may sound a little like science fiction and weird but that is just what our enemy wants us to think. We are in a spiritual battle every day for control of our thoughts. For some of us, the intensity of conflicting thoughts can torment us for hours. For others of us, we have learned such mind control that we do whatever it takes to press down those thoughts. Regardless of how we handle them, they are still there.

The only effective weapons that we can use are the ones given to us by God. The weapons of the flesh, such as mental imagery or positive thinking will not bring us the victory that God has for us. Many people use such tools and claim dramatic results, but the deceptions are dangerous. Why? Because the focus and the power are placed completely on us, not on the Lord. The more power we think we have in ourselves, the more power Satan has over us. Scary thought? Let's go for the real power and learn to bring our thoughts before Christ.

Are you in a battle today over things in your mind? There are two things you must incorporate into your daily routine. First, open your Bible and read at least one verse in the morning. Use a devotional or a Psalm or Proverb, whatever works for you. Second, take what you have read and ask the Lord to put His words into your mind throughout the day. If you will do this every day, in time you will begin to remember God's words and will think about them because His Spirit will bring them to your thoughts. For those things that you are struggling with in your mind, start talking to the Lord about them. Pray that His words will bring comfort and hope instead of torment and guilt. Let the Lord fight for you by letting Him have the control over your mind instead of trying to handle it yourself.

September 18

Today's Reading: Proverbs 30-31; 2 Corinthians 11:1-15
Today's Thoughts: When God Exalts

On that day the Lord exalted Joshua in the sight of all Israel; and they feared him, as they had feared Moses, all the days of his life.
Joshua 4:14

Joshua had already been appointed by the Lord and by Moses to lead the people into the promise land. But, the Lord waited until everyone (except the priests holding the Ark of the Covenant) had crossed the Jordan before He exalted Joshua in the eyes of the people. At that point, the Lord told Joshua what to command the people to do. Joshua went from being their leader to being their commander. The Lord knew what lay ahead of them and the people would need a strong, courageous leader to get them through the battles.

There comes a point in time when the Lord steps in and exalts His leaders, teachers, commanders, whomever He has chosen for His purposes. Often times, we want to help God in this way by exalting ourselves. We may know in our hearts that we are called for a specific purpose and we even know that He has gifted us for the task, but no one else will respect or submit to that calling until God reveals it to them. The people needed to fear Joshua and respect his authority but only God could instill that in them. The Lord will do the same today for His people that He has chosen to lead others, whether in teachers, leaders, and so many other areas of service. Wait on the Lord to let others know what He is doing with you.

Have you been called for a role in leadership, maybe in your church or in an area of your workplace? Whatever the role and wherever the place; if the Lord has placed that on your heart and is leading you in your role; trust Him to give you the authority to do the job. Be careful not to push your authority on others or do things to promote yourself. When the Lord exalts you, it is for His purpose and in His timing. People will see you differently, not because of you, but because of the Holy Spirit revealing those things to them. Take comfort in what God did with Joshua and be of good courage in all that the Lord is leading you to do.

September 19

Today's Reading: Ecclesiastes 1- 3; 2 Corinthians 11:16-33
Today's Thoughts: Things That Matter

***Then I looked on all the works that my hands had done And on the labor in which I had toiled; And indeed all was vanity and grasping for the wind. There was no profit under the sun.* Ecclesiastes 2:11**

It seems that time is passing so fast; birthdays come faster and even the holiday season comes sooner every year. But in reality, time is set at a fixed and constant rate and does not change. So I am not getting older faster than I used to, and holidays do not come sooner than they used to, but rather, time is passing at a constant speed. Why does time seem to be passing too fast?

The author of the book of Ecclesiastes had come to the point in his life in which he understood just how little time we really have. In the first chapter, the author identifies himself as David's son, the king of Jerusalem. He also calls himself the Preacher and he spends twelve chapters preaching on the meaning of life—the *meaningless* of life. His main point centers on how much time we spend on things that are "vanity." The word *vanity* means emptiness, unsatisfactory, and vapor. Too often, the things we chase after in life are done in vain, and end in emptiness. How many of us look back on things we chased in life and see the worthlessness of the time wasted and the end result? How many of us would love the chance to go back and do certain things over again? Let us go forward with a new mind, set on the things that matter.

Today, take time to stop, slow down and find the real meaning in your life. Our Savior was born to give the only gift that matters. A gift that is not bought, earned or labored for; it is totally free. Jesus has given the gift of everlasting life to all who believe in Him. In this gift, however, there is so much more than just what awaits us in eternity. Our gift starts today. Jesus promised that He would not leave us alone. He has given us His Holy Spirit who lives within us; and because of Him, we have love, joy and peace. Take the time to worship and praise Him for all that He has given you.

September 20

Today's Verses: Ecclesiastes 4-6; 2 Corinthians 12
Today's Thoughts: Friendship

Two are better than one, Because they have a good reward for their labor. For if they fall, one will lift up his companion. But woe to him who is alone when he falls, For he has no one to help him up. **Ecclesiastes 4:9-10**

We spoke at a church retreat where the theme was *Friendship*. Everyone who attended received a small card that had a picture of two little girls and the words, "Two are better than one...for if they fall, one will lift up his companion." It was very sweet and very cute, just like we expect little girls to be. However, shouldn't this verse apply to adults too? Running a large weekly women's study, we know that many come looking for friendship. But, as adults, friendships are not defined nor identified in the same terms as when we were young. We grow up, we grow distant, and we get too busy to spend time with, or even maintain, friends. That is not how Jesus intended friendships to be. He knows that we need each other, in times of prosperity and in times of failure.

Jesus had a way of putting things in perspective. When the disciples asked who was the greatest in the kingdom of heaven, Jesus showed them the little children. As adults, we need to become childlike, not childish, with others. Children do not worry about the boundaries or the real motive behind every spoken word. They just want a friend, someone to hang out with. Jesus also talked about friends. He said that a friend is someone willing to lay down his life for another. Maintaining friendships as adults is difficult because true friendship does "not [seek] its own" desires (1 Corinthians 13:4); true friendship is selfless and giving.

Jesus is our perfect example of friendship. He laid down His life just for us, because He is our true Friend. Today, we can take comfort in knowing that no matter what happens, Jesus is always with us. Do you have a friend or companion who is there for you? Regardless of how alone we may feel at times, we need to take heart in knowing that Jesus is always our Friend, and He is the best one of all. He will never leave us nor forsake us.

September 21

Today's Reading: Ecclesiastes 7-9; 2 Corinthians 13
Today's Thoughts: The Cross is Enough

"for all have sinned and fall short of the glory of God"
Romans 3:23

Where do you find a group of people who openly admit that they are sinners but do not go to church? Traffic School. A few years ago, I had the opportunity of being with a group of people who openly admitted their sins to each other as we went around the room explaining what *wrong* we had committed. These people may not have actually called themselves sinners but they all knew they were *wrong* with something in their driving. By showing up for traffic school, we were protected from insurance premium increases, points placed on our licenses and the DMV. While at traffic school, the instructor (who was a professional comedian—after all it's hard to admit you are wrong) sets the class up in three ways: what you want to know, what you need to know and what you should know in order to bypass future consequences. By the end of the day, I was filled with so many statistics of future repeat occurrences that I fully believed that I would break the law again—somehow, someway, some day. What are their motives for inducing such fear, guilt and inadequacy? Well, in this traffic school class, the organization was selling "lifelong memberships" to traffic school. I watched the whole class buy a membership, and then, he looked at me. And I said in front of everyone, "Is this legal? You are *soliciting a profit* off of our sins?" They are right that I will break the law again (it does not take much to realize that); however, their *intent* was wrong.

Man's ways are not God's ways. God demonstrated His love towards us in that while we were still sinners, Christ died for us. Jesus' motives and intentions are pure and selfless. He went to a cross, innocent and blameless, to become sin for us so we can be made righteous by His blood. Jesus paid the price of a debt we could never afford to pay. Every day we need to come to the cross of Christ to acknowledge our sins before Him. We need to confess them and believe that His sacrifice is enough to cleanse us of our sins. Every day we need to come to His feet and beg Him to use us and convict us of the wrongs we do. Every day we need a new filling of the Holy Spirit so that we can be used by Him to bring others into His freedom from old habits and behaviors. God has a plan for you. Jesus went to great lengths to help you to believe, for no other reason and for no personal profit, but because of His love for you. He was crucified and you are cleansed. Will we continue to sin? Absolutely. But do those sins now separate us from God? Never. All He wants from us is to believe. Do you believe? Is Jesus enough? Is Jesus all you want? Today, pray that God will open the door for you to share with someone else the hope you have found in Jesus.

September 22

Today's Reading: Ecclesiastes 10-12; Galatians 1
Today's Thoughts: His Everlasting Love

The LORD appeared to him from afar, saying, "I have loved you with an everlasting love; Therefore I have drawn you with lovingkindness." Jeremiah 31:3

God is love. You cannot separate Him from love and He works in our lives because of His love. Many of us know that God is love, but how many of us live like we believe it? Fear, doubt, guilt and thoughts of self-condemnation are all factors that wage against the love of God. So each time that we worry, we are not resting in the love of God. Each time we feel fearful regarding our finances or health, we are not resting in the love of God. What does it take to accept that God loves us unconditionally so that we can rest in His love?

For me, I understood my need for a savior far greater than my need for love. I was raised in a church that reminded me of my unworthiness weekly. Jesus was identified dying on the cross because of my sins much more than He was identified with the hope of His resurrection. As a result, I expected God to be angry with me each time I opened the Bible. Reading verses such as, "I have loved you with an everlasting love," seemed foreign, as if God wrote them to the patriarchs and not to me.

However, as I continued reading the Bible, I realized that these "beloved" patriarchs were very much sinners also. They each had shortcomings and each made mistakes. Slowly, I began to realize that God knows and accepts my sinful ways better than I do. That's why He made a way through Jesus to save me....because He loves me. The love came first!

God loves you so much that you cannot do anything to separate yourself from His love. He understands your struggles and limitations. But through it all, He wants you to find rest as you accept His love and love Him back. After all, 1 John 4:19 says, "We love, because He first loved us."

September 23

Today's Reading: Son of Solomon 1-3; Galatians 2
Today's Thoughts: Your Home and Family

"For this reason I bow my knees to the Father of our Lord Jesus Christ, from whom the whole family in heaven and earth is named"
Ephesians 3:14-15

I am going home for a few days to see my family. As I say the words "home" and "family," I catch myself wondering what my definition of these two words really are to me today. Technically, I have three families: one in my own home, the second in the family I grew up, and third is the family of God. The same context applies to with the word "home." For some of us, home is somewhere away from the family and vice versa. Confusing? Now let's add what Paul says about our "family." We are all one big family in heaven and earth named by our heavenly Father. Our real home awaits us in heaven. Until then, we all seek and search for that loving embrace and acceptance from our earthly homes and families.

So, what is the point? For me, every time I "go home," I end up dealing with certain emotions, feelings and thoughts. Some are pleasant and some are not. Death, divorce and disease are just a few of those unpleasant ones that have touched my family. However, there is also that family connection God has put within all of us of a desire to spend time together and have fellowship. For me though, I am most thankful for the family of Christ. As I look around my life today, I realize that the friendships I have represent the family that God has given me. I am so very thankful for the love and fellowship in the body of Christ.

Even if our earthly families are not the most stable and strong, we can be encouraged to know that our heavenly family is one in the Lord. It will not matter when we get to heaven who did what to whom or how much pain was involved. Our heavenly Father will take care of us all and home will be with Him. The old saying says that, "home is where the heart is," I believe that to be true. If my heart is with Jesus, then there my home is also, not just with Jesus in heaven but here on earth with His family. I love and appreciate my family very much but I take much comfort in knowing that I have the greatest family of all in the body of Christ.

September 24

Today's Reading: Son of Solomon 4-5; Galatians 3
Today's Thoughts: Are You So Foolish?

Are you so foolish? Having begun in the Spirit, are you now being made perfect by the flesh? **Galatians 3:3**

There are times that we all get discouraged when we try so hard in our homes and in our work. We want to do the best we can. We want to be giving and loving in our efforts. We want to know that we are pleasing the Lord. We equate blessings with God's favor, thinking that it is our motives and actions that are being rewarded. So when things don't go well, we wonder what we are doing wrong. It is a common misconception, but God doesn't work like that. His grace is not gained by our good behavior. Because we are under the covenant of grace, we are called into rest. That is why it is so important to hold on to His promises. He is always working in us and with us. His plan is that we almost become passive in the process. We get tired of trying as we get caught in the middle between His promises and our efforts. Our hearts long to spend time with the Lord but there are always too many other factors competing for our attention. These factors take away our energy, and at times, make us feel like we can't do it all or we didn't do a good job anyway.

If God has given you promises, then He is the One who is responsible to fulfill them. God does not need our help or efforts. What He has started in the Spirit will not ever get accomplished in our flesh. It is the enemy who tries to discourage you. Satan will put thoughts in your mind of doubt and self defeat. Out loud, tell those evil forces that God has a plan regardless of your abilities. You choose to rest in His promises instead of worrying and doubting God by looking at yourself. Tell the Lord that you are choosing to focus your eyes on Him. And then, get into His Word and claim those promises. His Word puts you on the Rock, and our only responsibility is to stand firm on that Rock.

September 25

Today's Reading: Son of Solomon 6-8; Galatians 4
Today's Thoughts: Song of Love

***Where has your lover gone, most beautiful of women? Which way did your lover turn, that we may look for him with you?* Song of Solomon 6:1**

While reading the Song of Solomon, I can almost blush. The book is so intense regarding the extent of a love between a man and woman. There are times that the dialogue between the man and woman shift to the woman and her friends. Both sets of conversations focus on love, not the kind of love we associate with Jesus, but sexual love. The Song of Solomon talks about breasts and passion. I read the verses almost forgetting that I was reading the Bible.

I have heard scholars say that the Song of Solomon should not be in the Bible because Solomon's counsel about women is wrong. After all, he had 700 wives and 300 concubines and his behavior was certainly not biblical. They continue to say that Solomon's attitude toward women turned him away from the pure worship of God, so his counsel is as a heathen not as a believer. I have a difficult time with this thought because God would not have allowed it to make the cut into His Holy Word if He did not value the book. I have heard others say that the Song of Solomon represents the love that Jesus has for the church. He loves us as His bride and this love is a passionate and a self-sacrificial love. Personally, this is difficult to accept because the book is about sex. Can we equate Jesus' love for the church as sexual?

I believe that the Song of Solomon is as inspired by the Holy Spirit as every other book in the Bible. I also believe that Jesus loves us more than anyone else can, and we can draw similarities between His love for the church as His bride. But I also believe that the sexual love between a man and woman in marriage is godly. Sex was in the Garden before sin. God made a woman's body in such a way that Adam wanted her sexually. We sometimes spiritualize our desires too much and presume that sex, passion or the instincts of the body are sinful. Not true. Sex between a man and woman is normal and spiritual. It is that union that keeps them one and brings them together in intimacy.

God is not afraid of our sexuality. He wrote a book about it—the Song of Solomon. If you have not read it, try to read it without blushing. But while you are reading it, allow the Lord to minister to you in the areas that you need Him to intervene in your opinion and views about your sexuality. Do not hide these thoughts from Him; He already knows them. Include Him and He will do amazing things because He loves you.

September 26

Today's Reading: Isaiah 1-2; Galatians 5
Today's Thoughts: Take Time to Stop

Be still, and know that I am God; I will be exalted among the nations, I will be exalted in the earth! **Psalm 46:10**

"Be still" is a phrase often commanded by parents. Children have an abundance of energy that seems to have no limits at times, regardless of where they may be. The command to "be still" usually comes when either the energy level has reached a fevered pitch or when the environment is not conducive to an overabundance of activity, such as riding in a car or sitting in church. However, getting a child to be still from just speaking the words is most often a futile attempt at best; sometimes, more serious action is required.

As children of God, we are not so different. We tend to lead overactive lives filled with busyness and activities that seem to have few boundaries themselves. The difference is that, as adults, we have no one telling us to "be still" and slow down. We think the busier we are, the better we are for us and for those around us. Most of us even feel better just by staying active and over stimulated. Our society not only endorses this behavior but also has, for the most part, created the standards for it. How many of us would like to be known as weak, feeble, or slack? We most likely would be insulted by such words. But God says that we are to "be still," which is defined in Hebrew with such words. God is the One who tells us not to put anything above or before Him in our lives, even our energy levels and activities.

Just as with a child, sometimes our attention must be captured by God using more drastic actions. I heard the Lord telling me to slow down but it took physical changes in my life before I began to listen and obey. All God wants is more of us. He wants to be with us and He wants our undivided attention when we are with Him. It is a command to "be still," to be weak and to be at rest, before our God. Take time today to stop and be still before the Lord. Pray that you can "know" that He is your Lord and Savior and that you can be with Him in stillness and peace. If you hear Him telling you to be still, then listen and obey and do not wait: start today.

September 27

Today's Reading: Isaiah 3-4; Galatians 6
Today's Thoughts: Do Not Give Up

And let us not grow weary while doing good, for in due season we shall reap if we do not lose heart. Galatians 6:9

Are you feeling "weary" these days? Do you ever feel as if "doing good" really gets you nowhere? I bet that if we were honest, most of us would admit to these feelings or thoughts at some points in our Christian life. If you spend much time at all reading Paul's letters in the New Testament, you hear him telling us to persevere, to run the race with endurance, and to not lose heart. The Apostle Paul understood the trials of suffering and how hard it can be just to keep going, day in and day out. Today's verse has a promise at the end, one that we can all claim in victory. If we do not grow weary and give up, there will be a time of reaping, a season of harvest.

The enemy will use every scheme he can against us to discourage us from moving forward. I wonder how many times we give up at the last minute. What would happen if we never gave up? When the Lord calls you to step out, you must go. Once you make that decision, you will be tested. You will be tempted to quit, especially in your thoughts. To combat the temptations to quit, we must stay in the Word of God to renew our minds and to surrender to His will over our natural-born desires. Through the Holy Spirit, we can overcome and live a life pleasing to God. Praise God, for He wants for you to overcome and live a life pleasing Him.

God is leading you, you must follow Him, step by step. Know that there will be times when you will feel weary and your heart wants to sink, but that is when you turn to the Lord for strength and encouragement. Do not give up! The victory may be a day away.

September 28

Today's Reading: Isaiah 5-6; Ephesians 1
Today's Thoughts: Here am I! Send Me.

Also I heard the voice of the Lord, saying: "Whom shall I send, And who will go for Us?" Then I said, "Here am I! Send me." Isaiah 6:8

The prophet, Isaiah, was blessed with an incredible experience...he found himself in the throne room of God. He saw the Lord sitting on a throne, high and lifted up and the train of His robe filled the temple. He saw glowing celestial beings, or seraphim, around and above the throne. And then, Isaiah heard the voice of the Lord say to him, *"Whom shall I send, And who will go for Us?"* Without hesitation, Isaiah shouted out, "Here am I! Send me." His life would be forever changed.

Isaiah had a tough ministry as a prophet. The people did not want to hear what he had to say. His prophecies foretold of destruction and doom. Why would anyone want to hear such dreadful predictions about their ultimate desolation? Isaiah did not have a glamorous life and he did not die in an honorable manner. Jewish tradition says that he was sawn in half when he was about 90 years old. He answered the call of God but the call on his life brought much pain and suffering. For people today to answer the call of God there must be an awareness and acceptance of the pain and suffering that comes with the call. It is a tough place to live, especially in a world engrossed in comforts and pleasures.

Many Christians today are afraid to say, "Here am I! Send me." They are afraid to surrender and to really take up their cross and follow Jesus. The price is high...very high. When the flame starts getting too hot to handle, they step back and cool off. I know because I have been there. So, why be an "Isaiah" if we do not have to be? We can be saved, live a nice life and wake up in heaven. Why go through all of this other stuff? There is only one reason, one answer, to that question: to have the experience of standing before the throne of God and hearing His voice. To experience His presence in such an intimate way that you feel as though you are melting at His feet. To see His glory and majesty as Isaiah did. Once you have tasted and seen, you will never ask "why" again. The pain and the suffering cannot compare to the glory of the Lord.

But first you must be willing to step out and be sent. Are you ready? Do you want more? If so, then start praying. Read the books of Isaiah and Jeremiah to learn more about their lives and how God called them. You will never know all God has for you until you let go and ask Him to send you out. As Christians, this is our great commission.

September 29

Today's Reading: Isaiah 7-8; Ephesians 2
Today's Thoughts: Hearing God's Voice

And when they say to you, "Consult the mediums and the spiritists who whisper and mutter," should not a people consult their God? Should they consult the dead on behalf of the living? Isaiah 8:19

We are doing a study right now on hearing God's voice. We didn't realize how many misconceptions there were and how much fear exists in the lives of the people who say that they hear the Lord. The fear mostly deals with wondering if it was really God who spoke. There is a definite problem if we question hearing God's voice because we would end up more double minded than if we didn't consult the Lord at all. If we just do what we want to do, we deal with the consequences of our own decisions and actions and don't have to get God involved. When we get God involved, even if our interpretation is wrong, we end up having an issue with Him.

When I see a large group of women all struggling with the same thing, I know that the enemy is having his way. Why? Because satan tries his hardest to distract the believer and make them doubt their faith. There is not very much victory in the church today, so if a group of people are dealing with the same issues without victory, satan is involved. How do we resolve the problem of doubt and fear when hearing God's voice?

The verse above shows us that God's people have been struggling with the same thing for a long time. It is easier to consult mediums and spiritualists who really don't know than to consult the Lord. A medium or spiritualist whispers and mutters and focuses on the dead while a Christian should be all about life. But to hear God's voice takes time, effort and the fighting of the flesh. To know the will of God means that we need to get into the Scriptures and acquaint ourselves with the character of God as well as our own desires before Him. Then we need to lay down our desires so that the Lord can work on our hearts from His perspective instead of our own. That's how His desires become ours. When we are not threatened by God's ways, we are able to hear His voice. We have to remember that God is on our side. We are His children; He does not want us walking around confused and frustrated. He wants us walking around sensing His love and His life. If you are a Christian who struggles with knowing the voice of God, talk to Him about it. Tell Him that you don't want to consult anyone or thing else because His will matters most to you. He is not going to reject a prayer like that. Jesus said that His sheep hear His voice and follow. Be sure that you are listening because you do hear if you have come to know Him. He will not lead you astray and His will is not to confuse you. Ask, seek and knock and you will find God. That's a promise.

September 30

Today's Reading: Isaiah 9-10; Ephesians 3
Your Thoughts on Today's Passage:

October 1

Today's Reading: Isaiah 11-13; Ephesians 4
Today's Thoughts: Repentance is First

***This is what the Sovereign Lord God, the Holy One of Israel, says "In repentance and rest is your salvation, in quietness and trust is your strength, but you would have none of it."* Isaiah 30:15**

Jesus says His yoke is easy and His burden is light. He has not called us to a life of burdens but to a place of peace. Our salvation is not found in a list of works or good behavior. Strength is not found through building our bodies or having all our bases covered. Over and over, the Lord calls us to a place of repentance and to rest in Him. "In repentance" means that we are in agreement with Him that our ways are not right. We fall short of His standards. Repentance also means to turn from our ways as we accept Jesus and accept that His sacrifice on the cross is all we need for salvation. There is nothing we can add. This leads to the next part of the verse: "rest" in His work. We develop inner strength by our stillness of soul (quietness of mind, emotion and will) as we are convinced that He will help us (trust).

Living out this verse goes completely against our nature. We want to do and to fix. We find comfort in feeling that we are in control. It is hard to accept that we have nothing to offer. That is why the verse concludes with, "but you would have none of it."

Isaiah 30:15 is a great verse to test your faith. Where are you today? Why do you do certain things? Is your mind at rest? Do you have a stillness of soul because you trust the Lord for your circumstances today? This is not easy. But that is why a repentant heart is needed every day, for we do fall so short of His standard.

Meditate on it: "Take My yoke upon you and learn from Me, for I am gentle and lowly in heart, and you will find rest for your souls. For My yoke is easy and My burden is light." Matthew 11:29-30

October 2

Today's Readings: Isaiah 14-16; Ephesians 5:1-16
Today's Thoughts: That you May Believe

And he who has seen has testified, and his testimony is true; and he knows that he is telling the truth, <u>so that you may believe</u>. John 19:35

Over 333 prophecies were fulfilled during the life of Christ. Twenty-eight of those prophecies were fulfilled while Jesus was on the cross. In this Scripture passage, John points out three times that certain events occurred for the Scriptures to be fulfilled, "so that you may believe."

In the Christian life, our salvation (our faith) is based upon one thing: our belief that Jesus Christ came to earth as the Son of God, fully God and fully man, and that He gave His life as a ransom for us. He was not just a good man or prophet or teacher—He was God incarnate. Jesus began His journey to earth back in the Garden. God told us that her Seed (the woman) would crush his head, the head of Satan. The rest of the Old Testament foretold of His coming and God's people knew to look for Him. Yet, when Jesus showed up on the scene, many of them did not believe. In the course of three years, He did more miracles than could be recorded. But still, many did not believe.

Jesus fulfilled the prophecies, just as had been told. He died on a cross for the sins of humanity and opened the way for all who look upon Him and believe in Him, to have everlasting life. But, He did not stay dead—He arose on the third day. Jesus conquered death so that we may live. One day the rest of the prophecies will be fulfilled, just as the Bible foretells. One day, Jesus will return and will reign as the King that many of these people were looking for when He came the first time. Until then, we wait and we eagerly look for Him to return. Do you believe?

The Bible was written so that we may believe. The Gospel of John was written by the disciple whom Jesus loved, so that we may believe. Regardless of life's trials and hardships, do not allow the hardness of this world to weaken your faith. Look up! Look to the Cross and be saved. Sometimes we need to be reminded to look up and see Jesus. He is with us right now—He is with you right now. Just believe.

Pray about it: Lord, I will never know how much it cost to see my sin upon that cross. Help me to never take for granted the forgiveness You have extended to a sinner like me. Help me walk in the freedom You gave to me as a result of Your perfect sacrifice.

October 3

Today's Reading: Isaiah 17-19; Ephesians 5:17-33
Today's Thoughts: Five and Two

***Then He took the five loaves and the two fish, and looking up to heaven, He blessed and broke them, and gave them to the disciples to set before the multitude.* Luke 9:16**

One day I was walking my dog and as I looked down at the ground, I saw a nickel and two pennies. At that moment, all I could think about was five and two. I picked them up and walked home. That evening, as I opened my daily Bible, the verses recounted the story of Jesus feeding the five thousand—with five loaves and two fish. I knew at that point that the Lord had a message for me personally, but what exactly was He saying? The numbers five and two kept coming to my mind.

Just recently, I was reading this same story, and once again the numbers five and two jumped out. At lunch on that same day, someone left a nickel and two pennies on the soda machine. Okay, Lord, what does this mean? Well, the message began to unfold in my heart and mind as I continued to mediate on the Word of God. The disciples gave Jesus all that they had, which was not enough; two fish and five loaves could not feed five thousand men. But, Jesus took what they had and did three things with it. He looked up to heaven and blessed it; He broke it; and He gave it back to the disciples to hand out. The men did nothing with the fish and the bread but handed them to Jesus. Jesus performed the miracle. Not only did they have enough to eat, but also they had twelve baskets left over—one for each disciple.

The nickel and the pennies reminded me that Jesus takes what seems so insufficient and works miracles in ways that I cannot. It is not up to me to make things work; it is up to Him. When I have less than enough, then He has an opportunity to be more than enough. When I hand things over to Him, He blesses, breaks and then hands back, in abundance. Isn't it interesting that when He broke the bread, it multiplied. We need to remember that sometimes when things seem to be breaking, maybe that is because Jesus is trying to bless and multiply His provisions for us. Ask the Lord today to help you see His provisions and may He bless you in abundance to meet your needs and desires.

October 4

Today's Reading: Isaiah 20-22; Ephesians 6
Today's Thoughts: Worth the Cost

And on the second day, at the banquet of wine, the king again said to Esther, "What is your petition, Queen Esther? It shall be granted you. And what is your request, up to half the kingdom? It shall be done!" Then Queen Esther answered and said, "If I have found favor in your sight, O king, and if it pleases the king, let my life be given me at my petition, and my people at my request. **Esther 7:2-3**

Rarely do we see such sacrifice as today's verses demonstrate. Queen Esther had hidden her true identity from her husband, the king. He did not know she was a Jew, especially in lieu of his signed proclamation that would lead to the annihilation of the Jews living in his providences. For Esther to make this petition, she had to be willing to lay down her life. Not knowing the outcome, she stepped out to do what she believed was their only chance for survival, risking her own life in the process. The result was better than expected. Her petition was granted, the enemy of the Jews was hanged and the Jews were saved. Truly, this story is inspiring, amazing and encouraging—do we not wish all endings were so happy?

How often do we find ourselves in such a situation? Maybe not in the same context as Esther, but we are challenged in society today to take a stand for what we believe. Jesus laid down His life for us. He said that a true friend lays down His life for another. But, how many of us live out that type of sacrificial commitment? Esther interceded on behalf of her people, the Jews, many of whom she would never personally meet. Jesus did the same for us. Today, we are called to bear witness of what Jesus did for us. For God so loved the world that He gave His only begotten Son that whosoever believes in Him should not perish but have everlasting life (John 3:16). Our job is to tell the world about Jesus, even if it costs us our lives.

Because of Jesus, we are not only offered "up to half the kingdom," but also we have been given the keys to the whole kingdom. We are heirs to the kingdom of God; we will inherit it all. Wow, what a thought! What else could be worth more than our eternity with Jesus? May we keep an eternal perspective. May we be willing to lose it all to win souls to Christ.

October 5

Today's Reading: Isaiah 23-25; Philippians 1
Today's Thoughts: Pursue Peace

***Seek peace and pursue it.* Psalms 34:14**

My computer crashed one night and oh, the feeling of anguish and loss that came over me. My husband and I tried hard to work on it, reboot it and reinstall the system, but to no avail. I was reading and studying Psalm 34:14 when the computer went down, "Seek peace and pursue it." Obviously the verse and my state of mind were not in the same place, so I started reasoning with the Lord over it.

To seek peace means to search for it like you are looking for a lost child. Have you ever had that happen to you? One minute your child is right next to you and the next minute, gone. Oh the feeling of panic that overwhelms you as your heart starts beating faster, a burning sensation moves throughout your body, your eyes become fixed and your voice raises as you firmly and seriously call out your child's name. Nothing is right until you find your child. That state of mind is not very peaceful and yet we are instructed to seek peace like we are seeking a lost child. What does that mean?

A missing child isn't normal for the parent as much as missing peace isn't normal for the Christian. Jesus left us His peace as if He left it in His will for us to claim, own and enjoy. However, it doesn't come naturally. Jesus says that He gives us His peace not the world's peace. If we don't have peace, we need to find out why and then pursue it.

I pursue peace by meditating on Scripture, and thinking on the verse over and over. I pray the verse back to the Lord asking Him to reveal His will for me from the verse. So when the computer crashed, I felt my peace being traded in for panic. Because the contrast was so evident, I sought for my peace first and then focused on the computer. You have to put yourself in that place to claim your rights to God's peace regardless of the circumstance at hand. Peace is yours. Seek for it, pursue it, claim it and own it!

October 6

Today's Reading: Isaiah 26-27; Philippians 2
Today's Thoughts: Name Above All Names

Therefore God also has highly exalted Him and given Him the name which is above every name, that at the name of Jesus every knee should bow, of those in heaven, and of those on earth, and of those under the earth, and that every tongue should confess that Jesus Christ is Lord, to the glory of God the Father. **Philippians 2:9-11**

Have you ever been in a position where you were the only Christian? Possibly now, you may be in the situation that your co-workers, neighbors or family members speak against Christians continually and frequently use His name as a curse word. What are you supposed to do? Our tendency is to try to maintain a position of peace (and so silence), especially where our paycheck is involved. We tend to back down, keep silent and ignore the comments. These actions reveal a person who is powerless and ashamed, even if that is not our intent. We should never appear powerless or ashamed of our Lord!

We need to stop and consider Who Our Lord Jesus is. One day all will bow before Jesus and confess that He is Lord. Every tongue, tribe and nation will be on their knees giving glory to God the Father, just by speaking the name of Jesus Christ. That day is coming and what a glorious day that will be! There will be no more denying Who Jesus is and no more discriminating against those who boldly speak His name. There will be no more hiding, embarrassment or guilt, only praise and pride in praising our Savior. And shame and sorrow, "gnashing of teeth," to those who did not praise Him now.

Most importantly, stop and consider Who He is to you personally today. Should anyone have the right to discredit Your Lord and Savior at any time or anywhere? Paul tells us in Romans 1:16 that he was not ashamed of the gospel of Christ for it is the power of God unto salvation for everyone who believes. By succumbing to fear and acting ashamed of the gospel, we fall right into the pit of the devil. Why? Because our power is stifled. To **not** be ashamed allows the power of God to be released through your words. We represent life. Those who are criticizing represent death. We have the responsibility and we have the power to share the gospel of Christ.

Today, ask the Lord to forgive you for not speaking up and for not speaking out for Him. Today, as you wait for Him to open the door, ask the Lord to give you an opportunity to speak about Him. Today, ask the Lord to teach you to be bold for Him because you do not ever want Him to be ashamed of you. And today, thank the Lord for saving you, forgiving you and giving you His Holy Spirit so His will.

October 7

Today's Reading: Isaiah 28-29; Philippians 3
Today's Thoughts: Hiding from God

***Woe to those who seek deep to hide their counsel far from the Lord, and their works are in the dark; they say, "Who sees us?" and, "Who knows us?"* Isaiah 29:15**

I watched my dog hide a bone the other day. Most dogs bury bones outdoors in the dirt, but my dog buries his bones in the house, preferably under my pillow on the bed. Fortunately, his "bone" is a kind of chewy rawhide so at least there is less mess to deal with (unless of course he has brought his "bone" once buried in the dirt into the house to then re-bury in the bed, not a pretty picture). The funny thing is that unless he knows I am watching him, he acts as though he is doing it in secret. In other words, if he does not *see me*, then I must not be seeing him. And that is how we act with the Lord at times. We somehow think that because we cannot see Him we can hide things from Him as if He were not watching.

God knows everything about us. He is omniscient (all-knowing), omnipotent (all-powerful) and omnipresent (ever-present). He has no boundaries or limits. He has known us from the beginning of time and already has seen our last day here on earth. We cannot hide from Him. Instead of trying to hide from God, we need to be open and honest with Him. Things done in the dark will ultimately come under His light. Others may never need to know what we have done, but Jesus always knows and He will eventually expose the truth to us. Why? Because He knows that unless we are made aware of these sins, we will continue down a path of destruction. Our flesh seeks darkness, not the Light. John 3:19 says, "And this is the condemnation, that the light has come into the world, and men loved darkness rather than light, because their deeds were evil." Jesus came as the Light to set us free from the darkness. Why would we want to stay in it?

Are you trying to hide from the Lord today? Are you engaging in activities of which you think He is unaware? If you are a child of God, then you know your sin because the Holy Spirit will convict you and impress you to turn and repent. The enemy will lie to you and tell you that you are hidden and no one will ever find out. But God already knows and you cannot live with the deception and condemnation. You will slowly self-destruct in the darkness. God's love beckons you back today. God's mercy and grace promise that He will forgive. You can start fresh just by asking for His light to shine on you and lift the darkness. Take the steps of obedience to turn and walk towards the Lord. Put away the sins done in secret, and embrace the freedom that comes when you throw off those things that bind you. Just ask for His help. He is waiting and longing to be gracious to you.

October 8

Today's Verses: Isaiah 30-31; Philippians 4
Today's Thoughts: Giving to God First

Not that I seek the gift, but I seek the fruit that abounds to your account. Indeed I have all and abound. I am full, having received from Epaphroditus the things sent from you, a sweet-smelling aroma, an acceptable sacrifice, well pleasing to God. And my God shall supply all your need according to His riches in glory by Christ Jesus. *Philippians 4:17-19*

We quote Philippians 4:19 frequently to ease our concerns about insufficient finances as we say, "I know that God will supply all my needs according to His riches in glory by Christ Jesus." However, did you know that the context of this verse is about giving, not receiving? According to the context of this Scripture, Paul was saying that we will receive from God if we are willing to give to God first. Paul was thanking the Philippi believers for giving to him. Paul also said that he had learned to be content in all circumstances and he was thankful that they gave, so that fruit may abound on their behalf.

Why is it so difficult to give first? How can we live with an eternal perspective when it comes to money? Because we need to pay real life monthly bills, make actual house payments and physically handle money, it is difficult to believe that our money will go the extra distance by giving tithes and offerings. We can give away older things to replace them with newer things, although that kind of giving tends to be to our personal advantage. We can give to our families and even to our own social life, but to tithe regularly seems impossible. The word "tithe" means a tenth or 10% and God specifies that the amount is subtracted from our first fruits (our gross salary). God also talks about "offerings." An offering is in addition to the tithe. This seems like it would require too much faith to even begin giving.

But God asks us to give so He can bless us, to give cheerfully, not with clenched fists. If we open our hands freely to the work of God, we are in a position that we can freely receive from His open hand toward us. Corrie ten Boom said, "I have learned to hold on to things loosely because it hurts when God pries them away." The interesting thing about clenched fists towards God is that we can never protect what we have from God any way. Remember: God will supply all of our needs when we supply to His first.

Oh Lord, help me to give to You without my left hand knowing what my right hand is doing. Let me be a blessing to You first and I will trust You to take care of me.

October 9

Today's Reading: Isaiah 32-33; Colossians 1
Today's Thoughts: Knowing the Real Jesus

***Now I rejoice in what was suffered for you, and I fill up in my flesh what is still lacking in regard to Christ's afflictions, for the sake of his body, which is the church.* Colossians 1:24**

My children took their achievement tests this past week. My daughter told me that there was a Bible achievement test also but the school opted to not have the children take it. I asked her if they told her why. She said yes and then gave me a sample question of what was on the test. She said, "Which characteristic does not belong to Jesus?

      a.   happiness

      b.   love

      c.   forgiveness

      d.   compassion

She told me that a question like that confused a lot of sixth graders because they equate Jesus with joy and their interpretation is that joy and happiness are the same things. I told her that a lot of adult Christians have the same interpretation. It is difficult for us to follow a King that was despised and rejected by men, a Man of sorrows and acquainted with grief (Isaiah 53:3). We don't want to believe that the Christian life is filled with those same sorrows and afflictions at times.

However, it is during those sad times in each of our lives that we understand the afflictions of Christ as He becomes our Sustainer, our Strength and our Peace. It is at those times that we discover joy from the depths of our hearts instead of happiness that is only superficial, changing with the circumstances. Our Christian life is not only about receiving from the Lord but also about enduring until the end.

If you are struggling with sorrow and grief, know that the Lord understands. Continue to come to Him. There is hope in every situation for we know the Lord will cause all things to work together for good. Just fix your eyes on Jesus.

October 10

Today's Reading: Isaiah 34-36; Colossians 2
Today's Thoughts: The "Why's" of Life

And at the ninth hour Jesus cried out with a loud voice, saying, "Eloi, Eloi, lama sabachthani?" which is translated, "My God, My God, why have You forsaken Me?" Mark 15:34

Isn't it wrong to ask God "why?" Isn't it wrong to question God? One day I was reading this verse, and suddenly I realized that Jesus asked God "why?" While Jesus was on the cross, dying for the sins of the world, He asked God, "Why have You forsaken Me?" We all know that God did not forsake Jesus. We also know that Jesus never sinned, so asking God "why" does not have to be wrong. Just by including the Lord in our thoughts shows a step of faith. It shows faith because we are praying, acknowledging that He is listening, and God will reward our faith.

By being honest with the Lord, our hearts are softened to receive His wisdom involving the things of the past, while helping us to move forward in faith for future decisions. God tells us in James 1:5-8:

> *If any of you lacks wisdom, let him ask of God, who gives to all liberally and without reproach, and it will be given to him. But let him ask in faith, with no doubting, for he who doubts is like a wave of the sea driven and tossed by the wind. For let not that man suppose that he will receive anything from the Lord; he is a double-minded man, unstable in all his ways.*

So the "why" questions are not the sin, but receiving an answer to those questions and then doubting the answer is the sin.

Today, start asking "why" about things you do not understand and ask the Lord to help you to go forward in faith. Ask the Lord to give you wisdom in how to receive His answers to the "why" questions in your life. Pray for faith, as we know that without faith, it is impossible to please God. As our faith grows, we will move forward to make decisions as the Lord leads us and we will trust in His guidance and direction for all things. The goal is to get past the "why's" and move on to the "what's"—trust, obey and get going.

October 11

Today's Reading: Isaiah 37-38; Colossians 3
Today's Thoughts: Not Our Favorite Word

***Wives, submit to your own husbands, as is fitting in the Lord.*
Colossians 3:18**

The word "submit" is not one of our favorite words in the English vocabulary. I actually heard that on certain computer programs and websites, the word "submit" will be removed (*Submit* is usually the button clicked on a website when someone agrees to a purchase, accepts contractual terms for a software product or approves other related entries to get to the next link). For wives, the word "submit" is often scoffed at. The thought of it makes many of us downright angry at times. Why is it so hard for us to submit, especially to our husbands? Oh, there are lots of reasons...too many to list here.

Here is the key for us to get past this issue: we must place our feelings about it before the Lord and ask Him to help us. Why? Because the Bible tells us to submit. But, let me comment on a couple of notes about this. First, submission involves respect and honor, not a doormat mentality. Second, submission is a place of strength, not weakness because it gives God the power to work on our behalf. Our first submission is to God, then the rest will follow. The enemy will blind you with your own pride and selfishness to keep you from submitting to the Lord. Our pride keeps us from humbling ourselves and serving others, even our own husbands. Let the Lord work these issues out with you and you will see the power of God displayed in amazing ways in your life.

I learn something every day about submission, whether I like it or not. Just the other day, my husband asked my help in something and then demanded of me what he wanted me to do. With this specific task, I was truly clueless. After unsuccessfully pleading my case with him, I submitted and vowed to do whatever I could to help him. Behind the scenes, I begged the Lord to help me. I repented of my rebellious heart and I gave up the fight. I saw the Lord not only rescue me from the task, but also show my husband what to do to find the answers himself. It is simply amazing to watch the Lord move like that! For me, my heart was tested. Was I willing to humble myself and submit? This time I did...but believe me, I have to pray and ask for the Lord's help every single time I get into these struggles with submission. God will get the glory as we submit to Him first and then ask Him for His help towards others.

October 12

Today's Reading: Isaiah 39-40; Colossians 4
Today's Thoughts: Patience with Perspective

But those who wait on the LORD shall renew their strength; they shall mount up with wings like eagles, they shall run and not be weary, they shall walk and not faint. Isaiah 40:31

Are there ever times when you become tired of trying? From work to your family, the list of needs seem endless with all fingers pointing to you. You know the promises of God and know that He is absolutely working in your life, but the "flesh" part of you becomes worn out and on edge.

The other night, at dinner time, I was trying to cook…and help with homework…and talk to my mother who was visiting. In addition, the phone was ringing, the news was on and my youngest child was singing for everyone to hear. I could sense the tension rising within me. The noise seemed too loud and the night was becoming too long. Because I recognized these feelings and knew how poorly I have handled them before, with all the self control I could muster, I prayed to the Lord and said, "Please help me do all this with an attitude and with words that would please You." Immediately I sensed the Lord say to me, "Enjoy it; find peace right now." That thought was the furthest thing from my mind but those words pierced my heart.

It just doesn't matter if everything gets done on time or in the right order. It just does not matter if there are grease stains on my daughter's homework paper and there are dishes left in the sink when I go to bed. What does matter is that I honor the Lord by pouring out His love through me especially in the times that push and pull on my nerves. What matters is that I find the joy of the Lord in everyday challenges and appreciate every passing moment of the day. It is in these most challenging times, when you feel as if everyone is looking to you, that you can only endure by continuing to look to the Lord. God hears and answers the smallest of prayers when they are dripping with honesty. He knows us and understands us. Today, when the anxiety is increasing and your patience is decreasing, turn to the Lord so that you can soar above the circumstances while renewing your perspective and strength.

October 13

Today's Reading: Isaiah 41-42; 1 Thessalonians 1
Today's Thoughts: Consider Your Calling

Paul, called to be an apostle of Jesus Christ through the will of God, and Sosthenes our brother, To the church of God which is at Corinth, to those who are sanctified in Christ Jesus, called to be saints, with all who in every place call on the name of Jesus Christ our Lord, both theirs and ours. 1 Corinthians 1:1-2

What do you picture when you hear the word "saint"? A deceased person, a statue, candles and incense? Years ago, my definition would have been, "One who is worthy to be prayed to." I also associated sainthood with death, as if it was an oxymoron to be a "living and breathing saint." But in Paul's opening to the Corinthian church, his definition is very different. First, he begins this letter by addressing himself and bringing attention to his calling as an apostle. An apostle means "a sent one." Paul then goes on to say that he was "sent" to the Corinthian church to remind them of their calling—<u>sainthood.</u> The Corinthian church seemed far from sainthood. As a church they struggled with pride, immorality, idolatry, envy, jealousy—just to name a few. Paul knew what they were like, which is why he wrote them this letter. So why did he tell them they were called to be saints?

According to the Bible, a saint is one separated from the world and consecrated to God; a believer in Christ (Psalm 16:3; Romans 1:7; Romans 8:27; Philippians 1:1; Hebrews 6:10). Becoming a saint starts right here on earth, the minute you accept Jesus as your Savior and Lord. Mans' ways of statues and candles and incense or even good works are not God's ways of being set apart to do His will. Paul was reminding the Corinthians that they are now saints because of Jesus, but they were not acting according to their calling. Their lifestyle contradicted their declaration of knowing, loving and living for God.

This contradiction happens to us too. We become saints because of believing in Jesus' work on the cross and then we have the ability to live for Him by the filling of the Holy Spirit. Our sainthood does not rest in our works but in our belief. Our beliefs should represent the way we live, not inconsistent like the Corinthians.

Today, ask the Lord if your calling in Christ (being called a saint) is consistent with your every day life. Are you living every day to please Him? Are your priorities consistent with a life of holiness? Do you really understand your calling? Consider the areas in your lifestyle that you know are not pleasing to the Lord. Then write a prayer specifically asking Him to help you be who He wants you to be.

October 14

Today's Reading: Isaiah 43-44; 1 Thessalonians 2
Today's Thoughts: Trust the One Who Knows You

But now, thus says the Lord, your Creator, O Jacob, And He who formed you, O Israel, "Do not fear, for I have redeemed you; I have called you by name; you are Mine!" Isaiah 43:1

God knows us by name. He calls us and says that we are His. We need to trust Him and give Him our hearts and lives. We need to be willing to come before Him, knowing that He alone cares about every hair on our head and every desire of our heart.

My husband and I went car shopping. While looking at a different make of car than either of us have ever owned, a salesman approached us. When he caught our attention, He called to us each by name. We had never been in this lot before and we had no idea how this man knew us. He then asked my husband about his job by title and place. Next, he asked about my husband's old car.

Finally I said to him, "How do we know you?" He replied, "Your last car, you bought from me." That was seven years ago! He then told us the kind of car we traded in for the new one. We were amazed. We literally met that man one time and it was to buy a car that same day. The car transfer was smooth, nothing out of the ordinary for him to remember us. I asked him, "How do you do that?" He said that he consistently and actively works on his memory.

I have to say that it was very eerie. My husband and I talked more about him than the car for which we were looking. When someone takes that kind of time to remember who you are, what you do and what you buy, you feel special but also humbled. We want him to sell us our next car, mostly because he gained our trust and respect.

This man remembered certain things about us but God knows all things about us. For some reason though, we do not give God our absolute trust. We second guess Him and wonder if He knows best. The Lord tells us that we are inscribed on the palms of His hands (Isaiah 49:16). His love for us is immeasurable and He knows every hair on our heads. He calls us His own and calls us to Himself. Oh, we need to be in awe of the Lord and completely trust Him. We need to cast ourselves down at His feet and be willing to let go. We need to thank Him and be all that He wants us to be. He not only knows our names but also He knit us inside our mother's womb. Life should be all about Him but for some reason, we make Him all about us. Come to Him with a heart of repentance and receive from His lot of love. He knows your name as well as your every thought and you can trust Him.

October 15

Today's Reading: Isaiah 45-46; 1 Thessalonians 3
Today's Thoughts: A Message from Isaiah

I will go before you and make the crooked places straight; I will break in pieces the gates of bronze and cut the bars of iron. I will give you the treasures of darkness and hidden riches of secret places, that you may know that I, the Lord, who call you by your name, am the God of Israel. Isaiah 45:2-3

One morning a few years ago, I woke up with the word "Isaiah" in my mind. I personally do not know anyone named Isaiah so I wondered why this name was on my heart. As I opened my Bible that morning, I literally turned to the first page of Isaiah. I immediately began wondering what message the Lord might have for me in this book. Since then, God has given me more messages from the book of Isaiah than I can count, and Isaiah now holds some of my most treasured sections of Scripture. One of my personal favorites is chapter 45.

When I first read that the Lord will make my "crooked places straight" and will "cut the bars of iron," I knew these were messages of freedom. Then to hear that the Lord calls me by my name, this made the message even more personal, as if He was speaking to me directly. I have come to understand that He *is* speaking to me directly. I took Him at His word and I began to pray:

> *Lord, please go before me and straighten out those parts of my life that are out of order or where I do not see the path before me. And please break the gates and cut the bars that keep me locked up, that have held me captive. I desire those treasures You have for me in the freedom that You want to give me. You are God and there is none other. I need You to be my God today and forever.*

These verses are just a snapshot of this great book. Through Isaiah I learned how to pray God's Word back to Him and through praying God's Word, I have seen His personal touch on my life in more ways than I could have ever imagined. Maybe today is the day for you to open your Bible to the book of Isaiah. Who knows how many messages God has for you there? He knows and you will soon find out. May God go before you and bless you.

October 16

Today's Reading: Isaiah 47-49; 1 Thessalonians 4
Today's Thoughts: Living through the San Diego Fires

When you walk through the fire, you shall not be burned, Nor shall the flame scorch you. Isaiah 43:3

During the worship session at a retreat, I showed Bobbye the verses from Isaiah 43:1-3. Well, the very next song that the worship leader played quoted the very same passage of Scripture in Isaiah. We knew that the Lord was getting our attention. God was speaking to us through the Scriptures. After the worship was over, Bobbye and I were next on the schedule to speak. We shared with the ladies the passage in Isaiah 43, knowing that "By the mouth of two or three witnesses every word shall be established." We did not realize the significance these verses would soon have. This was one of those times that we had ears to hear but it was difficult to discern if the message was literal or figurative. We all seemed to be content with believing that there are no coincidences with God and He uses people to speak and confirm His Words to others, through others. No one that I knew of, including myself, asked the Lord to clarify this message any further. In hindsight, we should have been asking, seeking and knocking to not only know His Word but to also know His Will. Why was He was giving these verses to us? Are the verses for our spiritual well being or are these verses God's Will to be accomplished on earth as it is in heaven…today? The reason for the verses became perfectly clear as we headed down the mountain.

On Sunday morning, the Santa Ana winds were so strong we had difficulty packing up the car. Little did we know that **Isaiah 43:2 was not only a promise but a prophecy** that was about to be fulfilled that day in all of our lives. Heading down the mountain, we heard that the fires had begun. We were about to be evacuated from our homes because of the fires and some of us would literally be "walk[ing] through the fire" as Isaiah 43:2 clearly states. At one given time, there were eight fires destroying the County of San Diego. Because fire is no respecter of persons, no one knew whose house would be next and no one could predict the path the fires would take. But we did know that God cared and that God had warned us up on the mountain. His desire was for us to know that He was with us, that these fires did not surprise Him, and He did not want us to fear. God speaks to us to comfort us and to bring peace regardless of the circumstance. The problem today is not that God is not speaking but that we are not listening. God wants us to know. He reveals His plans to His servants (Amos 3:7). God is willing to speak but are we ready to listen?

October 17

Today's Reading: Isaiah 50-52; 1 Thessalonians 5
Today's Thoughts: You Are Irresistible!

***And Stephen, full of faith and power, did great wonders and signs among the people. And they were not able to resist the wisdom and the Spirit by which he spoke.* Acts 6:8, 10**

Have you ever been with someone or just watched a person who was irresistible? Some people have a charisma that attracts others to them and the reasons for the attraction are not always obvious. I remember taking a Bowling class in college to fulfill one of my elective course requirements. On the first day I thought the instructor was a bit over-zealous as he enthusiastically explained the course curriculum. What is there to get excited about in a Bowling class? But to my surprise it was not long before I found this class to be a refreshing break and actually looked forward to going to it. Why? Not because I love bowling, but because the instructor shared his joy in such a way that I found myself attracted to his energy and enthusiasm.

We need to realize how often we affect those around us just by our expressions and attitude. It is amazing to think about how contagious our attitudes and behaviors can be to others. As Christians, we are to be lights in this dark world. We are called to share the good news of Jesus Christ and to proclaim His love, joy and peace to everyone. If we truly have His Spirit, as Stephen did, then we will be irresistible to those listening to us.

Think about your behavior and attitude today as you go about your activities, especially how you express yourself. A smile goes a long way, and there is something contagious about laughter. Before you start your day, stop and ask the Lord to fill you with His wisdom and Spirit. Ask for His joy to be your strength and for His peace to prevail over all of your circumstances. And, if you keep smiling long enough, eventually people will want to know your secret. Then you get to tell them all about Jesus because He is truly the only reason any of us can smile and have real joy.

October 18

Today's Reading: Isaiah 53-55; 2 Thessalonians 1
Today's Thoughts: God's Ways Can be Hard to Understand

"For My thoughts are not your thoughts, Nor are your ways My ways," says the Lord. "For as the heavens are higher than the earth, So are My ways higher than your ways, And My thoughts than your thoughts." **Isaiah 55:8-9**

I learned these verses in my early years as a Christian. I have heard them quoted often, and I have spoken them many times, either to others or to myself. I am not always sure if they are comforting words or if they are confounding words. Intellectually, I know that God's thoughts and ways are not like mine. Of course my ways are not like God's, God is God and I am not. But when I go down a path that seems to make so much sense to me, then out of nowhere, the course dramatically changes: I must admit that I am more confounded than comforted. There are times when we pray for things or people and our prayers are answered just as we prayed them. Then, there are times when we pray and our prayers seem to go unheard, with no answer at all. We can usually handle both extremes because God is God and He can act as He pleases

But what about those times when we pray about something specifically, and we clearly see God's hand in it? We pray, God leads and we are getting it. We can even see what is coming as God reveals things to us. Our faith is being increased and we are finally gaining some spiritual strength and understanding. Even though the course is tough, we are making steady progress. But... then... out of the blue, something happens that changes our course so drastically that we feel the air has been knocked out of us. All of sudden, we find ourselves in a place of desolation and despair. What happened? It seemed like everything was moving along so well.

It is in these moments that we all must glean a deeper understanding that God's thoughts and ways are not ours. God does not play by our rules and God is never in a rush. Keep one thing in mind today as you meditate on these verses. Our Lord loves us more than we can ever comprehend and everything He does is in our best interest. His main desire for us is that we are His, totally and completely. Do you feel as though God has abandoned you or maybe has just forgotten about your needs? Ask Him to help you understand more of His ways. Ask Him to open your mind to comprehend the Scriptures. Go to Him, repent if you have un-confessed sin and trust that His higher ways will always be what is best.

October 19

Today's Readings: Isaiah 56-58; 2 Thessalonians 2
Today's Thoughts: Highest of Heavens and the Humblest of Hearts

For this is what the high and lofty One says--he who lives forever, whose name is holy: "I live in a high and holy place, but also with him who is contrite and lowly in spirit, to revive the spirit of the lowly and to revive the heart of the contrite." **Isaiah 57:15 (NIV)**

How can it be that God dwells in the highest of the heavens but also in the humblest of hearts? How can the God of the Universe, who created all things and by Him everything exists, want to be a part of us?

From the beginning of creation, God created man to work and tend the Garden of Eden while in close relationship with Him. God then formed the animals and brought them to Adam so he could name them. God wanted Adam to have a partner, so He created Eve. Over and over in Genesis, we read about God's love, concern and willingness to have a relationship with man as well as God's desire to care for his needs. There was only one rule to follow. Everything else was burden free. God knew that man needed to have free will, for God wanted a relationship based on choice, not force. With the freedom to choose, Adam and Eve chose to disobey God. From that time moment on, Adam and Eve died in two ways: physically in time and spiritually that day. Sin entered into the heart of man that day and severed our relationship with Lord. Today, we are all born spiritually dead.

This broke God's heart. He created us to have a relationship with Him; He created us to dwell with Him. From Genesis chapter 3 until Revelation 21, the Bible is about God's plan to get the world back, reversing the destruction of sin. The answer was prophesied in Genesis 3:15 as being Jesus. Jesus is the One who came to restore sinful man back to a Holy God. John 3:16 says, "For God so loved the world that He gave His only begotten Son, that whoever believes in Him should not perish but have everlasting life." The wages of sin is death but Jesus came to bring life so we can live abundantly.

So whose heart does God dwell with today? He dwells with those humble of heart. He dwells with those who prayed and said,

> Lord, I know that I am a sinner and in need of a savior. I know that You sent Your Son to die on the cross, shedding His blood for my sins. I ask You, Jesus, to come into my heart and save me. I surrender control of my life to You. Please fill me with Your Holy Spirit and lead me in Your ways. Thank You for saving me."

October 20

Today's Reading: Isaiah 59-61; 2 Thessalonians 3
Today's Thoughts: Blessed Assurance

And having a High Priest over the house of God, let us draw near with a true heart in full assurance of faith, having our hearts sprinkled from an evil conscience and our bodies washed with pure water. **Hebrews 10:21-22**

The book of Hebrews instructs us to live by faith in God alone. The originator of our Christianity is Abraham, the father of faith. Because Abraham believed God, it was counted unto him as righteousness (Romans 4:3). His faith made him "right" before God. As a result of this righteousness, God protected his family as they grew into a nation, the Israelites. Moses was raised up as their first official leader to establish their worship, culture and Laws of the land. God spoke to Moses as he wrote down God's ways for the people, including the Ten Commandments. These rules were given to instruct the Israelites in how to live. These rules were enforced to protect them from walking away from God and to protect them from destroying themselves (through transmission of diseases) and each other. These ways were not given to replace "faith."

However, by the time Jesus came to earth, the rules were valued more than faith, mercy and kindness. The rules became more important than the people. For God so loved the world that He sent His Son, Jesus, to restore what was lost. Because of the love of God, our High Priest died on the cross and rose from the dead to allow us to live by faith, not by works.

Faith is not established by obeying the rules or the Laws of the land. Galatians 2:16 says, "a man is not justified by the works of the Law but through faith in Christ Jesus." Jesus did not abolish the law; He fulfilled it so that we can be accredited as righteous ("right" before God) through faith in Him alone. Christianity started through Abraham's acts of faith, and through our High Priest, we overcome the world through our faith. Jesus alone is the Author and Finisher of our faith (Hebrews 12:2). Today, let us draw near to the Lord with a true heart in full assurance of faith, knowing that we have a High Priest who understands our sufferings, forgives us of our sins and cleanses our conscience.

October 21

Today's Reading: Isaiah 62-64; 1Timothy 1
Today's Thoughts: God's Messengers Today

"Behold, I send My messenger, And he will prepare the way before Me. And the Lord, whom you seek, Will suddenly come to His temple, Even the Messenger of the covenant, In whom you delight. Behold, He is coming," Says the Lord of hosts. Malachi 3:1

A messenger simply stated is someone bringing a message to someone else. God used various means to get His message out to His people. In the Old Testament God used angels, prophets and other entities as His messengers to speak the words of God to the people. If we look more closely at the role of a messenger, we see that the message usually included one of two themes: encouragement or warning.

When the angel spoke with Hagar in the wilderness, he encouraged her to keep going and that Ishmael would be blessed. Haggai was sent to encourage the people to return to work on the temple. Jeremiah was sent to warn the people of what would happen if they did not turn from their evil ways. Moreover, we see how the prophet Malachi had a message of strong warning and rebuke to the priests and their attitude towards the things of God.

Is it not amazing that the Lord goes to such measures to make sure we get His messages? But, are we taking them to heart today? We often read of the Old Testament prophets and their predictions to the people and wonder why the people did not listen or take heed the message. We see clearly how the people were sinning against God and how their hearts were in rebellion, but seemed blind to the reality of their impending destruction. Are we so different today? We may not have Jeremiah or Malachi in physical form, but we have the Word of God. In fact, our accountability today is much greater because we do have the Bible.

When Jesus ascended into heaven, He gave us all a charge, a call to action. He said to go and make disciples of the nations. He told us to be His witnesses throughout the world. Being a messenger of the Lord is no easy task. The prophets of old had hard lives and most of them died without seeing the people change. It can be a thankless job. Jesus never promised that standing for Him would be easy. But, He did promise that He would always be with us. He also said that the Spirit would give us words to speak when we did not know what to say. Are we willing to be His mouthpiece today? Are we willing to stand for Him and speak what He tells us to say? Or, are we afraid of what others think and will say about us in return?

October 22

Today's Reading: Isaiah 65-66; 1Timothy 2
Today's Thoughts: Learning From Your Haircut

"I have stretched out My hands all day long to a rebellious people, Who walk in a way that is not good, According to their own thoughts" Isaiah 65:2

I went to a new hair stylist this past week. Have you ever had that feeling of vulnerability when meeting someone new who has scissors in his or her hands? I was anxious, but this stylist's reputation put my fears a little more at ease as I waited to get in the chair. To break the ice, he joked that he wanted to cut my hair really short. Everyone in the salon laughed including me, the "first time visitor" joke. However, by the time he was finished trimming my hair, he told me that my hair *should* be cut really short. It is a terrible feeling to get your hair cut just the way you requested, then have the person who cut it tell you that your hair would look better another way.

I left relieved and confused at the same time. Relieved that he did what I asked of him but confused that he did not like what he did. After talking and praying to the Lord about this experience, I came to the conclusion that it is a matter of professional opinion and training on the part of the hair stylist, but more of a matter of convenience and comfort on my part. There is no right or wrong answer, just preference.

With our issues in life, we pray and ask God for His guidance, but then, we go and do what we want. That works fine when you sit in a salon but not when dealing with the ways of God. We need to listen and obey what He says. God addresses many of these issues very clearly in the Bible. God's ways are not a matter of opinion or a strong suggestion; they are commands. We have to remember that those commands are written for our protection. Our Lord cares about everything we do and every detail of our lives (even how we get our hair cut). Regardless of the issues in your life, big or small, read God's Word every day for clear instruction and ask Him for guidance. You may not always agree with your hair stylist but that is okay. Agreeing with the Lord, however, brings a peace and joy that cannot even begin to compare to the best of haircuts.

October 23

Today's Reading: Jeremiah 1-2; 1 Timothy 3
Today's Thoughts: Worship Sets Us Free

And when they had laid many stripes on them, they threw them into prison, commanding the jailer to keep them securely. Having received such a charge, he put them into the inner prison and fastened their feet in the stocks. But at midnight Paul and Silas were praying and singing hymns to God, and the prisoners were listening to them. Suddenly there was a great earthquake, so that the foundations of the prison were shaken; and immediately all the doors were opened and everyone's chains were loosed. **Acts 16:23-26**

Worship sets people free. How much do you worship? When do you worship? Are you able to worship when your circumstances are the very worst, your patience has failed and your confusion is maxed out? Do praise songs come to your mind at the most stressful times? We need to assess our ability to worship because worship will change our lives. We need to train ourselves to worship in all circumstances.

Paul and Silas were beaten, thrown in the inner prison and their feet were fastened in stocks. They were probably in too much pain to sleep, so at midnight they broke out in worship. They did not have a worship leader, instruments, power point, or a radio to sing along to. They probably could not even tap their feet to keep the beat. Paul and Silas may not even have had good singing voices. But from the wellspring of their hearts, their mouths worshipped the Lord despite their circumstances. As a result, other prisoners listened to them and even better, everyone's chains were loosed. Worship set prisoners free.

In many ways, you may feel like a prisoner. You may have habits that you cannot break, you may find yourself in circumstances that have not changed, you may feel that God isn't answering your prayers. What are you to do? Worship. Worship when you are sad, feel bad, and after you were mad. Worship every day, at work and at play, worship any way. Worship will change your life and if you sing loud enough for others to listen, your worship may set them free as well.

Lord, teach me to worship. Lord, bring songs to my mind so that worship can be my way of life. Change me through worship and allow me to see the fruit of worship in the lives of others around me.

October 24

Today's Reading: Jeremiah 3-5; 1 Timothy 4
Today's Thoughts: The Valleys of Life

They went up over the mountains; They went down into the valleys, To the place which You founded for them. **Psalm 104:8**

In a physical definition, a valley is a low area between two hills, which can range from a few square miles to hundreds or even thousands of square miles in area. It is typically a low-lying area of land, surrounded by higher areas such as mountains or hills. It can also be a path between two mountains, or a *depression* in a single mountain. It is interesting to note that part of its physical landscape is a "depression." Metaphorically, we often compare the valleys of life to the more depressing times in our lives. Why do we associate valleys with dark, deep crevasses that we need to dig ourselves out of? Why do we associate the level plains of our days as times of depression?

We tend to long for the mountain top experiences of life. But even if we find them at times, we do not live there consistently. Let's face it, we all love our mountain top experiences with God, but they are usually moments in time. We spend most of our days in the valleys. It is a fact—we live in the valleys, day in and day out. And if the greatest amount of our time is spent in the valleys, we need to discover how the Word of God is a "lamp unto our feet and a light unto our path" as we walk through them, holding God's hand every step of the way. We can find great victory in the valleys when we keep our focus on the Lord and follow where He leads.

If you are in a valley today, pray this prayer:

> *Dear Lord, I know that You have created the valleys and the mountains. I know that You are with me and that You will never leave me. But, this place that I am in is dark, lonely and scary. I need Your touch, Your hand to guide me, and Your peace to surround me. Please help me to understand all that You have for me in this place and give me the strength to walk through it. Help me to wait on You and to not lose hope. I need Your help today. In Jesus name, Amen.*

October 25

Today's Reading: Jeremiah 6-8; 1 Timothy 5
Today's Thoughts: What Have I Done?

Why has this people slidden back, Jerusalem, in a perpetual backsliding? They hold fast to deceit, They refuse to return. I listened and heard, But they do not speak aright. No man repented of his wickedness, Saying, 'What have I done?' Everyone turned to his own course, As the horse rushes into the battle.
Jeremiah 8:5-6

The prophet Jeremiah warned the people to turn from their wickedness and return to the Lord. But Jerusalem carried out their temple rituals, and because of that they really thought they were doing only good things, and serving the Lord. *"What have I done?"* They were deceived in their thoughts and were perpetually slipping farther away from God. This happens when we start to really believe that we are not that sinful, just because we are doing religious activities. We believe we are getting closer to God but we can actually move away from Him. And the farther we get from God, the harder it is to repent and change.

We live in a world that is filled with deceit and wickedness. The world tells us what is acceptable. If it feels good, then do it. It you feel bad for doing it, then take a pill or do something to make yourself feel better. We have permission from the world to do whatever it takes to make ourselves feel better. Yet, so often, we end up feeling much worse. We can swing like a pendulum, from the heights of happiness to the depths of depression. And if we are not careful, we wake up one day wondering what happened. *"What have I done?"*

The Christian walk takes one day at a time. Some days are better than others, but every day belongs to the Lord. The only way to walk with the Lord consistently is to spend time with Him everyday. When times are hard and depression creeps in, do not look for the instant fix offered by the world. Turn to the Lord. When you feel good about yourself or things in your life, turn to the Lord and give Him thanks. None of us wants to feel badly, but when the Holy Spirit convicts us of sin, we must be willing to repent and to turn from our sin. As we walk daily with the Lord, these convictions keep us from falling into deceitful ways of thinking. Ask the Lord today to help you answer the question: *What have I done?* You may be surprised at the answer. Listen to the Lord and be willing to change as He guides you.

October 26

Today's Reading: Jeremiah 9-11; 1 Timothy 6
Today's Thoughts: Fight the Good Fight

"Fight the good fight of the faith." 1Timothy 6:12

Timothy was a young preacher, mentored by Paul and used in a mighty way to spread the Gospel of Christ. Paul knew how much Timothy's faith would be, and was being, tested. He would have to "fight" to keep going or his faith would be shaken. In our world, the word "fight" is heard so often. We hear of fighting actual wars in numerous areas of the world and we fight battles in our own lives every day. As Christians, we may not always be aware of the battles that we are fighting. Sometimes they are manifested in arguments, disputes and areas of stress in our homes. Sometimes they are internalized battles that rage in our minds, frustrating our thoughts and emotions. Ephesians 6 speaks of the battle as a spiritual war, one that tries and tests our faith to the limit, and to the end of our lives. We must learn how to fight to stay in the faith or we will find ourselves desperately weak and shaken.

So, how do we "fight" this "good fight of the faith?" Faith comes by hearing and hearing by the Word of God (Romans 10:17). If we are not in the Word on a daily basis, then deceiving thoughts and feelings have more opportunity to take root within us. The enemy roams about seeking whom he may devour and we are more easily devoured when we are not standing on the promises of God. But not only do we need to read God's word everyday; but also we need to learn how to *use* it. We must learn how to *fight* with it.

I challenge you today to read the Bible with a different perspective. Take certain verses and keep them in your thoughts. Speak them against the enemy forces that may attack you. For example, if a thought comes to you like, "God really doesn't care about what I am going through— where is He?" Read Psalm 139 and speak out the verses. Verses 7 & 8 speak about how God never leaves us; no matter where we go, He is there. Verse 17 tells us how much God thinks about us everyday. Take these verses and fight back. Let the Lord strengthen your faith through His word. You will gain strength spiritually, even when the battles rage. Ultimately you will win in the end. Our Lord and Savior Jesus has already won the war for us and He will be with us in every battle we fight. Just remember to take Him with you.

October 27

Today's Readings: Jeremiah 12-14; 2 Timothy 1
Today's Thoughts: A Four Letter Word

***For God has not given us a spirit of fear, but of power and of love and of a sound mind.* 2 Timothy 1:7**

More than once since I committed this verse to memory I have had to embrace it as truth and pray that I not give in to a "spirit of fear." I have come to believe that the word "fear" really is a four-letter word and should be considered profanity. Wouldn't it be great if we could somehow have it censored from all of our thoughts and feelings, like bleeping out a bad word? But instead, fear is very much a part of our reality, even more so than we realize or want to admit. I know there are times when being fearful can be a good thing, such as being afraid of doing something that might bring harm or danger. My husband bought me a motorcycle. I spent a significant amount of time preparing to ride it, taking courses, getting my license, but the bike still sat in the garage. I was hindered with fear from riding it.

The spirit of fear as mentioned in 2 Timothy 1:7 is a different kind of fear. This fear hinders what God has given us: His power, love and sound mind. This fear grips us with doubts and insecurities, waging war with our thoughts and feelings. This kind of fear is not of God but of the enemy, the devil. Fear is one of the greatest weapons Satan uses against us, as it attacks our minds. He tells us that we are no good, not worthy, unloved and useless. These thoughts affect us all, he picks on everyone, no one is spared. How can we fight against such thoughts? By believing the truth of the second part of the verse. God through His Holy Spirit has given us power, love and a sound mind, not in our strength but in His.

Are you dealing with fears today that are gripping you so tightly that you feel helpless and hopeless? How can these fears be conquered? Start by praying for the Lord to open your mind and help you understand how to use the power given by His Holy Spirit. The power precedes love, so next pray for your heart to be filled with His love, knowing that His perfect love casts out fear (1 John 4:18). What a great pair...power and love! Then, ask the Lord to clear your mind and bring to your thoughts His words, taking captive those thoughts not of Him.

> O, Lord, I pray that You teach us how to live victorious lives through the empowering love and strength of Your Holy Spirit. And in Jesus name we ask that You break those strongholds of fear that bind us up and keep us from experiencing the peace and joy that You desire for us. Amen.

October 28

Today's Readings: Jeremiah 15-17; 2 Timothy 2
Today's Thoughts: Why We Read the Bible

Your words were found, and I ate them, and Your word was to me the joy and rejoicing of my heart; for I am called by Your name, O Lord God of hosts. **Jeremiah 15:16**

The other day, my husband was asked a question by a friend of his at church. The question took him off guard which is why he asked me the same question. The question: *Why* do you read the Bible? At first I thought, "What kind of question is that? We all know why we read the Bible." As believers, we must know this answer or at least have an answer...right? Isn't the answer obvious, or is this a trick question? But my husband just kept looking at me and waiting on me to answer. As I thought about it, I realized the answer was not so obvious, not so easy.

So, I took a moment and asked the Lord to help me with my answer, not to sound haughty, but to glorify Him in my response. The verse above came to my mind, one of my life verses. "Your words were found and I ate them, And [they became] to me the joy and rejoicing of my heart." My response was fairly short as I quoted this verse and said that without God's Word, I would die. His words are my daily bread, my living water, and the air I breathe. I cannot live without the daily sustenance provided through His precious, living word. And yet I find myself so often wanting more, being hungry for more of Him and knowing how desperate I am without His continual filling. In reading His Word, I find strength to persevere throughout the day as God's child. I find wisdom in dealing with daily decisions. I find joy that there is more than what this world has to offer. And I find peace because I know that someone really loves me, despite all my shortcomings. When I discipline myself to read His Word, I not only find answers...I find God.

How would you answer this question? Why do you read the Bible? Go to the Lord today to find your answer. Ask Him to feed you with His Word, to write it on the tablets of your heart, and to fill your heart with joy and rejoicing. If you have not opened your Bible in a while, start today...you may be surprised at what you find. And you never know when someone may ask you, "Why do you read the Bible?" Think about, and pray for, your answer today.

October 29

Today's Reading: Jeremiah 18-19; 2 Timothy 3
Today's Thoughts: A New Start

Therefore, if anyone is in Christ, he is a new creation; old things have passed away; behold, all things have become new. **2 Corinthians 5:17**

The word "new" strikes a cord with most of us. There is something inspiring about getting a *new* anything. Shopping is one of our nation's favorite pastimes, bordering on obsession. We love to buy new things, go to new places, eat at new restaurants—you name it, if it is new, then we are interested. And if all of this is not enough, we can go a step further and get a brand new makeover for ourselves. Today, a person has several options from which to choose for their new look, from whiter teeth to a whole new face. Plastic surgeons are more popular than ever as they give hope of not only a new look, but a new life. People today are desperately seeking something new and are willing to do anything to re-create themselves.

I keep thinking of the phrases "beauty is only skin deep" and "the new will wear off." The basic point in these sayings is this: surface pleasures that bring joy and excitement do not last as they eventually fade away. Our joy is not on the outside but on our inside. Our happiness begins within us. If we are not fulfilled in who we are in ourselves then we will not ever be able to buy enough new things or make ourselves look good enough to fill the emptiness. How do we find this inner joy? How can we find true contentment? Our verse today is one that should give us all great encouragement and hope. Anyone who comes to Jesus Christ is given a brand new start. "All things have become new." What an amazing promise!

Nothing we could ever buy could compare to actually being given a new start in life. Jesus promises us that we not only get a new start, but we become a new creation. We are born again in His Spirit (John 3:3-8) and are forever changed. "I, even I, am He who blots out your transgressions for My own sake; and I will not remember your sins (Isaiah 43:25). Once we accept Jesus Christ as our Savior, He wipes all of our sins away and we are given a new start with a new life. If you are in Christ, take hold of this promise today. Do not believe the lies of the enemy who will tell you that you are still the same old sinner. Praise the Lord for His mercies and ask Him to teach you day by day how to live as His new creation. Ask the Lord to help you let go of the past and start living for Him today. Go forward, not backward. The Holy Spirit will lead you.

October 30

Today's Reading: Jeremiah 20-21; 2 Timothy 4
Your Thoughts on Today's Passage:

October 31

Today's Reading: Jeremiah 22-23; Titus 1
Today's Thoughts: The Lesser Blessed the Greater

Then Joseph brought in his father Jacob and set him before Pharaoh; and Jacob blessed Pharaoh. Pharaoh said to Jacob, "How old are you?" And Jacob said to Pharaoh, "The days of the years of my pilgrimage are one hundred and thirty years; few and evil have been the days of the years of my life, and they have not attained to the days of the years of the life of my fathers in the days of their pilgrimage." So Jacob blessed Pharaoh, and went out from before Pharaoh. **Genesis 47:7-10**

Joseph brought his family to Egypt and introduced his father to Pharaoh. Pharaoh is the head guy...he is the President, the Commander in Chief, the main man. In contrast, Jacob is a wanderer, a sojourner, the father of shepherds and a very old man who had to leave his homeland because he and his family would have died of starvation from the famine. But why do the Scriptures say two times: "Jacob blessed Pharaoh?" How does that happen? How does the empty bless the full? How does the lesser bless the greater? How does the poor possess more than the rich? How does the average overrule the superior? The answer—GOD.

God uses the foolish things of the world to confound the wise. God reveals His will and His blessings to the childlike. God empowers everyday people to fulfill abundant promises without money, without power and without titles. The prayers and praises of the saints have higher ratings and rewards than the most powerful, self sufficient persons on earth.

We have our logic all mixed up down here. If we *really* understood the ways of the Lord, we would not be held down by difficulty nor be lifted up with pride. When we fix our eyes on the things that matter, we understand that Jacob's "evil" years of pilgrimage have an eternal value that weighs more than this nameless Pharaoh who lived a life of luxury on earth.

Do you bless others? You are a child of God, just like Jacob. Jacob blessed Pharaoh because of Jacob's position in the heavenlies. He blessed Pharaoh because Jacob knew what being blessed really meant. Jacob blessed Pharaoh and today, his life still blesses me. Jacob has done more for me than Pharaoh because Jacob's family brought to us the greatest blessing of all...Jesus. When we really know Jesus, we truly understand what it is to be blessed so that we can freely bless others. Ask the Lord to use you to bless someone today.

November 1

Today's Reading: Jeremiah 24-26; Titus 2
Today's Thoughts: He is Worthy

And I saw a strong angel proclaiming with a loud voice, "Who is worthy to open the book and to break its seals?" **Revelation 5:2**

Who is worthy? Who is worthy to do the work of God? Who is worthy to fulfill the plans of God? John tells us in verse 3 that, "no one in heaven or on the earth or under the earth was able to open the book or to look into it." When I first read that verse I thought, "Where is Jesus? Isn't He worthy?" But I was wrong in comparing Jesus to any other created being. Those who are in heaven, on earth or under the earth are all created by God. No created thing is worthy or equal to God Himself.

Next we find John weeping, filled with sorrow that no one could open the book. Do you frequently find yourself weeping over your own inadequacies as well as over the shortcomings of others? It fills our hearts with grief that we can't do all that needs to be done, be all that we know we have the ability to be, or serve the Lord with limitless devotion. We are limited by physical boundaries: our flesh, our sin and time. So when one of the elders said to John, "Stop weeping; behold, the Lion that is from the tribe of Judah, the Root of David, has overcome so as to open the book and its seven seals" (verse 5), we find such hope and joy that Jesus is not limited. He has overcome, and, through the power of the Holy Spirit, He is helping us to overcome our shortcomings also. Jesus is God and He is worthy to accomplish all the works of God.

We stand before the Lord as a saint only because the blood of Christ covers our sin. The Lord Jesus gives us, unworthy sinful man, the strength to stand before a Holy God. On earth, we too cry with John but one day, we will stop crying. The Lord Jesus will come to our side to say to the Father, "They are worthy—because I died for them."

Oh, trust the Lord today. Allow His power to flow through you. Tell Him you are willing to be that vessel for Him to accomplish all His works.

He is worthy to accomplish all you need, if you are willing to allow Him to work.

November 2

Today's Verses: Jeremiah 27-29; Titus 3
Today's Thoughts: Get God's Attention

For I know the thoughts that I think toward you, says the Lord, thoughts of peace and not of evil, to give you a future and a hope. Then you will call upon Me and go and pray to Me, and I will listen to you. And you will seek Me and find Me, when you search for Me with all your heart. I will be found by you, says the Lord, and I will bring you back from your captivity. Jeremiah 29:11-14

The verse above from Jeremiah 29 may be one of the most popular sections of Scripture within the whole Bible. For me personally, it was one of the first that I ever committed to memory. So often, there are times in my life when I just need to hear that the Lord is thinking about me and that He wants to give me "a future and a hope." Many times, I have difficulty grasping the truth of that statement. The God of the universe, the Lord of heaven and earth, actually has time to think about me? Is that really possible? Not only is it possible but also there is so much more: He wants us to call upon Him, pray to Him, seek Him, find Him, search for Him. God Himself wants us to *want* Him.

One morning I got up early to pray and knew I needed to spend some time on my knees, literally. As I positioned myself on the floor, my dog soon discovered where I was and proceeded to demand my attention. He began nudging me, licking my face and pressing his body into a position where I could no longer deny his presence. He was insistent to the point that I continued in prayer while petting my dog. I must admit he is irresistible in his persistence and extremely affectionate. In those moments, I sensed the Lord impress upon my heart that He was doing the same thing to me. I knew in that moment that the Lord wanted me to give Him the same kind of attention. He wanted me to know that His love is unconditional, irresistible and affectionate. And He just wants to be loved back. He wants my prayers, but He also wants my never-ending desire to seek and search for Him, to never settle when I may not sense His presence. He is always with me, but sometimes I need to press into Him and not give up until I get His attention.

Just like my dog knew how to get my attention, knowing that I would respond with love and affection, the Lord knows how to get our attention and our response should be the same. But we need to learn how to respond to Him. We need to call upon Him, seek Him and press into Him. We need to give Him our love and affection. He just wants our hearts but He also wants us to sense His nudging. The next time your child or your pet demands your attention, especially in the areas of love and affection, just remember that the Lord is seeking the same kind of attention from you. And no one gives it back better than He does.

November 3

Today's Reading: Jeremiah 30-31; Philemon
Today's Thoughts: Seeing Jesus

When they had lifted up their eyes, they saw no one but Jesus only. **Matthew 17:8**

Have you ever gazed upon something so spectacular that you lost yourself in the moment? The beauty of God's creation can bring such moments. I remember the first time I stood on the ledge of the Grand Canyon. I could barely fathom what my eyes were seeing. I could only thank the Lord, not just for this awesome sight before me, but for the blessing of being able to see it with my own eyes. In itself, the human eye consists of some of the most amazing biological and physiological designs in all of creation. The intricate details of how the eye actually works are of continual study. Despite man's ever-increasing knowledge and technology, we are still limited in understanding the depths of this incredible sense. Only God knows how our eyes work because He is the One who made them, just for us.

What will it be like on that day when we stand before Jesus and we lift up our eyes to see Him? How can such beauty and glory even be imagined? As the song says, "I can only imagine." On earth we are limited in these physical bodies. There are those who have no sight, having never seen anything through their eyes. Countless others will lose their sight for various reasons. And, of the rest of us, many will lose components of our vision, whether near-sighted, far-sighted or color blind. Age has a way of dimming the lights of life. But one day, those of us who know Jesus Christ as Lord and Savior will see "no one but Jesus only." In that day, there will be no physical limitations to keep us from seeing our Jesus, in splendor and majesty more breathtaking than anything we have ever seen on this earth.

Today, take a moment to thank the Lord for what you can see here on earth. Ask the Lord to keep you from seeing the things that do not glorify Him. Our eyes are the windows to the world and we need to guard what we allow them to look upon. Open God's Word today, then read and meditate upon the Scriptures. Then close your eyes and ask the Lord to show you what He wants you to see. Spend time in prayer and lose yourself by being in His presence. Turn your eyes upon Jesus today and let Him show you what He wants you to see.

November 4

Today's Reading: Jeremiah 32-33; Hebrews 1
Today's Thoughts: Hope for the Promises

"Behold, the days are coming,' says the Lord, 'that I will perform that good thing which I have promised to the house of Israel and to the house of Judah." **Jeremiah 33:14**

There are times when I wonder if God will ever answer certain prayers of mine. There are prayers that I pray knowing that the Lord has put them in my heart. I have verses of Scripture and biblical illustrations to support these same prayers. I keep journals that detail the dates and descriptions of the promises the Lord has given me. But still I wait and wonder, not about the *what* but about the *when* and the *how*. Where is my faith?

These are the times when I stop and reflect more closely at what I know about my God. I must remind myself of His character traits. Unlike me, He is perfect in every way—so perfect that He is not capable of doing anything out of His character. For example, one of God's character traits is love. 1 John 4:7-8 says that "love is of God" and that "God is love." Therefore, everything God does is motivated by love. If the Lord has me waiting and has not answered certain prayers, then His reasons for waiting are based upon His love for me. Another example of God's character traits is that He is truth, perfect truth. Titus 1:2 says God cannot lie. He is incapable of lying; therefore, everything He has promised me will come to pass. God keeps His promises. But I must remember that He controls the timing, the process and the outcomes, for He is God.

When I step back, I can understand more clearly than when I am in the midst of being anxious over things. We all have seasons in life where we wonder if, and when, God will answer certain prayers. It is in these times that I find myself being tested most in my relationship with the Lord. Do I really know Him and believe Him and trust Him? I go to His Word and start asking for help and comfort. I confess my lack of faith. I am honest about my anxieties and worries. Then, I know I just need to be still and know that He is God. Before long, I am praising Him and my hope has returned. Don't give up and don't give in to the temptation to be discouraged and dismayed. Remind yourself of just who God is, what His true character is, and let His Word fill you with hope and joy for the promises that He has given you.

November 5

Today's Reading: Jeremiah 34-36; Hebrews 2
Today's Thoughts: Our Great High Priest

Seeing then that we have a great High Priest who has passed through the heavens, Jesus the Son of God, let us hold fast our confession. For we do not have a High Priest who cannot sympathize with our weaknesses, but was in all points tempted as we are, yet without sin. Hebrews 4:14-15

For the Jews, there was no higher office than that of the High Priest. The High Priest held a position of religious supremacy and honor. He was the one responsible for the cleansing of sins for the people on the Day of Atonement. On this holy day, the High Priest would enter the Holy of Holies. Through the acts passed down from Moses and Aaron (animal sacrifices, the sprinkling of blood, and prayers offered on behalf of the people), the High Priest would make atonement for their sins. In the book of Hebrews, we learn how Jesus Christ came to earth as the final High Priest. Not only was He the High Priest, but also He was also the Sacrifice. His death on the cross ended the need for the man-held position of High Priest, sacrifice and the Day of Atonement. Jesus took care of it all.

Not only was Jesus the High Priest, the Sacrifice and the Savior, but also He was still a person in flesh and blood. For us today, this point is so very important to grasp. Because Jesus was here on this earth in a human body, He understands everything we go through. Jesus experienced feelings, hurts, weaknesses and temptations, just as we do. He knows what it is to suffer, to hurt, to be rejected, ridiculed and mocked. Jesus got angry at the Pharisees, He cried over Lazarus and He showed great compassion to a prostitute. But, Jesus was God. He did not sin. He was the perfect, spotless Lamb—the ultimate and final Sacrifice. Why did Jesus go through all of this for us? Love. He loved us then. He loves us now. He has always loved us. He wants us to be with Him, now and forever. The only way to be with Jesus is to believe that He is the Son of God, the Son who came to earth as a Man, and who offered His life on our behalf as the final atonement for all sin.

Take time today to start reading the book of Hebrews. If parts of it seem hard to understand, pray for the Holy Spirit to open your minds to comprehend the Scriptures. Pray and meditate on those verses that stick out, or the ones that seem the hardest to grasp. If you will commit the time to the study of God's Word, you will grow closer to Jesus in more ways than you can imagine. You will see Him as One who understands you better than anyone else and who loves you more than anyone else can.

November 6

Today's Reading: Jeremiah 37-39; Hebrews 3
Today's Thoughts: Get out of the Boat

***Jesus said to them, "Come and eat breakfast." Yet none of the disciples dared ask Him, "Who are You?" knowing that it was the Lord.* John 21:12**

Do you ever find yourself wanting to ask a question when in reality you already know the answer? Jesus had already appeared to His disciples several times since His resurrection. But the disciples were still waiting for direction on how to go forward. So in the meantime, some of them decided to go fishing, their occupation before they followed Jesus. On this particular morning, as they were returning from their fishing, they looked up from their boat and saw a Man cooking breakfast on the shore. They did not need to ask who it was; they knew it was Jesus. I wonder if that moment punctuated their last fishing trip, as Peter jumped out of the boat to make his way towards Jesus. He did not have to ask *Who* was waiting on him; he just knew.

At times in our Christian lives, we look back at those things we used to do. We may even question whether we should go fishing again. Maybe we keep going back to old ways and habits, even though we know we should have gotten out of that boat a long time ago. If Peter and John had not stopped fishing and had stayed with their boat, what all would they have missed? They left it the first time Jesus asked them when He said, "Follow Me, and I will make you fishers of men." (Matthew 4:19) Now, they are leaving their boat for the final time, never to return to their former lives again. From this moment on, they would live and die for the message of the gospel of their Lord Jesus Christ.

Maybe today you need to look up from your boat. How far are you from the shore? You are never too far away from Jesus. If you look for Him, you will find Him. When you see that Person who seems familiar, you really do not need to ask who it is. Jesus is always with you, wanting to feed and comfort you. Maybe He is asking you to get out of the boat and come towards Him. Maybe He wants to make you "fishers of men." Is it a "maybe" or a "definitely"? I think we all know the answer to that question too. Today is the day to stop asking and start knowing and get moving.

November 7

Today's Reading: Jeremiah 40-42; Hebrews 4
Today's Thoughts: What Is Your Choice?

"And now look, I free you this day from the chains that were on your hand. If it seems good to you to come with me to Babylon, come, and I will look after you. But if it seems wrong for you to come with me to Babylon, remain here. See, all the land is before you; wherever it seems good and convenient for you to go, go there." **Jeremiah 40:4**

The prophet Jeremiah foretold the coming destruction of Jerusalem. The people hated him for what he predicted, but the Lord's words were true. The day came when the Babylonians invaded Jerusalem and Jeremiah was taken captive along with many others. But Nebuzaradan, the captain of the guard in the Babylonian army, said to Jeremiah: "The Lord your God has pronounced this doom on this place. Now the Lord has brought it, and has done just as He said. Because your people have sinned against the Lord, and not obeyed His voice, therefore this thing has come upon you" (Jeremiah 40:2-3). Even the enemy invaders saw the judgment of God upon His people. Jeremiah; however, was given a choice. He could go to Babylon or stay with the people left in their land. Jeremiah chose to stay.

Today, we have a choice as well. We can go with the Babylonians and pursue the things of the world, or we can remain in the place where the Lord has placed us. Sometimes the land we are living in is filled with hardships and afflictions. Sometimes we are tempted to choose the things that promise relief, even when they may not be God's will for us. Choosing the harder path does not come naturally for us, but it is usually what is best in the end. We must remember that our time here on earth is but a vapor, a brief moment. But eternity lasts forever and ever.

One of God's greatest leaders boldly proclaimed in Joshua 24:15, "And if it seems evil to you to serve the Lord, choose for yourselves this day whom you will serve, whether the gods which your fathers served that were on the other side of the River, or the gods of the Amorites, in whose land you dwell. **But as for me and my house, we will serve the Lord**." Make your choice today to serve the Lord. Do not stop serving Him during the hard times. Hold fast and firm to the choice you make today. Our Lord is faithful and just to get us through to the end.

November 8

Today's Readings: Jeremiah 43-45; Hebrews 5
Today's thoughts: The Peaceful Fruit of Righteousness

***Now no chastening seems to be joyful for the present, but painful; nevertheless, afterward it yields the peaceable fruit of righteousness to those who have been trained by it.* Hebrews 12:11**

We know that the Lord disciplines those He loves but sometimes I wonder why He has to go to such extremes to love me so much? I have told the Lord that it is a good thing He told us that discipline is based on love because, at times, I would misinterpret the pain as something else.

I have noticed that through the process of chastening, I probably will not remember the circumstance that brought the trial, but I will remember the intensity of the pain. When God really gets a hold of me and starts working on changing me, I feel hallow and empty. During those times, I feel like a pumpkin. God cuts around my handle and scrapes me clean of all the seeds and stringy things. But I liked all those seeds, stringy things and even the dingy environment; at first, I grieve over missing the "old me". But then, He goes a step further and takes the knife of the Word and starts carving at my hardened protected outside. Now my insides become exposed to the outside, and light has the ability to shine through. So often, I can barely handle such exposure and I feel embarrassment and vulnerability no matter what angle the light hits. During these times I cry, yell, pray and beg Him to stop, not understanding why I have to go through this painful process. But in His perfect timing, I realize that the peaceful fruit of righteousness has resulted in me. Finally, I pray for His light to be placed deep within me so others can see His good work.

We need the Lord. There should be nothing else we want to do but to love Him with all that we are and to serve Him with all of our lives. As we continue to walk with Him, we better understand that His purpose for us is **best for us**. We are called to endure these trials of life as we are becoming more like Christ, and being transformed by the work of the Holy Spirit who lives in us. Jesus needs to become our everything. We need to ask Him to help us yield to His will in the discipline process and to resolve ourselves with His timing, being content with His will. Though the process is painful, He knows what is best and we can trust Him to finish the work He started in us.

Are you dealing with something painful today? Maybe you are wondering why God has put you in such a place. Regardless of the reasons for our sufferings, we must remember that He is our answer. Go to Him today and ask for His strength and guidance to lead you through this time. Begin thanking Him for His purposes in the work He is doing in your life. He will never let you down.

November 9

Today's Reading: Jeremiah 46-47; Hebrews 6
Today's Thoughts: Down With a Shout

And the seventh time it happened, when the priests blew the trumpets, that Joshua said to the people: "Shout, for the Lord has given you the city!" **Joshua 6:16**

The story surrounding the city of Jericho is amazing. The Bible tells us that the walls came down with a shout. Just imagine the fear that already existed within the people in the city. For six straight days, the Israelites arose early in the morning and marched around the city of Jericho. A regiment of armed guards in the front, the priests carrying the Ark of the Covenant in the middle, and another regiment bringing up the rear marched around the city, never saying a word, just blowing their trumpets. Then, on the seventh day, they did it again, but not once, but seven times! And on the seventh time, at Joshua's command, they shouted with their voices...and the walls came down. What an awesome display of God's power!

We learn a lot from this story and this verse. We learn that God is sovereign and has a plan for all things. We learn that God's ways are not our ways and that His works seem mysterious at times. We learn to trust Him, obey Him and wait on His timing, despite what we think. We learn that the battle is His, not ours. We may want to use all of our weapons and all of our own strength, but God can strike the enemy with one word or one shout! We learn that we serve an awesome God, for He is God alone and there is none other. Let Him take down your walls today. Follow Him, trust Him and obey Him. Stop looking for worldly proof and evidence, just put your faith in His Word and start marching to the beat of His drum.

November 10

Today's Reading: Jeremiah 48-49; Hebrews 7
Today's Thoughts: Stop Talking to Yourself

You comprehend my path and my lying down, And are acquainted with all my ways. **Psalm 139:3**

Does God want to be involved in every decision we make? Is it necessary to pray about every little detail? Many would say that God doesn't care, He saved you and He gave you a mind to think things through for yourself. I would completely disagree. After spending years of praying over the details in life, I have come to appreciate all He does for me just by including Him and looking for His fingerprints in my life.

But how does this work? Well, for a great starter, stop talking to yourself and start talking to the Lord. If you are late, talk to God about it. If you are confused, include God in it. Talk to Him as a friend, a counselor and a companion. He says that He is all those traits so get to know Him in those ways. In every relationship, there needs to be two-way communication. Prayer obviously is your part in the communication process with the Lord. But the Bible is God's way of speaking back to you. So let's say it is on your heart to buy a new car. Start talking to Him about it. Ask Him to change your heart if it is not His will or to lead your heart in the car buying process. Next, go to the Bible and ask the Lord to give you verses to guide you in this decision. If some verse pops out to you, pray over it and ask God to make His message clearer through that verse. If He does give you the go ahead, do further research with that promise. Keep talking to Him and keep going back to the Word, asking for specific direction. Whatever you include Him in, the Lord will work with you from dealerships to your finances. As you grow in the knowledge of God's guidance, you will gain a peace in your decision of what to purchase. Regardless of the outcome, you will have assurance of God's hand in the process. Walking through these decisions in our lives with God is the only way to assure peace. The more we include Him in every decision, the more we grow in our knowledge of Him, which leads to a closer walk with the Lord.

Are you in the midst of a trial or major decision in your life? Are you searching for answers in various places? Knowledge can bring peace, but the knowledge of God brings everlasting peace that continues to surpass our understanding. Start out everyday in God's Word. Take time to put His Word in your heart and mind. Before the day is through, you will need to hear from the Lord on something that crosses your path. The knowledge of His Word will give you peace and assurance in ways you could never imagine. Try it and see what God will do.

November 11

Today's Reading: Jeremiah 50; Hebrews 8
Today's Thoughts: No Condemnation

"Most assuredly, I say to you, he who hears My word and believes in Him who sent Me has everlasting life, and shall not come into judgment [or condemnation], but has passed from death into life."
John 5:24

Although there is a huge difference between conviction and condemnation, they seem to be interchangeable in the Christian life at times. When Christians are convicted, often this leads them into feeling condemned. The thoughts and feelings associated with the word convicted suddenly become closely defined to the words "convict or a prisoner". When convicted, we act as if God wants us to wear prison clothes, eat prison food and wake up behind prison doors to remind us of how we have fallen short. God does not want that for us. Neither word: condemned nor convicted, applies to the believer negatively.

The Bible defines these words differently. Condemnation has to do with *who* you are, either in Christ or not. Conviction has to do with *what* you do, to bring you closer to Christ. If you do not know Jesus Christ as your Lord and Savior, if His blood does not cover your sins, you are condemned. You have not passed over from death to life and you remain in your sins. However, if you are a Christian, you will never be condemned because Jesus has given you eternal life. Thoughts of condemnation are not from the Lord, not ever. Condemnation is about your spiritual state for eternity, not your daily thoughts or actions.

For the Christian, there is a clear distinction between conviction and condemnation. Conviction leads us closer to Christ as we confess and repent to Him. Condemnation makes us feel as if we cannot come to God as we concentrate on our own shortcomings. Condemnation is a tactic of the enemy. Pray that you do not fall prey to his lies. Romans 8:1 tells us that there is no condemnation for those who are in Christ Jesus.

If you know Jesus, He is for you. He is always on your side and He wants you to succeed at everything you do. Stay on His team by thinking His thoughts about you. Work with Him to change. There is no condemnation for those who are in Christ because He has come to give you life in abundance.

November 12

Today's Reading: Jeremiah 51-52 Hebrews 9
Today's Thoughts: A Good Sharpening

As iron sharpens iron, So a man sharpens the countenance of his friend. **Proverbs 27:17**

Today's verse is one that is quoted often when referring to healthy and productive friendships. Just as iron is used to sharpen iron, a man or woman is used to sharpen his or her friend. What is this verse really saying? How do we apply this wise proverb to our friendships?

To sharpen a knife (even the basic kitchen knife) we must use a metal tool designed to sharpen the blade. Wood will not do the job, nor will plastic or glass. Certain types of stones were used in ancient times but many of the knives were also composed of stones or metallic substances. It takes a substance of similar strength and durability to handle the friction required to sharpen the object. Sharpening in itself is a process of fine grinding, requiring just enough to remove the dull edges. How does this compare to sharpening a friend? The knife is without a will of its own. It is in complete submission to the person doing the sharpening. Friends, however, have wills and opinions, and are much less likely to submit.

To sharpen the countenance of a friend we must first *love* them. When our motives are founded on love, our words and actions will reflect the love in our hearts. The Bible tells us to speak the truth in love and that without love, nothing we do really matters anyway. It is when we truly love a friend that we are willing to take the steps necessary to *sharpen* them. Just as with a knife, the process requires friction, some grinding and the removal of bad edges. And, just as with a knife, a person of similar strength and durability is required to handle the job. Sharpening involves heat, intensity and duration. A loving friend knows when heat and friction are needed, despite how long or intense the process. A loving friend is honest, even to the point of pain. If we are willing to submit to that friend, then we will be sharpened. Friends willing to work through these things in the Lord will bear much fruit together.

Do you have a friend who sharpens you? Are you the kind of friend who sharpens others? God has given us friends for His purposes. We have the greatest Friend of all as Jesus calls us His friends. He will sharpen us if we submit to Him. Friendships based on prayer and love will result in growth and maturity. Pray for your friends today. Ask the Lord to lead you in how to sharpen them. Also, ask the Lord to help you receive their counsel. If we place everything in God's hands, He will work through us in mighty ways. And one of the greatest ways to see God work is through our friendships.

November 13

Today's Reading: Lamentations 1-2; Hebrews 10:1-18
Today's Thoughts: A Blessing For Others

And Laban said to him, "Please stay, if I have found favor in your eyes, for I have learned by experience that the Lord has blessed me for your sake." Then he said, "Name me your wages, and I will give it." So Jacob said to him, "You know how I have served you and how your livestock has been with me. For what you had before I came was little, and it has increased to a great amount; the Lord has blessed you since my coming. And now, when shall I also provide for my own house?" **Genesis 30:27-30**

Jacob had two wives and eleven children and he wanted to return to his own land. The time had come for him to leave his father-in-law, Laban, but Laban was not ready to let Jacob go. Both men acknowledged the blessings upon Laban's house as being from the Lord. Jacob lived with and worked for Laban for over 20 years and both men prospered greatly, but the blessings came to Laban because the Lord's hand was upon Jacob.

The same should be true of the Christian today. Because we know the Lord Jesus, our lives should bring blessings to others. The blessings come in many ways, such as recognizable calmness in our presence and stability. Many times, others do not want to admit that their blessings are a result of the Christian's convictions and prayers. The unbeliever's pride causes them to take the glory for themselves. But as Christians, our lives do bless others. In time, God will get the glory as the fruit of our lives, as well as the testimony of our mouths, will clearly point to His intervention

Are others blessed because of you? Even if you are the only Christian in your family, God's hand is on you and your home will be covered with your prayers. Continue to pray for your family and seek the Lord's blessings over those who do not yet know Him. Press on to not grow weary in doing good. Wherever we go, we bring the Lord's presence because His Holy Spirit lives within us. We need to acknowledge the Lord's blessings not only in our own lives but also in the lives of those around us. Let your light shine to those around you and give God the glory for all things.

November 14

Today's Reading: Lamentations 3-5; Hebrews 10:19-39
Today's Thoughts: *Staying Spiritual During the Holidays*

"And you will have joy and gladness and many will rejoice at His birth." Luke 1:14

If the holidays are about our faith, family and food feasts, then why do we struggle the most with staying spiritual during these times? Shouldn't these times bring us peace on earth and joy to the world? But the reality for many of us is that we become emotionally stressed out, financially maxed out and physically burned out trying to keep up and keep it all together. Day after day, we become burned out and stressed out and now, our faith becomes affected as we start questioning God, "Are You hearing me?" and "Why won't You help me?" Like Eve in the garden, we question God's goodness and His basic command. Eve exchanged the peace of God for a piece of fruit but we exchange His presence for presents. The serpent tempted Eve to compromise and we know that Jesus called him a murderer from the beginning. He is still up to the same old tricks. We are allowing the enemy to have too much influence at times. He kills our dreams, steals our peace and destroys our hope for the future. Jesus came to earth to reverse the process of death and yet our undisciplined lifestyle of choosing not to obey can lead to negative consequences in our lives.

To stay spiritually minded during the holidays, we need to keep our eyes focused on the Lord and include Him in our thoughts. Instead of talking to ourselves, we need to start talking to Him and then give Him the time to talk back. The enemy will use every scheme he can against us to distract us from maintaining a walk with the Lord. It has been said, "If the devil cannot make you bad, he will make you busy." To combat the temptations to over commit in your time and finances, we must stay in the Word of God to renew our minds and to surrender to His will over our natural-born desires. Through the Holy Spirit, we can overcome and live a life pleasing to God that brings peace in the midst of chaos.

Joy to the world begins with each of us individually. Joy cannot come without the conscious awareness of the Lord's promised presence wherever we go and whatever we do. By allowing our thoughts to think about God in the daily routines of life, we allow the Lord to have the ability to move through our actions. Only then can we truly maintain the Spirit of Christmas and sing the song, "Let there be peace on earth and let it begin with me."

November 15

Today's Reading: Ezekiel 1-2; Hebrews 11:1-19
Today's Thoughts: Faith and Prayer

And without faith it is impossible to please God, because anyone who comes to him must believe that he exists and that he rewards those who earnestly seek him. **Hebrews 11:6**

God rewards faith. Christianity started with the faith of Abraham. Faith means that we do not see to believe. So, praying is an act of faith. When we pray, we are acknowledging that God exists and listens even though we do not see Him. God rewards our faith by answering our prayers.

I have noticed that when I first came to Christ, He answered my prayers pretty quickly. It was as if I was in "obedience school" and He wanted to reward my good behavior of including Him. As I grew in my faith, I next noticed that God answered some prayers quickly but other prayers, He did not answer at all. He was then training me to learn how to pray and ask for certain things differently. When I studied the Word to see God's heart on the matter or asked God to clarify my motives of why I was asking, He would answer those prayers as well as what I asked.

Now, after having walked with the Lord for many years, my prayer life has changed again. I have more of a conversational prayer life than an "ask and believe and receive" prayer life. Now, I talk with the Lord about my requests. I ask Him to reveal His heart to me about the matter and I wait to see if there is any change in my heart as I also search the Scriptures to hear His heart. The requests themselves have now become smaller in comparison to the relationship of talking and listening and learning from Him. There is no request or unanswered prayer that becomes more important than my ongoing relationship with Him. So my faith is rewarded in a completely different way than answered prayer; my faith is rewarded by enjoying His presence in the process.

God is good. Learn to seek Him in faith. It starts with prayer requests and continues until you are praying without ceasing. You will not be disappointed. How could we ever be more satisfied than with the very presence of God Himself?

November 16

Today's Reading: Ezekiel 3-4; Hebrews 11:20-40
Today's Thoughts: Our Real Hope

***By faith Moses, when he became of age, refused to be called the son of Pharaoh's daughter, choosing rather to suffer affliction with the people of God than to enjoy the passing pleasures of sin, esteeming the reproach of Christ greater riches than the treasures in Egypt; for he looked to the reward.* Hebrews 11:24-26**

The world news is depressing. There seems to be so much suffering in the world. *Why must we suffer?* As children of God, why must we go through such painful trials and tribulations?

When we are going through painful times, there is no easy answer to those questions. But let's take a moment and just remind ourselves of what the Bible tells us. First of all, we are told throughout the pages of God's Word that we *will* suffer. Suffering is not an option. It is a fact of life. First Peter 4:12 tells us to not be surprised by the fiery trials. In other words, be ready for them. Do not be caught off guard when they happen. Moses chose to leave the palace and suffer the same types of afflictions that he had watched his people endure. Hebrews 4:15 comforts us by saying that we have a High Priest who understands our sufferings. No one suffered more than Jesus. No one understands our pain more than Him. We are never alone and never forsaken.

The next point for us to remember is this: Philippians 4:7 says, "the peace of God, which surpasses all understanding, will guard your hearts and minds through Christ Jesus." The peace that Jesus promises is not always consistent with our definition of peace. The peace of God is a feeling of inner comfort and joy, which cannot be explained or humanized. How do we get such peace? We ask for it in faith. We pray. We seek Jesus. We continually lay our cares at His feet. We ask the Lord to give us an eternal perspective so that we can keep our focus on His purposes and not our own. Romans 8:17-25 is a great passage of Scripture that reflects this message of an eternal hope. The Apostle Paul speaks of our suffering with Christ, a suffering that all of creation must endure because of sin. *"We ourselves groan within ourselves, eagerly waiting for the adoption, the redemption of our body."* (Romans 8:23)

For us today, our only hope is in Jesus. His living Word is our eternal food. His Holy Spirit is our living water. His very presence in our lives will change our hearts and minds with a wisdom and understanding that goes beyond our earthly desires. Jesus never promised that we would not suffer, but He did tell us to be of "good cheer." Even though we will have tribulation in the world, we have His peace and He has overcome the world. (John 16:33)

November 17

Today's Reading: Ezekiel 5-7; Hebrews 12
Today's Thoughts: Miracles

Gideon said to Him, "O my lord, if the Lord is with us, why then has all this happened to us? And where are all His miracles which our fathers told us about, saying, 'Did not the Lord bring us up from Egypt?' But now the Lord has forsaken us and delivered us into the hands of the Midianites."
Judges 6:13

This past weekend we hosted a radio talk show. We were told to pick whatever topic the Lord put on our hearts and then spend the hour discussing it with our audience. The Lord impressed upon us the message of miracles. How does God work miracles in our lives today? Does He do the same kinds of miracles as He did in the Bible? We asked these types of questions and discussed the answers. We pray that we shed some light on God's desire to perform miracles in our lives everyday. We need to recognize miracles and give glory to the Lord when things happen that we tend to call *"just a God thing."*

I gave an illustration that seemed to strike a cord with people. I talked about missing a freeway exit. I kind of spaced out and perked up past my exit. I got off the exit, turned around, and then missed the street I was to take. After missing both exits, I said, "Ok, Lord, You must have a reason for this. I thank You for whatever reason that may be." I did not know what was going on but I knew it was not a coincidence. Someone called in and gave a story of getting off on an exit that they had not planned to get off on. When they returned to the freeway, an automobile matching their description almost exactly had just been totaled in an accident. This person knew the hand of God had just saved them. Miracle or coincidence? We believe that God is working in our lives on a daily basis, some things we see clearly, others we may never know about.

We need to believe in miracles because we need to be encouraged and have hope that our God is with us, protecting us, and doing supernatural wonders in our lives. We do not need an explanation for all that happens. Just give glory to the Lord and watch what He does in your life. Gideon would soon learn of God's miraculous power as the Lord would use him to win the battle against the Midianites. Sometimes, we need to just ask and believe. God will respond.

November 18

Today's Reading: Ezekiel 8-10; Hebrews 13
Today's Thoughts: I Will Not Fear

So we may boldly say: "The Lord is my helper; I will not fear. What can man do to me?" **Hebrews 13:6**

I have noticed recently that my dog has developed an unusual kind of fear. He hesitates at times as he walks over thresholds. He scopes out the door frame and the floor in front of him, especially when he is going outside. It all started when something jumped at him one day when he was going out the back door. To be honest, I think it was a small insect that startled him (yes, he is a big baby.). I watch him now as he cautiously looks at the ground, suspicious of danger lurking about. The other day, he stopped in front of a twig. I think he is becoming neurotic about this fear thing. I wonder if the Lord looks at us and how we fear things and thinks we are also a little neurotic at times.

My dog has no ability to understand and cannot be reasoned with, regardless of how much I try. After all, he is a dog; but what about us? God has not only created us in His image, but also He has given us His Holy Spirit who lives within us. But fear has a way of creeping into our lives and before we know it, we fear situations, circumstances or people that are no real threat to us at all. Sometimes we are startled or caught off guard in a situation, and moments of fear can be a normal response. The problem lies in allowing the fear to remain. It can then take over, control the situation and even enter other areas of our life. Fear will keep us in a place of bondage, instead of freedom.

As I watch my dog tiptoe over a twig, I am reminded of how silly some of my fears look to God. What about your fears? Have you thought about the things or even the people who bring fear to your life? Maybe it is time to step back and really look at the cause of the fear. When and where did it begin? Has it gotten blown out of proportion? Are you hindered or even crippled by it? If so, please know that the Lord wants to set you free, beginning today. Ask the Lord to help you pray through it. Take a bold stand and say, "Enough!" Life is too short and Jesus wants to give you freedom, liberty and victory. *"For God has not given us a spirit of fear, but of power and of love and of a sound mind."* (2 Timothy 1:7)

November 19

Today's Reading: Ezekiel 11-13; James 1
Today's Thoughts: Count It All Joy

My brethren, count it all joy when you fall into various trials, knowing that the testing of your faith produces patience. **James 1:2-3**

In our Discipleship Program, we closely examine the biblical teaching on trials. How often do we find joy in the midst of our trials? Can we realistically count it all joy today? Does this statement by James even make sense? Let us examine these verses more closely and see if we can better understand what James is saying. One of the main reasons for us to read and study the Bible is not just to learn meaning and content, but also to learn how to apply its principles to our lives. So, what is God saying to us about trials, faith and patience from these verses?

Since this first statement comes right after the greetings in verse 1, we can assuredly note its importance. In these two verses, there are three main points: we will fall into different types of trials, our faith will be tested, and patience will be produced from the testing. James is not trying to be politically correct and he is not concerned about offending anyone. He is stating the facts. Today, we deal with all kinds of trials in our lives. The issue is not *if* we will have them, it is *how* we will get through them. When we feel as though the battle is too much and the struggles are too intense, how do we respond? Difficulties with marriage, kids, finances, health and daily issues of life can seem overwhelming at times. But James says to "count it all joy" when these problems come. We can only begin to understand the meaning of this statement if we look at the results in our own lives.

God's goal in allowing us to experience various trials is to make us grow stronger, to learn patience, and to develop a deeper faith in our walk with God. We can choose to "count it all joy" as we go through them because we have the assurance in God's word that the end result will benefit our faith. How many times have you looked back after a trial has passed and recognized the benefits it produced in your life? So often, we get the point after the fact. We tend to acknowledge the blessings after the trial is over. "Oh, now I see what God was doing." Trials and testing can produce patience. There is not much we can do about avoiding such things in life; however, we can decide how to get through them.

If you find yourself in trials today, ask the Lord to help you find joy through His Holy Spirit so that you will come through the testing with an increased patience, strength and faith.

November 20

Today's Reading: Ezekiel 14-15; James 2
Today's Thoughts: Who Do You Say That I Am?

And it happened, as He was alone praying, that His disciples joined Him, and He asked them, saying, "Who do the crowds say that I am?" So they answered and said, "John the Baptist, but some say Elijah; and others say that one of the old prophets has risen again." He said to them, "But who do you say that I am?" Peter answered and said, "The Christ of God."
Luke 9:18-20

"But who do you say that I am?" As many people began to discuss and debate who Jesus really was, Jesus asked His disciples this question. Did the people closest to Jesus even know or believe in Him? Peter's answer was correct. Jesus was, and is, the Christ, the Son of God. Jesus is the Messiah, the One who came to earth to redeem us from sin and save us from eternal darkness. But even though Peter confesses rather boldly in this statement who he believes Jesus to be, not too long after this, Peter would also deny Jesus three times (Luke 22:54-62). In Jesus' greatest time of need, His closest friends deserted Him. So it is not just in words that we need to know and confess who Jesus is, but also our actions testify of our belief in Him. Do your actions line up with your words?

Today, take time and pray for situations that you may encounter with others who have questions about your faith. Pray for strength and courage to speak up and stand up for Jesus. Pray that you can be His witness. Only the Holy Spirit reveals the truth of Jesus Christ to our hearts. Pray for the Holy Spirit to reveal Jesus to those you love and to use you as a testimony as to His saving grace and mercy. Pray that your words and actions will line up with His Spirit.

November 21

Today's Reading: Ezekiel 16-17; James 3
Today's Thoughts: The Roman Road

"Do not fear, nor be afraid; Have I not told you from that time, and declared it? You are My <u>witnesses</u>. Is there a God besides Me? Indeed there is no other Rock; I know not one." Isaiah 44:8

Christians seem to have a lot of difficulty witnessing to unbelievers. It is easier to talk about the church we attend than to talk about Christ. The most common excuse that I have heard from Christians is that they do not know enough Scripture verses. This misconception is a lie from hell itself. From personal experience, I have not led anyone to Christ because of my Scripture knowledge, but I have led many to Christ because of my passion. Sharing the gospel is a verbal expression of your heart. If I am absolutely convinced that Jesus Christ is real and has changed my life, I will reflect that in my words and it will be evident in my actions. No one can challenge my own personal experience and testimony because it is mine. People hear with their eyes but no one comes to faith in Jesus Christ without hearing about Him. The Holy Spirit only needs a willing heart and an open mouth. His words will not return void because witnessing pleases the Lord and witnessing is God's will for our lives. Just say it.

However, with that being said, there are four scripture verses in the Book of Romans that have been very helpful to me while witnessing. These verses have been called the "Romans Road" implying that *All Roads lead to Rome*! And the "Romans Road" leads to God. These are great verses to become familiar with and to memorize:

Romans 3:23 "for all have sinned and fall short of the glory of God"

Romans 5:8 "But God demonstrates His own love toward us, in that while we were still sinners, Christ died for us."

Romans 6:23 "For the wages of sin *is* death, but the gift of God *is* eternal life in Christ Jesus our Lord."

Romans 10:9-10 "that if you confess with your mouth the Lord Jesus and believe in your heart that God has raised Him from the dead, you will be saved. For with the heart one believes unto righteousness, and with the mouth confession is made unto salvation."

Oh Lord, give me an opportunity to share Your love with someone today. Open the doors for me to witness and give me the boldness to open my mouth to share about You.

November 22

Today's Reading: Ezekiel 18-19; James 4
Today's Thoughts: Resist the Devil

***Submit to God, resist the devil and he will flee from you.* James 4:7**

As believers, we face so many obstacles that can take us away from serving the Lord. We have thoughts that control us, wants that consume us and feelings that move us. Because none of us were born understanding God's ways, our natural inclinations work against living out a Spirit-filled life. Coming to Christ means that we lay aside our natural inclinations and ask the Lord to fill us with His wisdom instead. That process is known as submitting. To submit means to be under obedience. So we take what we want to do, what we feel is right to do and what we think is best and tell the Lord that we want His Will instead. Waiting on God and trusting Him begins.

To resist the devil means that, after we have submitted to God, we go to the Word of God to counteract our thoughts, feelings and wants. The devil is fully aware of who we are and how we think. As a result, he uses our desires to entice us to keep from submitting. We start questioning the love of God and His motivation. We begin to fear and hold on to our thoughts, feelings and wants, instead of trusting and letting go. We have to realize that the devil is alive and well. His desire is to use your fleshly desires against you so that you sin against God.

Jesus resisted the devil by the Word of God. Now, if the Living Word, Jesus, needed to speak forth the written Word, the Bible, to resist the devil and have him flee, how much more do we need to submit to God in this pattern also? Trust Him; in faith speak forth and pray the Word of God. Believe Him, submit to Him and the devil will flee from you. It is a promise—God's promise!

November 23

Today's Reading: Ezekiel 20-21; James 5
Today's Thoughts: Daily Victory in Battles

For though we walk in the flesh, we do not war according to the flesh. For the weapons of our warfare are not carnal but mighty in God for pulling down strongholds. **2 Corinthians 10:3-4**

To understand the battle, we need to begin with acknowledging that we are in a war. Battles make up smaller components of the bigger picture. By definition, battles involve combat between two persons, between factions, between armies and they consist of any type of "extended contest, struggle, or controversy" (Webster-Merriam). As Christians, we are in a spiritual battle of some sort on a daily basis. In warfare, battles are fought on different fronts, for different reasons, and with varying degrees of intensity. The same is true in spiritual warfare. Our spiritual battles are real, even though we cannot physically see the attacker. But, we can educate ourselves on how the battles are fought and how they impact our lives on a daily basis.

We have to ask ourselves, "Why do we even want to fight?" It will do us no good to educate ourselves on the battle if we see no reason for the fight. War is very controversial today in the physical realm. Those attitudes, beliefs and convictions will transfer over to the spiritual realm. However, in the spiritual realm, there is a battle going on regardless of our opinion. We are either victors or victims. Jesus has come and conquered. The war is already won in the heavens. Jesus told us in Matthew 28:18 that, "*All authority has been given to Me in heaven and on earth.*" We now have the privilege of having an eternal relationship with God. Many of us enter into that covenant of salvation by grace. But Matthew 28:18 is not only about our salvation; but also it is about our every day victory, which adds up to victorious living in Christ. Every day victory is achieved by knowing, believing and understanding the battles that we are enduring daily, regardless if we are passive or active in the battles. Do you want all that God has for you on this earth right now, or do you want to wait until you get to heaven to receive the victory and blessings?

We can only receive the abundant life by understanding the battles that we are facing every day. We need to pray for spiritual eyes and we need to understand what weapons the Lord has given us to not only fight the battles but to win the war.

Yes, Lord, I want all that You have for me on earth as it is in heaven. Give me spiritual eyes to see the world as You see it and do not let anything stop me from receiving good gifts from You. Educate me to live a victorious life in Christ because of Christ. In His name I pray, Amen.

November 24

Today's Reading: Ezekiel 22-23; 1 Peter 1
Today's Thoughts: Submitting to God

"Surely not, Lord!" Peter replied. "I have never eaten anything impure or unclean." The voice spoke to him a second time, "Do not call anything impure that God has made clean." **Acts 10:14-15**

Do you feel comfortable saying "no" to God? We might not have the ability to admit it but we each say "no" to God in some way. Sometimes we say "no" by choosing to not witness. Sometimes we say "no" by refusing to volunteer. Sometimes we say "no" by ignoring the prompting of our heart. But most of the time we say "no" because we refuse to accept His will for our lives.

In Matthew 16:13, Jesus asked His disciples, "Who do men say that I, the Son of Man, am?" The disciples answered His question by stating what others think about Him. "But who do you say that I am?" Simon Peter is the only one who answered and said, "You are the Christ, the Son of the living God." Jesus acknowledges to Peter that God, Himself, has revealed this to him. A few passages later, Jesus begins explaining that He will have to suffer. Peter again is the only one who speaks up as he disagrees with God's plan, taking the liberty to rebuke the Lord. Matthew 16:22-23 says, "Then Peter took Him aside and began to rebuke Him, saying, 'Far be it from You, Lord; this shall not happen to You!' Jesus turned and said to Peter, 'Get behind Me, Satan! You are an offense to Me, for you are not mindful of the things of God, but the things of men.'" Peter has gone from a place of esteem to a place of rebuke. In Acts 10, we find Peter disagreeing with God again as he says, "Surely not, Lord."

We are flesh and our natural born nature rebels against the Lord at times. God can handle whatever position we take and He wants to reason with us on our position. Talk to Him and be honest. The problem comes when we place our thoughts and desires over God's, to a point of rebuking the Lord. We may not have the boldness to tell Him we think He is wrong, but our actions show the thoughts and intents of our hearts as we choose not to submit or yield or even listen to what He is trying to say. God's ways do not make sense to us and His thoughts are not our thoughts.. Peter was a man who had an honest relationship with God. His rebukes may not seem respectful but he kept himself in a position to keep listening until his will lined up with God's will.

What is the Lord trying to tell you today? Are you going through a set of circumstances that you refuse to acknowledge *is* God's will for your life? Are you fighting or submitting? Are you willing to listen or are you running? Let us all pray that we are on God's side. When it comes down to it, He is going to win. Don't you want to be on His team?

November 25

Today's Reading: Ezekiel 24-26; 1 Peter 2
Today's Thoughts: Be on God's Side

The word of the LORD came to me: "Son of man, set your face against the Ammonites and prophesy against them. Say to them 'Hear the word of the Sovereign LORD. This is what the Sovereign LORD says: Because you said "Aha!" over my sanctuary when it was desecrated and over the land of Israel when it was laid waste and over the people of Judah when they went into exile, therefore I am going to give you to the people of the East as a possession.'"
Ezekiel 25:1-4

In the first half of the book of Ezekiel, God rebukes His people for their attitudes toward Him. The Lord uses words like stubborn, hard-hearted and rebellious to describe His people. The Lord speaks through Ezekiel and tells the people to understand that their choices come with consequences. If they make good and righteous choices that honor God, they will be blessed. If they make poor and wicked choices that honor themselves, they will be punished. As the story continues, the reader knows what choice the Israelites made because God starts talking about the judgment that is to come upon them.

In Chapters 25-27, God redirects His message and begins proclaiming judgment on the nations that punished the Israelites. God warned His people to turn from their sins and to turn back to Him. They did not heed the warnings and God allowed the other nations to invade Israel (just as He said would happen). But then God judged the nations that came against His people because they did not show respect towards the Lord or towards His people during their captivity. That is amazing.

If you are God's child, He desires that you walk with Him and love Him with your whole heart. If you do not, you will not lose your salvation but you will have to live with the consequences of making bad choices. Those consequences are usually inflicted upon us by other people. But those people will not get off easy in their unjust deeds and prideful attitudes towards you. They, too, will have to suffer the consequences for how they deal with you, God's child. God is always on your side. He wants you to walk with Him and He will go out of His way to warn you. God is amazing. I always want to be on His side for He always gets the last word.

November 26

Today's Reading: Ezekiel 27-29; 1 Peter 3
Today's Thoughts: In Whose Strength?

"...then you say in your heart, 'My power and the might of my hand have gained me this wealth'" **Deuteronomy 8:17**

I was talking to someone the other day who has been up and down financially more than I could personally handle. He has been very successful in more than one business venture, only to lose it all, and then start over again. This last venture; however, has cost him greatly. Great amounts of money have gone into something that now sits in a warehouse, with little hope of distribution. What is he going to do? In his own words, "I am going back to church." The key point here is the word *back*. At one time, he and his wife were involved in their church, gave to missions, and looked for ways to use their financial blessings to bless others. Then, as the business grew, they got busier, and had less time for the things of God. Before long, they even stopped attending church.

The Lord warned the Israelites to be very careful when they entered the new land of promise and prosperity. They would be tempted to think that their success and wealth were of their own doing. Or they would forget God when they felt satisfied with their prosperity. They would serve mammon rather than God. But the Lord wanted them to remember that He was the One who brought them up from the land of Egypt. He was the One who delivered them from slavery and bondage. He was the One who brought them to this land of blessings. We can be so arrogant as humans. Not only do we start to think that we have made our own way by our own strength, but also we turn away from the One who blessed us so wonderfully. That is what happened to the Israelites and that is what happened to my friend.

When times are hard, we cry out to God for help. We press into Him. He is our only hope. We tend to be less attentive to God when we are doing well and life is cruising along. Regardless of where you are today, remember one thing: You have a Lord and Savior who only wants you. We have a Lord who desires to be with us, have a relationship with us, and wants us to desire Him above anything else in the world. He is a jealous God and He will do whatever it takes to get our attention. Draw closer to Him today. Pray that you do not fall away. If you have, then come back to Him. Do not get caught up in the things of life and lose sight of your relationship with Jesus. Stay with Him so He does not have to come get you *back*.

November 27

Today's Reading: Ezekiel 30-32; 1 Peter 4
Today's Thoughts: Hypocrisy

Now when Peter had come to Antioch, I withstood him to his face, because he was to be blamed; for before certain men came from James, he would eat with the Gentiles; but when they came, he withdrew and separated himself, fearing those who were of the circumcision. And the rest of the Jews also played the hypocrite with him, so that even Barnabas was carried away with their hypocrisy. **Galatians 2:11-13**

The Apostle Paul was a bold man in his faith. He seemed to have no problem confronting Peter about his hypocrisy. At the time this letter was written, the Gentiles were new converts into the salvation of Jesus Christ. The Jews had always been God's chosen people, circumcised and set apart, and were never to associate (especially to eat) with the heathen Gentiles. The Jewish prominence and exclusivity was abolished when Jesus died on the cross for all sins, for all people. Paul was the apostle to the Gentiles and he would not tolerate the contradictory behavior of his fellow Jews, especially from their leader, Peter. The leader has tremendous influence on the others, as seen in Barnabas' behavior in following Peter as well.

This holiday season will bring many families together for meals and fellowship. How will you behave in front of your un-believing family member? Will you stand firm as Paul and make your faith in Jesus Christ your standard of behavior? Or, will you go with the flow of the others and return to past traditions like Peter? A change is especially hard in families where traditions do not glorify God. If you are new to Christ, you may have a hard time with the old traditions, but will have a harder time going against them. Fear of ridicule and rejection are common to many of us who have had to go through these changes. It is much easier to just go with the family consensus and keep the peace. But is that what God wants you to do?

Regardless of where you find yourself in these situations, please remember to pray about them. Holiday gatherings tend to be times of deep-rooted family rituals—too much food, too much drink, too much indulgence of many kinds. As a child of God, pray for wisdom and discernment in these situations. God may not be telling you to be a Paul and confront all your family members. Paul confronted his brother in Christ, Peter. You, however, can set an example by simply not partaking in the behaviors that you know would not glorify Jesus. You never know who else will follow your lead. Do not criticize or judge. Hypocrites are those who judge others while committing the same acts. Let your actions be without hypocrisy. Pray in advance.

November 28

Today's Reading: Ezekiel 33-34; 1 Peter 5
Today's Thoughts: Perfect in Suffering

But may the God of all grace, who called us to His eternal glory by Christ Jesus, after you have suffered a while, perfect, establish, strengthen, and settle you. **1 Peter 5:10**

The word "suffer" is not a word that brings comfort or peace. When we hear someone is suffering, we feel concern, sympathy and even sadness for that person. The definition for suffering means "to *experience* a sensation or impression (usually painful)." The other fact that we as Christians must deal with is that God allows suffering. Nothing on this planet or in the universe happens without God's approval.

Why does this loving, gracious God allow tragedies that bring such suffering? Why does evil persist, even in homes and families who seek the Lord the most? These and many more questions are asked when we try to make sense of God's ways. Satan has taken full advantage of these events in our lives; at least he has tried to. Have you ever had thoughts like: "Why should I believe that doing things God's way brings me any assurance of protection or blessings?" Or, "If God loves me then why is He letting this happen to me?" The same logic applies when we turn those same thoughts to others. People hurt us, so let's not trust people anymore. Circumstances hurt us, so let's control everything we possibly can to prevent being hurt (Hurt, by the way, is just another description for suffering.).

The bottom-line answer is that we will never fully understand why God does what He does. We must live as He has told us to live—by faith. We must believe and trust that the Lord has a reason for everything that happens to us. When we dig into His Word and study His teachings, then we will have more peace in those times of confusion and pain. Today's verse is an example of how we can find peace and hope in times of suffering. We should pray this verse and meditate on it as we are going through those times of suffering. If we can look beyond our pain for the moment and pray to see the bigger picture that God has for us, then we can find hope and even peace in what we are going through. Even in the midst of great tragedies that tempt us to ask God "why," we can trust that He has a much greater plan than what we see in the short term. The question for us is whether or not we will trust Him or turn from Him.

Take time today to really meditate on the verse above. Take comfort and trust the Lord to "perfect, establish, strengthen and settle you" after you have suffered a while. Only God knows the timing of such things. Pray for the perseverance to endure.

November 29

Today's Reading: Ezekiel 35-36; 2 Peter 1
Your Thoughts on Today's Passage:

November 30

Today's Reading: Ezekiel 37-39; 1 Peter 2
Today's Thoughts: The Holy Spirit in You

I will put my Spirit in you and you will live, and I will settle you in your own land. Then you will know that I the LORD have spoken, and I have done it, declares the LORD. Ezekiel 37:14

When we accept Jesus Christ as our Lord and Savior, we receive His Holy Spirit. The Holy Spirit is God and He is a gift from God the Father that allows us to have a relationship with the Lord. The Spirit of God intercedes for us in prayer as He uses words that enter into the throne room of God when we can only moan and groan. He is our Counselor, our Teacher, our Comfort and our Guide. But the most important eternal promise we receive when the Spirit of God is placed in us is the seal of our salvation as we come alive spiritually for eternity. Amazing!

Do you realize the power you have living inside you? That power raised Lazarus from the dead, healed all kinds of diseases and delivered souls from bondage. That same power is found in you through Jesus. Jesus shows us this power through His indwelling peace. When we receive the Spirit of God, we have peace with God, not peace as the world gives, but the peace of God that surpasses understanding. It is peace that "settles" you. No matter where you are or what you are struggling with, God has settled your soul to have peace with Him.

The peace of God is a promise for you today. Claim it, speak it and live in it. Throw off the distractions of illness, financial loss, negative comments, guilt, and self-condemnation. And let the peace of Christ dwell in you richly. Nothing can interfere with the love Christ has for you and the peace that is yours by knowing Him. All these things on earth will pass away, but your security for eternity is fixed. Fix your eyes on Jesus by reading His Word. Get a verse today that you can claim as a promise and live beyond the struggles. Cast your cares upon the Lord for He cares for you.

December 1

Today's Reading: Ezekiel 40-41; 1 Peter 3
Today's Thoughts: Great Treasure

I rejoice at Your word as one who finds great treasure. **Psalm 119:162**

The Psalmist who wrote this verse knew where to find real treasure. He rejoiced at the Word of God. He considered the Word as precious as treasure. The prophet Jeremiah declared a similar expression of his heart when he said, "Your words were found, and I ate them, And Your word was to me the joy and rejoicing of my heart." (Jeremiah 15:16) As Christians, we should all consider the Word of God, the Bible, a treasure to rejoice over. But do we?

I was raised in church, went to Sunday school, and accepted Jesus as my Savior when I was very young. I read the Bible, knew most of the stories, and really loved the Gospels in the New Testament. I bought tapes of my favorite sermons and watched programs that touched me personally, all in addition to going to church. But did I *rejoice* over the Bible? Honestly, I must answer a definitive "no." To be even more honest, I had trouble just sitting down and reading the Bible. I would rather have someone "talk" to me and teach me what it says. I did not know how to read it for myself and get a message that applied directly to my own life.

One day I knew that I needed more in my life and I said to God, "I need more of You. I need to know who You really are in my life." Very quietly but clearly, the Lord said to me that it was time to start spending time with Him. I must start in His Word. I must make a choice to turn off the other things and start to listen to Him. I picked up the Bible and began reading from the beginning. Six months later, I finished it. I cannot begin to tell you the change that began in my life and has continued to this day. By the time I came to the verse in Jeremiah 15:16, I knew that if I could truly eat the Word, I would. I could not get enough. The Word of God became the greatest treasure I have ever found. Nothing compares.

If you want to find a treasure worth rejoicing over, then pick up the precious Word of God and actually read it for yourself. Read it every day and keep going. You will be changed in ways you cannot imagine. Another of my favorite verses says just that: "Now to Him who is able to do immeasurably more than all we ask or imagine, according to His power that is at work within us." (Ephesians 3:20) Get to know the Lord through His Word, start today.

December 2

Today's Reading: Ezekiel 42-44; 1 John 1
Today's Thoughts: There is Only One Way

This is the message we have heard from him and declare to you: God is light; in him there is no darkness at all. **1 John 1:5**

A few years ago, the Lord put a burden on my heart for someone who worked for my husband. She was a loyal employee, a hard worker and a mature manager. Her ethics and values made her trustworthy and honest. Why would I have a burden for such a good, well-balanced person? She began attending a church and her life began to change. Her church was a Self-Realization Fellowship temple. She learned from them that her good works and good life would gain her much favor when she was reincarnated. If she continued to improve in her good works each time that she returned to earth (in whatever form that might be), then eventually she would reach the pinnacle of her life's goal: she would become god. Believe it or not, this is the teaching of the New Age movement. She bought in to their lies and deceptions.

The Lord gave me the verse listed above and told me to tell her that He is the only Light. I remember the morning that I asked her to come over and have a cup of coffee with me. I told her that the Lord loved her and that He wanted her to know Him. I read 1 John 1:5 to her and said that in God, there is no darkness. She began to cry and she shared with me her fears of darkness. She was struggling and hurting. I talked with her about Jesus and that there is only one true God. Jesus is the only Way. There is no alternative path to everlasting life. The hard part in witnessing to someone in this position is that they will often agree with you. She agreed with me in everything I said about God and that she most certainly believes in God. But, she was not so sure about Jesus being God. She believed in Jesus but she had made Him out to be just a really good man, a prophet and a teacher. She could not accept His sacrifice for her sin, because she convinced herself that, in the end, her good works were all that really mattered. She would earn her way.

She would later leave our company and move on. The differences in our beliefs really stressed our relationship at times. Without realizing it, she changed in ways that made her hard-hearted and opinionated. In denying Jesus, she was continuing to bring increasing darkness into her life. The "good life" is deceptive and blinding. It simply is not possible for us to be "good" people. I still pray for her when the Lord brings her to my mind and I hope that one day she will come to the true Light. For me personally, I learned a lot from going through the experience of witnessing to her. Despite my burden for her soul, I know that only the Lord can save her. If you know someone in this type of cult or other false religion, do not give up praying for them. There is always hope—until their very last breath.

December 3

Today's Reading: Ezekiel 45-46; 1 John 2
Today's Thoughts: Our Perspective Matters

The righteous perishes, and no man takes it to heart; Merciful men are taken away, while no one considers that the righteous is taken away from evil. He shall enter into peace; they shall rest in their beds, each one walking in his uprightness. Isaiah 57:1-2

A friend of ours went home to be with the Lord recently. God had blessed her with a heart of wisdom as she served Him in ministry as well as in the company of those in the world. We, as a church, prayed fervently for her healing and for extended life but the Lord took her home a few months after her diagnosis.

It is difficult to understand why the Lord would want to take someone home to be with Him when she was serving Him so faithfully and fully here. Thinking in the earthly perspective, she was too young to die, still had too much left she could do and was too much of a godly model. When I think about her from this perspective, I am stumbled by God's will and confused by God's ways. And I just want her back. However, when I pull myself back enough to see it from an eternal perspective, I am sincerely envious that I am not in the presence of the Lord with her right now too. Our hearts should groan for heaven even though our flesh refuses to accept death. The dichotomy of life and death is a continual reminder that earth is not our home. We are just passing through. The goal of each Christian should be that our walk with Him doesn't change from here to there. The place changes but not the relationship.

Our perspective matters. If we chose to continually see things from man's perspective, we will be discouraged and disappointed with the Lord. If we choose to get into the Word and see life from His perspective, we find joy peace and excitement by drawing closer to Him. I am saddened by my loss of a sister in Christ but I am also thankful that we will remain friends for eternity.

Oh Lord, bring us home as dedicated and committed to You as our friend was.

December 4

Today's Reading: Ezekiel 47-48; 1 John 3
Today's Thoughts: In Your Sight

Then Moses said to the Lord, "See, You say to me, 'Bring up this people.' But You have not let me know whom You will send with me. Yet You have said, 'I know you by name, and you have also found grace in My sight.' Now therefore, I pray, if I have found grace in Your sight, show me now Your way, that I may know You and that I may find grace in Your sight. And consider that this nation is Your people." Exodus 33:12-13

My family and I went skiing for a few days. The first day, we were in a blizzard and I quickly discovered that my goggles were broken. At times, I thought I would have skied better if my eyes were closed because my goggles made the weather conditions even worse. The second day, the sun was out shining and the conditions were beautiful. I didn't need goggles, sunglasses would work just fine. Well, I pulled out my sunglasses on the first chairlift. To my surprise, one of the lenses fell into my hand. When instances like this happen two times in a row, I know that the Lord wants my attention. While staring at the lens, I immediately said to the Lord, "What are you trying to teach me about my eyesight?"

What came to my mind at the end of that prayer was the Scripture that I had been meditating on from Exodus 33: "found grace in Your sight." Three times Moses said to the Lord that he knew he found grace in God's sight. I was meditating on why Moses added in "Your sight." I know that I have also found grace but what does the "in Your sight" mean to me? God used my goggles and broken sunglasses to explain. We are very dependent on what we can see or what we perceive. We make judgment calls and assessments through our eyes. When something impairs our ability to see, we compensate by using our other senses, and in those times I do a lot more praying for wisdom. I realized how much I judge by what I see. I depend on my sight (or perception) more at times than on how God see things. But it cannot be about how I see things, but what the Lord sees in me. Like Moses, I know that I have found grace in His sight but am I living to be pleasing in His sight every day, with every decision? Do I value the things the Lord sees more than the things I see?

The Lord wants us to live with spiritual eyes. He wants us to fix our eyes on Jesus so that our decisions and choices may be pleasing in His sight. We have already found grace because we know Jesus, but we also need to live a life that is pleasing in His sight daily. I just wish I could have learned this lesson safely at home instead of on top of a mountain. But of course, it was up on a mountain that Moses learned the most too.

December 5

Today's Reading: Daniel 1-2; 1 John 4
Today's Thoughts: God's Favor

Now God had brought Daniel into the favor and goodwill of the chief of the eunuchs. Daniel 1:9

I love the Old Testament stories that illustrate how God brought someone into favor in an unfavorable situation. Daniel was given favor in the king's palace, even though he was a captive and taken prisoner. Joseph was given favor with the guards as a prisoner in Egypt. And even David, before he was king, was given favor when he sought refuge in enemy territory with the Philistines. Why such favor with these people in these situations? Was it because of what they had done to earn it? No. It was because of *who* God is.

Even today, we see God's favor upon us. As I opened my hometown's newspaper last week, I saw a picture of my nephew being awarded "Student of the Month." Knowing that he is not a stellar student but wondering if he had suddenly had a major change, I called my sister to congratulate him. When I asked her why he received the award, she said, "Well, for some reason the teacher just picked him once before, and she picked him again for this month." My sister had no real answer as to why he got the award the first time, much less the second. I knew that my nephew had found favor with his teacher. Because of what he had done? I doubt it. But because of *who God is.*

There are many times that I ask God to give me favor in certain situations or with certain people. Sometimes we hold back from asking for such things because we feel unworthy or undeserving. But the Lord wants us to ask because He loves to answer. I have been given favor in situations where I deserved nothing but wrath. When we truly embrace God's grace, then we can begin to understand a snippet of how much He loves us.

December 6

Today's Reading: Daniel 3-4; 1 John 5
Today's Thoughts: Be Aware

Now Dinah the daughter of Leah, whom she had borne to Jacob, went out to see the daughters of the land. And when Shechem the son of Hamor the Hivite, prince of the country, saw her, he took her and lay with her, and violated her. **Genesis 34:1-2**

The Lord had a plan for His people as He was beginning to develop His own nation. But long before the children of Israel would fully understand what that meant, God began to teach them the importance of being set apart from the Gentile (or pagan) nations. When Dinah "went out to see the daughters of the land," she may not have realized the potential danger in mingling with the foreigners. This danger became real when Hamor "violated her."

As Christians, we must be aware of the danger of mingling in the world's activities. There must be a distinction between us and the world. There are times to witness and to shine the Light of Jesus Christ into the world, but we need wisdom to know when to be careful of dangerous places. Jesus prayed for us to be *in* the world but not *of* the world. God has a purpose for us in how we live and also in the activities of which we participate. We must train ourselves to be aware of the enemy's schemes to lure us into trouble, which often happens because we are in a place where we should not be.

Dinah found herself in trouble by visiting the women of the land. This trouble led to even greater problems for the family. Let us receive this message and apply it to our own lives as we pray to be aware of these types of dangers. What can seem harm-*less* could actually be harm-*ful*. Get into the habit of bringing every decision to the Lord first and then asking for the wisdom and guidance in what to do next.

December 7

Today's Reading: Daniel 5-7; 2 John
Today's Thoughts: What is Double Minded? Part 1

"he is a double-minded man, unstable in all his ways." **James 1:8**

James tells us not to be double-minded because we will be unstable in all our ways. No one wants to be unstable, so how can we prevent ourselves from being double-minded?

First, we have to define what it means to be double-minded. Thinking through decisions and weighing out the cost/benefit ratio is not being double-minded. That, instead, is called "counting the cost." In every decision we make, we need to count the cost; it is important to gather information and know the options first. This process can take some time, and for some of us, it takes too long when our thoughts become consumed by our choices. It is important to include the Lord during the "count the cost" process because if we have missed something, the Lord will bring to mind what we need to remember. But there is only so much information we can gather and there is only so much time that can be allotted for each decision. Continuing to extend the process; leads to second guessing (or being double-minded). Set a time limit to gather information and then make a decision. No looking back and no regrets. It is better to make a wrong decision than to make no decision. God can work together anything for good and we learn from every decision, the right and the wrong ones.

Another way to prevent double-mindedness is by asking the Lord for wisdom. Your goal should be to make a choice that will bless the Lord as well as be best for you. At the outset, do not jump to conclusions of what you think is best in pleasing Him. When I have done this in the past, I usually ended up being wrong. For some odd reason, I tend to believe that whatever is best for God is not what is best for me. WRONG! Or I want to believe that what is best for me has to be best for God—wrong again. Be open, be willing, and be in the Word. When we ask for wisdom, we are really asking for God to make His choices known to us. We have to be willing to listen. The only way I have learned to be willing to listen is by understanding that God's ways are best for me, regardless of what those "ways" may look like. He is a God of love and purpose, and His purposes are filled with love. We can trust Him. Pray and believe in faith that God hears you. James tells us to not doubt that it is God answering. Pray for faith to do what He has led you to do, despite your doubts.

(Part 2 in tomorrow's reading)

December 8

Today's Reading: Daniel 8-10; 3 John
Today's Thoughts: Being Double Minded Part 2

"he is a double-minded man, unstable in all his ways." James 1:8 (Part 2)

After we ask for His wisdom, we may still be apprehensive to step forward in faith. We might need to look for further confirmation on the matter. Now, we need to make sure that we are sensitive to His confirmation, assuring His Will over our own will. We are instructed in Matthew 18:16 that at the testimony of two or more witnesses, every Word may be established. He will confirm His Will to you. God can use various means to confirm His message to our hearts; always start in His Word then continue to ask Him to bring things to your mind. When you have received the same message several times and you know in your spirit that the Lord is impressing His Word upon your heart, then rest in His counsel and have confidence in knowing that He is speaking to you, to guide you in His ways.

But we still have the choice to choose His ways or not. James 1:5-8 says, "*If any of you lacks wisdom, let him ask of God, who gives to all liberally and without reproach, and it will be given to him. But let him ask in faith, with no doubting, for he who doubts is like a wave of the sea driven and tossed by the wind. For let not that man suppose that he will receive anything from the Lord; he is a double-minded man, unstable in all his ways.*"

If we know how God is leading and we still chose to wrestle, struggle, doubt, worry and fret—we become double-minded. It would have been better to just make your own decision and live with the consequences. Instead, ask for wisdom, believe He gives it without reproach and receive His answers in faith.

Let's say something goes terribly wrong after going through this process. I have seen this happen also. We need to remember that God rewards those who diligently seek Him and God rewards faith. If you went through this process of seeking His will over your own, and you stepped out in faith, then rest with the results. Those results are now God's burden and we all know that His burden is easy. Always remember: God works all things together for the good of those who love Him and are called according to His purpose.

Oh Lord, thank You for allowing me to find You when I search for You with all my heart. Please give me wisdom, knowledge and discernment as I desire to do Your will over my own. And I will trust You with the results.

December 9

Today's Reading: Daniel 11-12; Jude
Today's Thoughts: In Time

Blessed is he who reads and those who hear the words of this prophecy, and keep those things which are written in it; for the time is near. **Revelation 1:3**

The key word in today's verse is the word "time." When the Lord says, "the time is near," He is referring to time from His perspective, not from the perspective of time that we have. Our minds cannot comprehend God's concept of time, which puts us in a place of trusting Him for His perfect timing in all things.

Time seems so finite to us on earth, but in the kingdom of God, time is very different from what we know today. Even setting the clocks back one hour throws our schedules and internal systems off. It takes many of us all week to get adjusted to the time change. How can we begin to even imagine time being eternal? But it will be eternal for our spirits, and the choice of where we will spend it is up to us. For those in Christ, eternity in Heaven awaits us and the concept of time will no longer matter. For those who die without knowing Christ as their Savior, eternity consists of a Hell too terrible to comprehend.

Today, we should value time from the perspective that it is too short and is running out fast. Let's get out and spread the Good News before time runs out. The book of Revelation is written so that we will know of what is to come and to believe that the "time is near." We must never become complacent or apathetic regarding the Lord's prophecies and promises. His words will come to pass…in time.

December 10

Today's Reading: Hosea 1-4; Revelation 1
Today's Thoughts: Watch Your Stubbornness

"For Israel is stubborn like a stubborn calf; Now the Lord will let them forage like a lamb in open country." Hosea 4:16

Giving my dog a bath gives me a good glimpse at a stubborn animal. He instinctively knows when I am getting ready to bathe him. Once he is convinced that I am coming to get him, he tries to hide. At that point, talking to him will not get him to change his mind. He must be dragged by the collar to the hose outside (hot and cold water, of course). It is not like he isn't spoiled and usually gets what he wants, but a bath is what is best for him, and for those around him. When the bath is over, he is the happiest, most energetic dog I have ever seen. He seems to like the after effect much more than the process he has to go through to get there.

The Lord compared Israel to a stubborn calf. They would not listen to the Lord's commands and would not obey His Word. Regardless of how much God blessed them, they still returned to their wicked ways. Regardless of how much they saw God's hand taking care of them, rescuing them, and providing for them, they had a short memory of those things. They were constantly returning to their sinful practices of worshiping other gods and pursuing their fleshly desires. Stubborn and stiff-necked is how God referred to them as He looked upon their rebellious ways.

I wonder how often the Lord looks at us the same way. How often do we rebel against what God has for us because we do not want to go through the process (or the bath we need)? For me, even though I revel in the after effects of God's touch, I still resist that necessary cleansing of the sin in my life. If we are not careful, we will find ourselves too comfortable with being unclean. Once that begins to happen, we may also find ourselves slipping into practices that do not glorify God, just as the Israelites did. When you know the Lord is coming to get you for that special time of bathing, just let Him have His way with you. The process will go much faster and you will feel so much better when you surrender to Him. I keep giving my dog the same advice even though he is not catching on. **But we should.**

December 11

Today's Reading: Hosea 5-8; Revelation 2
Today's Thoughts: The Weather Man

Let us acknowledge the Lord; let us press on to acknowledge Him. As surely as the sun rises, He will appear; He will come to us like the winter rains, like the spring rains that water the earth. **Hosea 6:3**

Wherever you live, rain is a big deal. If we have little rain or a lot of rain within the year, the weather makes a huge impact on us. The weather conditions command our daily attention as we graph and chart the amount of rain. We have entire television stations dedicated to the weather around the world. When a natural weather disaster hits, all other news stops to report the extent of the damage. We pay a lot of attention to the weather as we depend on this information to plan our day.

There is nothing wrong with following the weather patterns, but Hosea is stating that as surely as we acknowledge the sun and rain coming, we need to acknowledge the coming of the Lord. Jesus rebuked the people for acknowledging the signs of the weather but not acknowledging the signs of the Lord. In Luke 12:54-56, Jesus said to the crowd: "When you see a cloud rising in the west, immediately you say, 'It's going to rain,' and it does. And when the south wind blows, you say, 'It's going to be hot,' and it is. Hypocrites! You know how to interpret the appearance of the earth and the sky. How is it that you don't know how to interpret this present time?"

Are we sensitive to His ways? Are we in tune with His will? Do we study His Word daily for direction and guidance in our day as much as we tune into the weather? Could Jesus rebuke us today for the same things? Even though the rain seems to be later than expected some years, we do get rain. We might think that Jesus is late in His coming too. But surely He will come. If we are not acknowledging Him, He may not acknowledge us. And even if Jesus doesn't come here during our life time, it is inevitable that we will go to Him some day.

Who do you worship? Who do you serve? Just by examining your need to know the weather in comparison to your need to know the Lord should shed some light on what your priorities are. When we acknowledge the Lord in everything, we are able to interpret His signs because we become sensitive to His ways over our own. We give so much attention to the weather but Jesus truly is the ultimate Weather Man, *whether we acknowledge Him or not.*

December 12

Today's Reading: Hosea 9-11; Revelation 3
Today's Thoughts: Consider Your Deeds

"I know your deeds, that you have a name that you are alive, but you are dead." Revelation 3:1

Companies hire consultants to come into their business to assess and evaluate the services they provide. The goal is to become more efficient, make more product and have better outcomes. Hiring an outsider allows an unbiased opinion without any emotional attachments to the employees involved or the personal dynamics that make up each "excuse" of why the employees do not achieve the company's goal.

In the Book of Revelation, Jesus does not ask for a consultant, like a prophet or angel, to come in to evaluate His church and His people. He clearly and lovingly assesses each church Himself. He starts each letter with the words, "I know your deeds" and then to six out of the seven churches, He says, "But". Deed is also another word for work, which in the Greek means an act, deed, work that is done or anything accomplished by hand, art, industry or mind. These churches know Jesus because they each have a saving faith to be called and hear what He is saying. So, Jesus speaks to each one about their works, not their faith. And we know that faith without works (or deeds) is dead. However, the Overseer of each church rebukes their works and deeds that follow their faith. The stakes are high. The rebuke is not about losing your job but the shaking of your eternal foundation.

Jesus writes these letters explaining that He understands who they are, where they live, what they do and to six of the churches, how they fall short. Then, He lovingly tells them how to return or be restored. God is also for us. He wants us to be our best and live abundantly. How we live our life here day by day does matter for eternity. For me personally, I do not want any "Buts" when standing before Jesus face-to-face. I want to hear, "Well done, good and faithful servant. Enter into My rest." I want all He has for me here and I want to do it and live it His way.

Today, read over Revelation 2-3. Meditate on the verses; ponder His descriptions in your mind. Then ask the Lord which description looks a lot like you. He will honestly and clearly answer and then help you become all He wants you to be. The stakes are too high—take His words seriously.

December 13

Today's Reading: Hosea 12-14; Revelation 4
Today's Thoughts: Return To The Lord

"Their deeds do not permit them to return to their God. A spirit of prostitution is in their heart; they do not knowledge the LORD."
Hosea 5:4

During Hosea's time, the people would not return to the Lord. They continued to do what they thought was right in their own eyes, not acknowledging God's standard for their lives. As a result, they lived to please their king, and themselves, to their own destruction. However, the people were still trying to appease God by following in the things that God had established for them, like the rituals and sacrifices. It was easier for them to "do" for God than to "be" with God. God told them that He would rather have them acknowledge Him than to burn sacrifices for Him. The people's sins became such a habit and place of bondage that their deeds prevented them from returning to the Lord.

The Book of Hosea is written because God wanted them to return to Him. Their deeds did not have to separate Him from them, but their deeds kept their hearts from turning to God. If you find yourself in sinful habits and patterns that are keeping you from returning to the Lord, today is the day to come back to Him. The Lord's love cannot ever be separated from you, but you have the ability to separate yourself from Him. Start being honest with Him today. Tell Him the truth—He already knows everything about you. Let the secret out so that He may come in. Do not beat yourself up during the restoration process, it will take time and change often comes slowly. Praise God that He is never in a rush. Relax. He is on your side.

December 14

Today's Reading: Joel; Revelation 5
Today's Thoughts: No such thing as wasted years

"So I will restore to you the years that the swarming locust has eaten..." **Joel 2:25**

The book of Joel was written by the prophet of whom we know the least. His book is only three chapters in length but is packed with sincere pleas for repentance for the nation of Israel. Joel knew the Lord's Day was coming but he also knew and spoke of God's endless mercy and grace. One of my favorite verses in the Bible is Joel 2:25. For me, this verse represents great hope and encouragement for the future. Not only is our God a God of second chances, but also He is a God who restores our past. I have witnessed this restoration first hand in my own life.

I was raised in church, accepted Jesus as my Savior and was baptized, all before the age of 10. I can honestly say I have known the Lord most of my life, but I have not always chosen to walk with Him. There is a big difference in *knowing* Him versus *living* for Him. There is also a big difference in knowing *about* Him versus knowing Him personally. I spent many years as an adult choosing to live my life my way. I still read the Bible, went to church and prayed...prayed a lot for the things I wanted. But my life was my own. Then one day, something changed in my heart. My eyes were opened in a very different way. I realized who was really on the throne of my life and it was not Jesus. I began to see how empty my life had become and that I wanted more. More meant more of Him. It was time to live for Jesus. It was time to change my course. But what about those wasted years? I had such sadness realizing the time I had wasted. If only I had done things differently.

Then one day, I come across this verse in Joel. And the Lord Himself said to me that He would restore all of those years that the locust had eaten. He has a plan for my life and His plans and callings are irrevocable. Not only is God with me as I go forward but also He has never left me. I look back and see God's tremendous grace. And I have seen God do immeasurably more than I could ever ask or imagine in just a few short years. Do you have regrets about wasted years? Know today that God will restore them all. If you have been away from Him, come back today. Repent and ask the Lord to sit on the throne of your life and carry out His plans and callings for you. Let this verse give you hope and encouragement to begin a new day.

December 15

Today's Reading: Amos 1-3; Revelation 6
Today's Thoughts: Willing To Walk

Can two walk together, unless they are agreed? **Amos 3:3**

What does it mean to walk with someone? If you walk with someone, you keep the same pace or stride. You walk beside them, close enough to see and hear them clearly. Walking with someone in the physical sense represents fellowship and synchronicity, where neither one is moving ahead or lagging behind. To walk with someone requires a willingness to move ahead together in the same direction and for the same duration. As the verse says, for *two* to walk *together*, they must *agree*—on quite a few things.

The Bible is filled with "two [who] walk together." The Lord gave Moses a partner in Aaron. Naomi had Ruth. David had Jonathan. Even Jesus sent His disciples out in pairs, two by two. Peter and John would continue as friends and partners as they started the first church. Ecclesiastes 4:9-11 says that two are better than one because they are there to help each other, pick each other up, and even help keep each other warm. And, of course, from the beginning, God put man and woman together: to walk together and to become one in marriage. But to truly walk together, we must agree with our partner, have common goals, be willing to submit, and work together for the purposes of God.

God puts us together because He knows the value of fellowship and friendship. He has made it a necessity to the point that if we do not have fellowship with others, we will struggle with loneliness and depression. Jesus wants us to walk with Him in fellowship and friendship. He desires for us to agree with Him, submit to Him and allow Him to set the course. If we can truly learn to walk in agreement with the Lord, then we will successfully walk together in our marriages and other friendships. We cannot walk with someone and be at odds, eventually the walk will end.

If you are struggling today in your walk with someone, a marriage, friendship or partnership, you must first get your walk back in agreement with the Lord. Ask the Lord to help you walk with Him in those areas in which you are struggling. Maybe you need to repent from rebellious ways or attitudes or maybe you need to submit to going in a direction you have not wanted to go. Once you learn to walk with God first, then you will be so much better in walking with others.

December 16

Today's Reading: Amos 4-6; Revelation 7
Today's Thoughts: Weary People with No Holiday Cheer

You, O God, sent a plentiful rain, whereby You confirmed Your Inheritance, when it was wear. Psalm 68:9

I was shopping in Costco during the summer months and ran into an old acquaintance from a Bible study class we attended together. I asked her how she was doing and her response was, "Well, I have the most difficult time keeping the faith in the summer and during the holidays." Her words were absolutely honest and sadly enough very true. We grow weary in doing good at different seasons of life. The holidays tend to be one of those seasons.

After talking with so many women at retreats, Bible studies, live radio programs, women's events and even through emails, we have found that many Christians are basically weary (or at least the people who talk to us). Circumstances, people, finances, negative words and thoughts have all contributed to not having a desire to persevere. We know that the right thing is to persevere, run the race with endurance, and not lose heart, but sometimes it seems too difficult to just keep going, day in and day out.

If we were really honest, most of us would admit to having thoughts like, "How can I do it all? How is this going to work out? Why does life have to be this hard?" The added pressure of the holiday season demands more of our time and attention. Suddenly we find ourselves having no time to pray or seek the Lord. Instead, we try to spiritualize and justify our position by saying we are doing all these things for others as if we are living a sacrificial life. However, we can't give others something we do not have. We can become unspiritual in the process. My father frequently said, "When your output exceeds your intake, your upkeep is your downfall." It is too easy to fall down in the holiday seasons.

Why? Because God does not ask us to sacrifice but to obey. Jesus defined obedience in John 15 as abiding in Him. We need to take the time to be with the Lord. Lift your eyes toward Him, open your heart in prayer and seek the Lord. He will help you and He is able to provide rest for the weary soul.

December 17

Today's Reading: Amos 7-9; Revelation 8
Today's Thoughts: Whose Problem is It?

See if there be any hurtful way in me and lead me in the everlasting way. **Psalm 139:24 (NASB)**

One of the roles of the Holy Spirit is to convict us of wrongdoing. In my life, there have been things I have been doing for years and years, never thinking much about it. All of a sudden, out of nowhere, I start getting convicted. When the Lord starts working with us on these behaviors, thoughts come to mind like, "Maybe I should not do that any more" and "Why do I do that?" Personally, I do not like these thoughts because I know how long I have been on this path and it is embarrassing to think that I might have been wrong all those years. Who wants to be wrong?

At these times, the Holy Spirit is the One who is working with you. God's standards are different than man's. Behavior that might be completely acceptable to others may not be acceptable for you in the Lord. However, it is the Lord's responsibility to change you. He is just letting you in on what He is doing. Our responsibility is not to suddenly change our behavior (as if it were that easy), but to yield to what the Spirit wants to do with your behavior.

If there is something in your life which you believe the Holy Spirit is working on, be thankful. It is confirmation that you are God's child. The Lord will not force His will or His best on you. But He wants you to be willing to yield to His leading in that behavior. I pray things like, "Lord, if this is a conviction from You, I ask that You will remind me before I do it again." The Lord hears all your prayers and He wants to work with you. He will not convict you and let you fix it by yourself. He wants to do the work by His Spirit, not by you in your flesh. It is in the Spirit that God gets the glory. If He led you into conviction, He will be responsible for leading you into a different behavior. Rest, Trust and Relax. You will like who you are when He is done.

December 18

Today's Reading: Obadiah; Revelation 9
Today's Thoughts: Jesus with Skin On

Praise be to the God and Father of our Lord Jesus Christ, the Father of compassion and the God of all comfort, who comforts us in all our troubles, so that we can comfort those in any trouble with the comfort we ourselves have received from God. For just as the sufferings of Christ flow over into our lives, so also through Christ our comfort overflows. If we are distressed, it is for your comfort and salvation; if we are comforted, it is for your comfort, which produces in you patient endurance of the same sufferings we suffer. And our hope for you is firm, because we know that just as you share in our sufferings, so also you share in our comfort.
2 Corinthians 1:3-7

There is a story about a little girl who was being tucked in to bed by her mommy. The little girl was scared to be left alone and asked her mom to stay in the room with her. The mother kindly answered, "Honey, it's okay. Jesus is with you." The little girl said back, "But Mom, I need someone with skin on."

People need to see and feel Jesus through us. We have heard repeatedly that God loves us. However, sometimes, we do not sense that love. God does not use His arms to physically touch us because it is our spirit that responds and testifies to the love of God. But sometimes, others need to see Jesus and feel Jesus to accept His love. God does that by using us. Think of Moses when He was called by God in Exodus 3:7-10:

> "And the Lord said: "I have surely seen the oppression of My people who are in Egypt, and have heard their cry because of their taskmasters, for I know their sorrows. Therefore, I have come down to deliver them out of the hand of the Egyptians. Come now, therefore, and I will send you to Pharaoh that you may bring My people, the children of Israel, out of Egypt."

God was saying to Moses that He was going to answer the Israelites' prayers by sending Moses to free them. Moses became "Jesus with skin on." He is asking us to do the same thing today.

December 19

Today's Reading: Jonah; Revelation 10
Today's Thoughts: Are You Going or Running?

Then the men were exceedingly afraid, and said to him, "Why have you done this?" For the men knew that he fled from the presence of the Lord, because he had told them. **Jonah 1:10**

The Lord told Jonah to go to Nineveh and preach repentance. Jonah took off in the opposite direction, running from the Lord. Nineveh was a city of wickedness, but the Lord wanted to give the Ninevites a chance to repent from their evil ways and turn to Him. Jonah did not want anything to do with this mission and instead decided not to do what the Lord had told him. As a result, everyone on the boat that Jonah had boarded (as a way of escape) was in fear in a terrible storm that threatened their lives. They were bearing the consequences of Jonah's disobedience.

In the end, God had His way with Jonah and he went to Nineveh. The people on the boat were spared and the city of Nineveh was spared, at least for a time. Because Jonah refused to obey the Lord the first time, he put others in jeopardy. By God's grace, no one was hurt by Jonah's decisions. God dealt directly with him (The big fish that swallowed Jonah readily took care of any misconceptions about Who was really in charge.). Make no mistake about it: God is in charge. We may delay things for a period of time, but ultimately God always gets His way. The question for us is: Are we putting others in danger of being harmed by the consequences of our decisions to disobey God? If so, then we are held accountable for that too.

Has God been telling you to do something and you just keep putting it off? Are others being affected by your decision? Maybe you have jumped on a boat heading out to sea, trying to escape the Lord's words. God may let you get away with it for a season, but in time He will take control; He will have His way with you. If something is gnawing at you, or if you have any doubts about what God is telling you, then go to Him immediately and ask Him what He wants you to do. If you are afraid, He will help and comfort you. God never asks us to do anything that He has not already prepared and planned out for us. Do not delay: You never know when there is a big fish lurking about. God bless.

December 20

Today's Reading: Micah 1-3; Revelation 11
Today's Thoughts: He Equips the Called

***And the L**ORD** said to Joshua, "This day I will begin to exalt you in the sight of all Israel, that they may know that, as I was with Moses, so I will be with you."* Joshua 3:7**

It is clear that *God was with Moses*. No one understood that more than Joshua, Moses' assistant. But Moses struggled with his call of *being with God*. He understood that he was just a man with limitations and shortcomings. Moses had committed murder, attempting to defend his fellow Israelite, but his actions caused great consequences as he became an outcast from both sides, the Jews and the Egyptians. Even though Moses was raised as royalty in a palace the first forty years of his life, he ended up in the desert shepherding someone else's sheep for the next forty years. So when God called him, Moses wasn't expecting it and questioned God's will in choosing him. Exodus 3:11 says "But Moses said to God, 'Who am I, that I should go to Pharaoh and bring the Israelites out of Egypt?'" and then Moses told the Lord in Exodus 4:10-13, "O Lord, I have never been eloquent, neither in the past nor since you have spoken to your servant. I am slow of speech and tongue." The Lord said to him, "Who gave man his mouth? Who makes him deaf or mute? Who gives him sight or makes him blind? Is it not I, the Lord? Now go; I will help you speak and will teach you what to say." But Moses said, "O Lord, please send someone else to do it."

God heard the fear and insecurity in Moses' words. But God had placed this desire in Moses' heart forty years prior and God was not going to take no for an answer. He instead provided Moses' brother to be his partner and then gave Joshua to Moses as his assistant. Moses was able to accomplish all that God intended for him to do. Since Joshua was faithful, Moses trained him up to be his replacement.

We read promises like this, "***This day I will begin to exalt you in the sight of all Israel, that they may know that, as I was with Moses, so I will be with you***" and understand what an honor that was to Joshua. However Moses and Joshua both had it hard before receiving such an honor. Both were tested and tried, both needed to submit and be willing to obey. Becoming God's leader doesn't come naturally or easily. It comes through faith and obedience. And it took both Moses and Joshua forty years of preparation.

God doesn't call the equipped; He equips the called as He answers the desires of your heart. Our response needs to be, "Here I am Lord. Send me."

December 21

Today's Reading: Micah 4-5; Revelation 12
Today's Thoughts: Is Fasting Necessary?

But this kind does not go out except by prayer and fasting.
Matthew 17:21

Is fasting necessary? And what does prayer with fasting do that prayer alone can't accomplish? I don't know about you, but fasting is difficult for me. My flesh does not like anything being withheld from it. If my flesh could talk, it would make comments to my spirit like, "You don't really love me. Why are you holding out on me? This fasting thing just hurts, there isn't anything really to it."

Well, fasting without prayer is called "being hungry." But fasting with prayer is called "feasting in a spiritual sense." When we choose to make a conscious effort to put aside our needs of the flesh to concentrate instead on the spiritual issues that are nagging our hearts, God moves. Sometimes He moves the person who is fasting, by changing their desires, stilling their storms and flattening their prideful attitudes. And sometimes, He moves in the circumstances and moves mountains. Other times, He is able to work in others because of our ability to intercede while fasting.

Fasting with prayer is important because it causes us to focus while reminding us to pray. Every time we want to pick up something to eat, we pray instead. Fasting does not have to be related to food though. We could fast from watching television, listening to the radio, wearing jewelry or even lipstick. We can fast from anything that we are in the habit of doing, which then causes us to crucify the flesh and to concentrate on the spiritual realm.

As for me, I need to "almost fast" before I really fast. I need to prepare myself spiritually before jumping into this commitment. Crucifying the flesh isn't easy and unless I am led spiritually into a fast, there are too many reasons to break it. But God is so willing to reward us when we fast, that the pain of denial is nothing compared to His blessings in abundance. The disciples once asked Jesus why they could not cast out a demon, He replied: "This kind can come out only by prayer and fasting."

If you haven't ever fasted, pray about it; God will lead you, and if and when He leads you, get ready - for He will move strongly in your life too!

December 22

Today's Reading: Micah 6-7; Revelation 13
Today's Thoughts: Who is Like Our God?

Who is a God like You, Pardoning iniquity And passing over the transgression of the remnant of His heritage? He does not retain His anger forever, Because He delights in mercy. He will again have compassion on us, And will subdue our iniquities. You will cast all our sins into the depths of the sea. **Micah 7:18-19**

Have you ever known anyone who truly delights in mercy? Think about it for a moment. How many of us really enjoy extending mercy to others? I am not talking about the kind of mercy that we give in times of need or extend to someone because we feel sorry for them. I am talking about the kind of mercy that the Lord extends to us on a continual basis. And not only does the Lord grant us this mercy, but also the Bible says He loves to do it. Isaiah 30:18 says the Lord waits (or longs) to be gracious to us. Sometimes we need to be reminded of how good God is to us. There is no one like our God.

We also need to be reminded of something else concerning our awesome God: He pardons our iniquities and casts our sins into the depths of the sea. He remembers them no more (Isaiah 43:25). So often, we are the ones who hold on to our sins, not God. We are the ones who condemn ourselves, not God. We are the ones who judge ourselves harshly, not God. And, we are the ones who cannot forgive ourselves, but God does. Jesus says that we are to come to His throne of grace with boldness and to take His mercy with the love to which it is extended to us. Jesus died for us so that we may live in freedom, knowing that all of our sins are forgiven, even the ones we have yet to commit.

To our friends, We pray that today can be a new day for you. We pray that you will accept God's mercy and grace with open arms. We pray that you, in turn, will begin to delight in doing the same for others. We pray that you will fall in love with this awesome God who truly does delight in forgiving all of your sins and waits just to be kind and gracious to you. Ask the Lord to open your heart to receive His mercy and take hold of all that He has for you.

December 23

Today's Reading: Nahum; Revelation 14
Today's Thoughts: Praying God's Will

But shun profane and idle babblings, for they will increase to more ungodliness. And their message will spread like cancer. Hymenaeus and Philetus are of this sort, who have strayed concerning the truth, saying that the resurrection is already past; and they overthrow the faith of some. Nevertheless the solid foundation of God stands, having this seal: "The Lord knows those who are His," and, "Let everyone who names the name of Christ depart from iniquity." But in a great house there are not only vessels of gold and silver, but also of wood and clay, some for honor and some for dishonor. Therefore if anyone cleanses himself from the latter, he will be a vessel for honor, sanctified and useful for the Master, prepared for every good work. Flee also youthful lusts; but pursue righteousness, faith, love, peace with those who call on the Lord out of a pure heart. But avoid foolish and ignorant disputes, knowing that they generate strife. And a servant of the Lord must not quarrel but be gentle to all, able to teach, patient, in humility correcting those who are in opposition, if God perhaps will grant them repentance, so that they may know the truth, and that they may come to their senses and escape the snare of the devil, having been taken captive by him to do his will.
2 Timothy 2:16-26

Can you honestly say that you know what God's will is for your life, or a specific area of your life? Most Christians today do want to know God's will but many of us are not sure how to find it. The answer lies in His Word, the Bible. It sounds simple, yet so many of us make it more complicated than it needs to be. We need to open the Bible and ask the Lord to speak to us. We need to start praying God's Word back to Him. It is in praying God's Word that we begin to truly understand His will for us.

Our Prayer revised from 2 Timothy 2:16-26:
Dear Heavenly Father, I pray that I will "shun profane and idle babblings" that increase ungodliness and "spread like cancer." Lord, I pray that I "will be a vessel for honor, sanctified and useful" to You, my Master. I pray for Your guidance in leading me to "pursue righteousness, faith, love, peace with those who call on the Lord out of a pure heart." Dear Father, lead me away from "foolish and ignorant disputes" that "generate strife." And, Lord, give me a servant's heart that does not "quarrel but (is) gentle to all, able to teach, patient, in humility correcting those who are in opposition." Use me Lord in those whose lives You want to touch "that they may know the truth," "come to their senses," and "escape the snare of the devil." In Jesus name, Amen.

December 24

Today's Reading: Habakkuk; Revelation 15
Today's Thoughts: Being Bold in Prayer

***I will stand my watch And set myself on the rampart, And watch to see what He will say to me, And what I will answer when I am corrected.* Habakkuk 2:1**

How many of us open our Bibles and search for the writings of Habakkuk? The most likely answer: not too many of us and not very often. I remember the first time I actually studied this short, three-chapter book. To this day I still quote chapter two, verse one, back to the Lord. Habakkuk tells the story of how he questioned what the Lord was doing with His people. It is a great illustration of how we can seek out the Lord's plans, reason with Him, and then fall at His feet when He answers us personally. However, in this case, the answer was not in favor of the people, Habakkuk stood in awe of God's awesome power and mercy.

As Christians we tend to think too much about what God wants from us. We spend a lot of time praying for the things that we think He wants us to pray. We hope that He will not be upset with us if we pray for the wrong things. And then there are times when we are just amazed that He answered our prayers at all. It is one thing to be thankful and grateful for the Lord's gracious gifts, but we need to consider and question why we are just surprised that He answered in the first place. Our relationship with the Lord should be stronger than that. The Lord wants us to not only pray for the things we want and need, but also He wants us to *reason* with Him, wrestle with Him, and be honest in our prayers. In Isaiah 1:18 the Lord says, "come and let us reason together." Habakkuk had questions and he was not afraid to reason with the Lord, even to the point that he states in this verse that he will stand, wait and watch for God to do something. Talk about boldness!

For each of us today, God wants us to be bold. He wants us to come to His throne of grace with boldness and receive the mercy He has freely given through His Son Jesus. God wants us to pray with that same boldness, asking, seeking and knocking to know His will. Is something on your heart today about which you are not being completely honest with the Lord? Do you have questions or maybe you are upset with the way things are going? Go to the Lord in boldness. Tell Him your thoughts and feelings then tell Him that you will "sit on the ramparts [watchtower]" and wait for His answer. Do not be afraid. You will see God's goodness, just as Habakkuk did. "Yet I will rejoice in the Lord, I will joy in the God of my salvation. The Lord God is my strength; He will make my feet like deer's feet, And He will make me walk on my high hills." (Habakkuk 3:18-19) Now that is a great promise!

December 25

Today's Reading: Zephaniah; Revelation 16
Today's Thoughts: *Merry Christmas!*

For unto us a Child is born, Unto us a Son is given; And the government will be upon His shoulder. And His name will be called Wonderful, Counselor, Mighty God, Everlasting Father, Prince of Peace. Isaiah 9:6

If you look at today's verse, you see how it beautifully proclaims the birth of Jesus, our Savior. But if you look at its reference, you see that it comes from the Old Testament book of Isaiah. So often, we tend to focus only on the Gospels to learn about Jesus, His life and His purpose, but the coming of Jesus was foretold from the beginning.

Many people, even Christians, tend to focus on the New Testament writings more than the Old Testament. One year, we studied the book of Exodus in our women's Bible study and the impact for most women was amazing. They were able to really see how God worked through His people. They learned new insights to His sovereignty, His mercy, His power, and His amazing love for us. And every week, we pointed out the message of Jesus. We found Jesus in every chapter. The study of the tabernacle went from words describing seemingly endless details of clasps and boards to a powerful understanding of how God just wants to dwell with His people. We looked at how that foretold of what Jesus would do when He came to earth. People who never really had an interest in studying the Old Testament left with a desire to study even further. I think many of them went into the books of Leviticus and Numbers. They discovered the awesome power of the Word of God...in the Old Testament.

The birth of Jesus was not just mentioned in the New Testament Gospels. If we see it only from that perspective, we miss the awesome sovereignty and love of God. God had a plan from the beginning. This plan was not begun in the New Testament, it was completed there. We need to be challenged and encouraged to read and study the whole Bible. As the year end approaches, now is the time to make some new goals for next year. Prayerfully consider beginning a reading program that takes you through the whole Bible in one year. If you have never done that before, you are in for a blessing of unimaginable value. It will change your life. The enemy will do everything he can to thwart these goals, but if you are persistent, God will prevail. And when you get to the book of Isaiah sometime next year, maybe you will remember today's verse. If today's verse and this reading stir your heart to make this commitment, then give God the glory when you get to it next year. We pray that you make this resolution and that you allow the Lord to truly bless you through His word. **There is no greater Christmas gift than Jesus Christ, the living Word.**

December 26

Today's Reading: Haggai; Revelation 17
Today's Thoughts: Consider Your Ways

Then the word of the Lord came by Haggai the prophet, saying, "Is it time for you yourselves to dwell in your paneled houses, and this temple to lie in ruins?" Now therefore, thus says the Lord of hosts: "Consider your ways!" You have sown much, and bring in little; you eat, but do not have enough; you drink, but you are not filled with drink; you clothe yourselves, but no one is warm; and he who earns wages, earns wages to put into a bag with holes." **Haggai 1:3-6**

The prophet Haggai was sent to speak to the group who had come back to Jerusalem to rebuild the temple of the Lord. The Book of Ezra explains how the first group came back to begin the rebuilding process. Due to great opposition, the work came to a stop over the course of time. Haggai was sent by the Lord to re-start the work and get the people going again. The Lord was telling them to get their priorities straight. They had settled into their lives and homes but the temple was still not finished. They lived in paneled houses while God's house lay in ruins (Haggai 1:4). If they took a close look at their lives, they would see how their efforts had produced little.

Do you ever feel as though you work really hard, try to do the right things, yet nothing seems to come from it? Are you blessed? Do you see the Lord's hand in your work? If you feel as though you work to put your wages in a bag with holes, then maybe you need to "consider your ways." Check your priorities. Are you giving your first fruits unto the Lord? It is not just about your work for God that matters, it is about your heart before the Lord. Is He truly first in your life? Do you do things for Him because you love Him, or is it for other reasons?

These are the questions that we all must ask ourselves. Sometimes we get so focused on our own lives that we lose God's perspective. When that happens, we often begin to sense that things are not quite right. We do not see the blessings as we would expect or hope to see. As the Lord continued to tell His people through the prophets, "Repent and Return." The message is the same for us today. Maybe it is time to put some new priorities in your life. Pray about what the Lord wants you to re-prioritize and ask Him to help you to make the changes He wants. Instead of pockets with holes, you will see storehouses that overflow with His blessings.

December 27

Today's Reading: Zechariah 1- 4; Revelation 18
Today's Thoughts: Knowing God Loves You

***The Lord your God in your midst, the Mighty One, will save; He will rejoice over you with gladness, He will quiet you with His love, He will rejoice over you with singing.* Zephaniah 3:17**

I remember a day a few years ago when I was talking with someone who kept quoting Zephaniah 3:17. He was deeply impacted by the words in this verse to the point that he could not withhold his emotions. What was interesting about this conversation was that this man had spent his life in the ministry, serving in various capacities from missionary to seminary professor. He knew the Lord, walked with Him, talked with Him and loved Him with all of his heart, but did he really understand how much God loved Him?

The same is true for many of us today. We proclaim our love for the Lord. We pray to know Him more. We deeply desire a life that is fruitful and pleasing to Him. And for the most part, I believe that we spend more time thinking and praying about how to please God and how to live our lives in accordance with His will, than we spend thinking about how much He loves us. How often do you personally stop and think, "God loves me...He is rejoicing over me right now"? We tend to think that God is disappointed or upset with us instead of thinking about how much He loves and adores us. Why? Because so much of our thought processes are shaped by the world, not by the love of God. And we have an enemy who feeds right into those negative thoughts and will do anything he can to keep us from experiencing that peace and joy that comes from Jesus.

I encourage you today to write down this verse, meditate on it, and memorize it. Keep repeating it over in your heart and mind and let the Lord Himself speak these words into the depths of your soul. The Lord God is with you and will save you. He rejoices over you, quiets you with His love and sings to you. Can you believe it? Yes! God's Word is truth and you can trust that these words are spoken just for you. The man who spent his life in ministry was on the brink of retirement before he finally realized the truth of these words. Jesus wants you to know this truth today. Experience His love, peace and joy right now.

December 28

Today's Reading: Zechariah 5-8; Revelation 19
Today's Thoughts: Our Prophet, Priest and King

"as His divine power has given to us all things that pertain to life and godliness, through the knowledge of Him who called us by glory and virtue" **2 Peter 1:3**

The other day, I was teaching second graders about the reason for the crucifixion. It was a two week lesson so I started with explaining why God set up a Temple in the Old Testament. That led to the explanation as to why there were prophets, priests and kings. In second grade language, I told them that a prophet has a big mouth because he listens to God and then *tells* the people. But the priests have big ears because they *listen* to the people and then tell God. Prophets represent God's will for the people. The Priests represent the issues of the people to keep them in right standing with God. A king's job was different from both because a king uses his authority to establish what he thought was best for the people and then had the people do the work. Frequently, the king's commands were very different from God's commands, which led the people into a whole bunch of trouble. They seemed to understand all this so I moved on to Jesus.

Jesus came to earth as our Prophet, Priest and King. Jesus fulfilled all three roles because of His death and resurrection. He is our Prophet because He told us and showed us all about God. He is our Priest because He listens to us and intercedes at the Father's right hand today. Jesus is our King because all authority has been given to Him to now use us to do His will on earth. When I started to explain that the temple sacrifices are no longer needed because of Jesus' blood that was shed on the cross, they were very relieved. One of the little girls said, "Oh, that is so good for all of the animals." When I told them that prophets and priests are no longer needed because Jesus is everything we need on earth to have a relationship with God, they were so happy that God did that for them. Six of those little second graders accepted Jesus as their Prophet, Priest, King, Savior and Lord that day. Praise God that the Gospel is simple enough for children and yet, confounds the wise. Be sure to thank the Lord for all He has done for you to have a relationship with Him for He is worthy of our praise!

December 29

Today's Reading: Zechariah 9-12; Revelation 20
Your Thoughts on Today's Passage:

December 30

Today's Reading: Zechariah 13- 14; Revelation 21
Today's Thoughts: Refined and Tested

This third I will bring into the fire; I will refine them like silver and test them like gold. They will call on my name and I will answer them; I will say, "They are my people, and they will say, The LORD is our God." **Zechariah 13:9**

I read a verse like this and ask the Lord, "Why do the ones that you choose and love need to be refined and tested?" In some strange way, I look at love so much differently.

It is not natural in the flesh to look at my children and think that I willingly plan to discipline them. I often succumb to discipline when their actions are not socially acceptable and it embarrasses me. I really want to believe that they don't need me to discipline them and they'll grow out of this stage. I want to believe that I can show love by having the self-control to not discipline them and patiently wait for them to grow up. This is what my flesh wants to believe, and this is absolutely wrong because God is perfect love and He does not think that way. So, I need to align myself up with Him; both as a parent and as a child of God.

God says in this verse that He brings His own people into the fire. It is in the fire that we are refined like silver and tested like gold. If He is comparing us to precious metals, then we have to assume that the fire is pretty hot! The Lord does this for our good. He wants us to call on His name and know that He will answer because we are His people. His people have to be trained, disciplined, tested and refined to think His thoughts and to live in His ways. Because our flesh is so strong, we resist His work and end up turning from the Lord or blaming Him. We accuse Him of not being loving or caring. We wonder why He isn't answering our prayers or rescuing us from those hot circumstances, without realizing that He is the One who has placed us there.

God loves you. He is all powerful and can absolutely rescue you but He cares more about you than you can care about yourself. He needs you to change to be more like Him. Unfortunately that takes time in the fire. Trust Him because this is only for a season. For when you are trained, you will produce a harvest of righteousness and a deep peace knowing that He is your God.

December 31

Today's Reading: Malachi; Revelation 22
Today's Thoughts: Honoring the Lord

"For the lips of a priest should keep knowledge, And people should seek the law from his mouth; For he is the messenger of the Lord of hosts." **Malachi 2:7**

The magazines, newspapers and front page of the websites are filled with the most memorable events and significant people that have affected our lives in the past year. We can look back over the last year and see certain markers that changed our course and shaped our lives individually. Now with the New Year here, we not only reflect on the past, but also attempt to set goals for the future.

While being in this frame of mind, I started my devotionals this morning reading the book of Malachi. Malachi has four chapters with themes that are easy to remember: mouth, marriage and money. God is rebuking His priests because of the words of their mouths, their attitudes toward their marriages and their motives toward money. The priests were to be an upright and holy group of men representing the Lord. Outwardly, the priests did carry on their functions and requirements. But inwardly, their hearts were far from the Lord.

Considering the past year, we have to take the time to reflect on what we did and why. "Were my motives pure? Did others look at my actions and wonder why I was different? Did I give my best to the Lord or did I justify my offerings in giving the leftovers? Did my attitude in my marriage honor the Lord? What kind of words did I speak? Did my mouth glorify Jesus and edify others around me?"

After examining my heart in asking these questions, I found myself in repentance. How can anyone be that good? But that's not the point. None of us will ever be sinless but each of us needs to be above reproach in our hearts before God. God does not hold our behaviors against us if our hearts remain right before Him. It's when we think our behaviors justify our hearts that God rebukes us. The Lord is more than willing to work with a repentant heart.

In this New Year, be sure your heart is willing to be obedient. As you open your heart to confess, seek, yield and love the Lord, He will work with your behaviors that follow your yielded heart. We all want the Lord to bless us this New Year. We all pray that His face may shine upon us and give us peace. No matter how short we fall, His mercies are new every morning and His desire is to bless us indeed. As you reflect on the past and fix your eyes on the future, remember to consider the Lord. He has a plan for you as He watches over your coming and going, not only today but also for the this coming year.

This book was produced by Daily Disciples Ministries, Inc. We hope it has been life-changing and deeply enriching to your daily walk with Jesus Christ. The power of God's Word is the power that changes lives through the wisdom that comes in spending time with Him intimately. Daily Disciples Ministries, Inc. provides resources that reach the world for Christ including radio programs, videos, Bible studies, live webcasts, women's events, books and daily devotionals. For more information, please visit us online at www.dailydisciples.org or write us at:

Daily Disciples Ministries, Inc.
PO Box 131780
Carlsbad, California 92013
888-727-7206
Email: info@dailydisciples.org
Website: dailydisciples.org

Bible Study resources available include:

TOPICAL BIBLE STUDIES

Life in the Spirit

    4 lessons include the topics of *Believe & Receive, Power, Fruits of the Spirit and Gifts & Callings*

Practicing the Presence of God

    4 lessons include the topics of *Worship, Prayer, Hearing God's Voice I & Hearing God's Voice II*

Spiritual Warfare (also available in DVD)

    4 lessons include *Understanding the Battle, The Strategies of Satan, Our Weapons of Warfare, & Living in Victory*

Victory in the Valleys

    4 lessons based upon actual valleys from the Old Testament designed to teach us how to live in victory in our daily lives

My Identity in Christ (6 lesson study in Ephesians) *This study is available in DVD*

Checking Your Motives (5 lesson study in Malachi)

    Master, Mouth, Marriage, Messenger & Money

BIBLE BOOK STUDIES

Exodus (22 lessons)

Joshua (22 lessons)

Esther (4 lessons)

Malachi (5 lessons)

The Gospel of John (23 lessons)

Romans (8 lessons)

1 Corinthians (21 lessons)

Ephesians (6 lessons)

James (6 lessons)

Building Divine Intimacy
Tonilee Adamson & Bobbye Brooks

Have you ever wondered what was missing in your life?
Do you want more of what God has for you?

Take time and get to know God intimately. Your life will change forever!

Grow deeper in your personal relationship with the Lord through intimate worship, a powerful time talking to God in prayer, and knowing how to recognize God's voice speaking to you personally. This book will lead you step by step down a path of intimacy with the God of the universe.

Trade Paper ISBN 0-9788-726-4-9

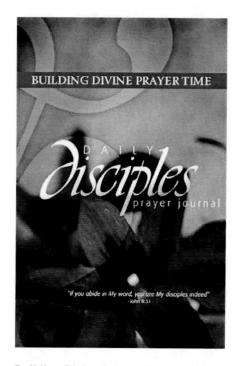

Building Divine Prayer Time: A Daily Disciples Prayer Journal
Tonilee Adamson & Bobbye Brooks

As Christians, we realize that we need Jesus for salvation, but do we realize that we need Him every day?

Learning to practice His presence daily, you will discover that God really is intimately acquainted with all your ways!

We have a deep down desire to know that God is working in our lives. This Prayer Journal is a tool designed to teach you how to apply the word of God to your life every day. The discipline of seeking Him daily will lead you into an exciting and victorious life in Christ.

An instructional CD and DVD are also available.

Trade Number ISBN 0-9840913-1-7